Religion and
Sustainable
Agriculture

RELIGION AND SUSTAINABLE AGRICULTURE

WORLD SPIRITUAL TRADITIONS AND FOOD ETHICS

Edited by
Todd LeVasseur, Pramod Parajuli,
and Norman Wirzba

Foreword by Vandana Shiva

Copyright © 2016 by The University Press of Kentucky

Scholarly publisher for the Commonwealth,
serving Bellarmine University, Berea College, Centre College of Kentucky,
Eastern Kentucky University, The Filson Historical Society, Georgetown
College, Kentucky Historical Society, Kentucky State University, Morehead
State University, Murray State University, Northern Kentucky University,
Transylvania University, University of Kentucky, University of Louisville,
and Western Kentucky University.
All rights reserved.

Editorial and Sales Offices: The University Press of Kentucky
663 South Limestone Street, Lexington, Kentucky 40508-4008
www.kentuckypress.com

Maps and figures by Richard A. Gilbreath, University of Kentucky Cartography
Lab.

Library of Congress Cataloging-in-Publication Data

Names: LeVasseur, Todd, editor. | Parajuli, Pramod, editor. | Wirzba, Norman,
  editor. | Shiva, Vandana, writer of foreword.
Title: Religion and sustainable agriculture : world spiritual traditions and
  food ethics / edited by Todd LeVasseur, Pramod Parajuli, and Norman Wirzba;
  foreword by Vandana Shiva.
Other titles: Culture of the land.
Description: Lexington, Kentucky : University Press of Kentucky, [2016] |
  Series: Culture of the land | Includes index.
Identifiers: LCCN 2016026839| ISBN 9780813167978 (hardcover : alk. paper) |
  ISBN 9780813167985 (pdf) | ISBN 9780813167992 (epub)
Subjects: LCSH: Sustainable agriculture—Religious aspects. | Food—Religious
  aspects.
Classification: LCC S494.5.S86 R45 2016 | DDC 338.1—dc23
LC record available at https://lccn.loc.gov/2016026839

This book is printed on acid-free paper meeting
the requirements of the American National Standard
for Permanence in Paper for Printed Library Materials.

Manufactured in the United States of America.

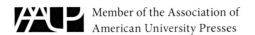 Member of the Association of
American University Presses

Contents

Foreword vii
  Vandana Shiva

Introduction: Religion, Agriculture, and Sustainability 1
  Todd LeVasseur

1. Our Flesh Was Made from Corn 25
  Leonor Hurtado Paz y Paz and Cristóbal Cojtí García

2. Soils, Spirits, and the Cosmocentric Economy: Re-creating Amazonian Dark Earth in Peru 45
  Frédérique Apffel-Marglin

3. Renewal of Non-Western Methods for Sustainable Living 71
  Eston Dickson Pembamoyo

4. Nature Spirituality, Sustainable Agriculture, and the Nature/Culture Paradox: The Permaculture Scene in Lower Puna, Big Island of Hawaii 99
  Michael Lemons

5. Hindu Traditions and Peasant Farming in the Himalayan Foothills of Nepal 121
  Jagannath Adhikari

6. *Dharma* for the Earth, Water, and Agriculture: Perspectives from the Swadhyaya 139
  Pankaj Jain

7. Gandhi's Agrarian Legacy: Practicing Food, Justice, and Sustainability in India 153
  A. Whitney Sanford

8. Thailand's Moral Rice Revolution: Cultivating a Collective Ecological Consciousness 173
  Alexander Harrow Kaufman

9. The Seven Species and Their Relevance to Sustainable Agriculture in Israel Today 195
  Elaine Solowey

10. Tending the Garden of Eden: Sacred Jewish Agricultural Traditions   211
    Yigal Deutscher

11. Religion, Local Community, and Sustainable Agriculture   233
    Anna Peterson

12. Heideggerian Reflections on Three Mennonite Cookbooks and a Mennonite Farm in Northwest Ohio   251
    Raymond F. Person Jr. and Mark H. Dixon

13. Steward or Priest? The Possibilities of a Christian Chicken Farmer   277
    Ragan Sutterfield

14. Religion and Agriculture: How Islam Forms the Moral Core of SEKEM's Holistic Development Approach in Egypt   295
    Maximilian Abouleish-Boes

15. Tohono O'odham *Himdag* and Agri/Culture   315
    Tristan Reader and Terrol Dew Johnson

Conclusion: Searching for Annapurna; or, Cultivating Earthbound Regenerative Abundance in the Anthropocene   337
    Pramod Parajuli

Acknowledgments   361
List of Contributors   363
Index   369

Foreword

The Dharma of Food

The seed, the soil, the water, and our food are sacred gifts of the earth.

Cultures and civilizations that have revered these gifts, and grown food in partnership with the earth, have nourished and sustained themselves, and lasted. Cultures that exploited and destroyed the biodiversity, the soil, and the water have perished. As the ancient Vedas tell us, "Upon this handful of soil our survival depends. Husband it, and it will grow our food, our fuel, and our shelter and surround us with beauty. Abuse it, and the soil will collapse and die, taking humanity with it."

Over the last three decades I have learned that agriculture is primarily about caring for the earth. The growing and sharing of food is therefore a spiritual act, as the contributors to *Religion and Sustainable Agriculture* remind us. This book is therefore an important contribution to sustainability. But it is also a contributor to peace in these times of religious and cultural polarization, and hate and fear of the "other." All faiths are in unity when it comes to sacred seed, sacred soil, and sacred, sacred food. No faith says to destroy the earth and starve your neighbors.

When we realize that the sacred duty to care for the soil, to save and share seeds, to share food and let no one in your sphere of influence go hungry, to conserve water, to give back to the earth in gratitude—through what Albert Howard called "The Law of Return"—is part of every religion, as this book records in such detail, then we will be able to start seeing how we are similar, instead of how we differ. We will start becoming aware that loving and caring for the earth as an intrinsic part of sustainable agriculture includes loving and caring for one another because we are part of Earth's family.

I started to work on ecological issues of agriculture in 1984 because of two deep tragedies in India—the violence in Punjab, and the Union Carbide gas tragedy in Bhopal. My study on Punjab, published as *The Violence of the Green Revolution*, revealed to me that what are called agricultural technologies in the industrial system are primarily tools of war. And as they destroy the earth, conflicts increase. Tragically, conflicts rooted in nonsustainability are often given a religious hue. The assumption is growing that religions are at the root of the escalating social conflicts and violence we are currently witnessing around the world. Yet when we look deeper, we find societal conflict growing from a destructive relationship to the earth. The Seed Freedom organization reports, for example, that such conflicts are occurring in both Syria and Nigeria. (See http://seedfreedom.info/campaign/terra-viva-our-soil-our-commons-our-future/.)

My own approach to the sacredness of food and farming has been informed by my culture. *Dharma* is the unique gift of Indian civilization to humanity. It has provided the compass for right action and right livelihood. There is no equivalent word in English. Dharma is not reducible to religion, as has often been erroneously done. It is the "right way of living" and "path of righteousness." All religions that grew from Indian soil—Hinduism, Buddhism, Sikhism, and Jainism—refer to dharma.

In *Hinduism*, dharma signifies the "right way of living" aligned with *rta*, the right order that makes life and the universe possible. In *Buddhism, dharma/dhamma* means "cosmic law and order." In *Jainism*, dharma refers to the teachings of the *Jinas* (deities) for the moral transformation of human beings. For Sikhs, the word *dharm* means the "path of righteousness." In hymn 1353 Guru Granth Sahib refers to dharma as duty.

The etymological root of the word *dharma* is *dhr*—to hold, bear, support, maintain, keep, carry—and its meaning is "that which holds," "that which sustains." Dharma encompasses the universe and the whole creation, the earth and the entire earth family (*Vasudhaiva Kutumbakam*)—from the microcosm to the macrocosm, from the tiniest microbe to the largest mammal—all of humanity from individual people to their relations with other human beings and nature. Because dharma holds and sustains the earth community and all of humanity it embodies the principle of unity—of humans with the rest of nature, and of humans across our diversities. Dharma arises from the interconnectedness of all life and from our duty to care for all humans and all species. The opposite of dharma is *adharma*, the violation of rta, of the ecological laws of the planet, and of the care for fellow humans. Whatever separates us from nature and from one another, whichever actions lead to the disintegration of ecosystems and societies, is adharma.

Dharma guides us in choosing between right and wrong actions. Into it are built ethical assessment, environmental assessment, technology assessment, socioeconomic assessment—all measuring our impact on nature and society by our choices of tools and instruments. If our choices contribute to "holding together," to integration, then they are dharmic. If they lead to ecological destruction and social displacement, then they are adharmic.

The great *Mahabharata,* sacred Hindu texts, have at their core the land and earth that sustain us, the food that nourishes us, and the systems of production that provide our livelihoods and basic needs. Whether we tread lightly on the earth or not, how we grow and produce our food, and how we define the economy (Aristotle called it *oikonomia*—the art of living) are all intimately intertwined with right living and right livelihood.

The growing and giving of food in abundance is the highest dharma. The dharma and discipline of food in India is in the *langars* (Sikh kitchens) at the *gurdwaras* (Sikh temples), and in the feeding of the poor found in every culture and tradition. My colleague Jatindar Bajaj, with whom I studied particle physics at Punjab University, along with M. D. Srinivas, another particle physicist, has written the most comprehensive book we have today on the dharma of food—*Annam Bahu Kurvita: Recollecting the Indian Discipline of Growing and Sharing Food in Plenty*. Their citations from ancient texts resonate with the latest in the science of agroecology, that everything is related and that the web of life is the food web. Food is life; therefore it is sacred. It is not just a commodity: "Food is indeed the preserver of life, and food is the source of creation. . . . All beings are formed of *anna* [food]; anna arises from the rains, the rains arise from *yajna* [ritual offering], and yajana arises from karma action. . . . Out of the contented Earth grow the food crops, which sustain all life. Flesh, fat, bone and marrow are formed of these alone" (5–6). Moreover "it is of anna that indeed all beings are born, it is from anna that they obtain the necessary sustenance for living, and having lived, it is into anna that they merge at the end" (161).

From the Upanishads we learn that good food is the best medicine. As Bajaj and Srinivas relate: "All that is born is born of anna indeed.

Whatever exists on earth is born of anna, and in the end merges into anna. Anna indeed is the first born amongst all beings; that is why anna is called *sarvausadha*, the medicine that relieves the bodily discomforts of all" (152). They also tell us that the growing and giving of good food is the highest dharma: "Do not look down upon anna, food. That is the inviolable discipline for the one who knows. *Prana*, the winds of life, are indeed anna, and *sarira*, the body, is the partaker of anna. Sarira is secured in prana, and prana is enshrined in sharira. Being dependent on each other, the two are anna for each other, and thus indeed anna itself is secured and enshrined in anna. . . . Do not neglect food. That is the inviolable discipline for the one who knows. Because, water is indeed anna, and *jyoti*, fire, is partaker of anna. Fire is enshrined in water and water in fire. Thus it is that anna itself is enshrined in anna" (165–66).

The growing and serving of good food means the knowledge, intelligence, and capacity to know the difference between good food and bad—between systems of growing food that hold the web of life together and sustain the earth, and those that push species to extinction and cause ecosystems to collapse by transgressing planetary boundaries due to ignorant, limitless greed. It means knowing which farming system will improve the well-being of farmers, and which system will disintegrate agrarian societies, pushing our farmers into debt, distress, and suicide. It means knowing and assessing which culture of eating will improve our health and well-being, and which culture will spread hunger, malnutrition, and disease epidemics.

The dharma of food guides all these plural, diverse assessments of food, agriculture, seeds, and biodiversity, as well as our relationship with the soil and the land. The contributors to this book have articulated the dharma of food across various religious traditions to help readers understand both food and religion at a deeper level—and through a deeper understanding, sow the seeds of sustainability, justice, and peace.

Vandana Shiva

Introduction

Religion, Agriculture, and Sustainability

Todd LeVasseur

Distinct practices of eating are at the heart of many of the world's faith traditions. In Western traditions, Jews observe kashrut, Christians gather around a Eucharistic table, and Muslims observe a monthlong season of Ramadan. In Eastern traditions, Hinduism and Buddhism offer rich symbols and practices for relating to the earth, each other, and the cosmos through practices of *puja* (offerings to deities, often in the form of food), vegetarian, and ascetic eating. How we eat matters because eating is an intimate act for human beings, establishing our connections to natural, social, and ultimate reality. What we eat, how we eat, whom we eat with express—often more clearly than our verbal piety—what we value most and think of supreme significance.

Religion and Sustainable Agriculture: World Spiritual Traditions and Food Ethics pulls back the curtain on food's *consumption* and examines more closely food's *production*. Religious traditions have much to say about what and how we eat. Do they also have anything to say about what we grow and raise and how we go about doing that? This line of inquiry has not been very well developed in research and writing. This book shows that practitioners of religious traditions care about food's production because they care about the world that is the source of our daily sustenance. In other words, agriculture is never simply a technical exercise for food's procurement. It is also a daily and necessary exercise in which we learn to nurture, receive, and share the gifts of life and world. Agriculture is, in the many senses of the term, an act of faith and an embodied and economic expression of what people care most about.

This volume presents the work of scholars, theologians, food activists, and lay farmers, all of whom explore how religious beliefs, values, teachings, and practices influence and are influenced by the goals, practices, and values of sustainable agriculture. This collection of essays, global in scope, shows the diverse potential and challenges emerging within what can be called a green/sustainable religious farming milieu. It takes up the issues of postcolonialism, indigenous rights to land and seed, the transmigration of environmentally sustainable religious values and beliefs, the polysemous meaning of "sustainable agriculture," the appropriation of green philosophies, and the complex demands of mindful and compassionate eating.

The impetus for this book came in 2009 while I was undertaking research for my dissertation on what I term *religious agrarianism*.[1] I noticed huge gaps in the extant literature; there was minimal mention of contemporary religion/agriculture issues, including sustainable agriculture issues, in leading journals and books.[2] It seems that the majority of scholars producing monographs or studies in the humanities are extremely competent in talking about religion, but not necessarily about agriculture (although, thankfully, this is quickly changing, as seen in the creation of the "Religion and Food" group at the American Academy of Religion). Or, as members of the growing subfield of "religion and nature/ecology," scholars offer sophisticated studies of religion/environment issues, but within this subfield there is still little work on religion and sustainable agriculture. The same is true of social science literature; here, many studies exist about agriculture, and sustainable agriculture, but only a minority attempt to tackle the tougher "values" element by researching why farmers, consumers, or legislators advocate for sustainable farming. When the values element is incorporated into social science studies, it tends to focus on ethics or political values—very rarely on religious values and identities.

The time in which most people could hunt and gather their food is long gone. To eat, we must now grow the bulk of our food. Given a growing global population, that means questions of agriculture will continue to be central to the economies and cultures of people around the world. Given the widespread degradation of lands and waters, that means questions of *sustainable* agriculture will have to assume central importance for all people. This book deepens our understanding of the diverse ways in which food's production and consumption can be sacred acts and practices that contribute to the healing of the world. In so doing, the volume helps fill in some of the gaps in current academic literature about religion and sustainable agriculture.

The following questions are central guides to each individual contribution to the book:

- How are the religions of the world contributing to a maturing sustainable agriculture movement?
- How are sustainable agriculture projects unique to both their bioregion and to the nations within which they are located? And, conversely, how are these projects similar on a global level?
- How and why are farmers and citizens motivated by religious belief, faith, ritual, and/or teaching to either undertake or participate in sustainable agricultural practices?
- What criticisms of global trade agreements, agricultural subsidies, and corporate farming practices are inherent within faith-based sustainable agriculture initiatives?
- How do such farmers compete with conventional farmers for market shares and profitability?

Because people must eat to live, there is no such thing as a postagrarian society. That raises two questions: what are the indispensable ecological and agricultural conditions necessary for good food to be produced over the long term, and how might religion play a role in creating, nurturing, and sustaining such conditions?

Though there are remnant pockets of gatherer-hunters, and an even larger aggregate of nomadic pastoralists, a perhaps surprisingly great number of humans are peasant farmers: Jan Douwe van der Ploeg estimates that there are roughly 1.2 billion peasants farming small plots of land.[3] But a majority of the world's population is dependent upon industrial agriculture products, either in full or in part, for its daily caloric needs. This majority includes the powerful elite of our globalized world and almost all the citizens of the Global North who, given their rates of consumption, bear a disproportionate duty to ameliorate the emerging crises caused by industrial agriculture: the harmful environmental impacts; the abuse of animals, soil, and farm workers; the major contributions to human-induced climate destabilization. Indeed, the impact of industrial agriculture on the biosocial planetary commons precipitates one of the goals of this book: to situate these effects within the emerging ecocrisis of the Anthropocene. Another premise immediately follows: that by better understanding alternative agricultural models and the religious worldviews that underpin these models, policy makers, religious communities, and farmers can move toward adopting sustainable farming techniques, as these are inherently better for the planet and society, based on a variety of metrics (see below).

We need not belabor the astounding amount of evidence that has accumulated in the past few decades in regard to the human impact upon the earth's various biomes and biogeochemical processes. In fact, it is studying the social and environmental impacts of humans, coupled with the normative goal of sustainability (and especially agricultural sustainability), that unites our scholarship and that in large part stimulates the publication of this book. Nonetheless, a quick list will suffice to once again remind readers of our perilous situation: the sixth-largest extinction process in the history of our planet; massive deforestation; loss of freshwater sources; changing weather patterns (including rain and snow, a huge issue for agriculture); overpopulation; overconsumption; loss of genetic diversity of wild plant and animal species; the emergence of failed states; terrorism; and the seeming inability to regulate international finance/globalization. All of these issues are anthropogenic in origin, and they add up to create the even larger issue of climate destabilization. Most climate specialists, and those involved in climate policy and those who work in the insurance industry, readily accept that the planet is on track to become two degrees Celsius warmer within this century, regardless of what we do to mitigate climate change from here on out. Worst-case scenarios predict a rise of five degrees.

The impacts of these nested and interlinked processes are being felt by all life forms the world over. In terms of Homo sapiens' need to imbibe calories, the implications are startling. The oceans are overharvested and becoming more acidic. The latter will most likely cause the collapse of the remaining aquatic food chain, at least in terms of viable scales for providing humans with calories. As for terrestrial food chains, and here we are most concerned with agriculture, since most arable land is used now either for urban living or growing the food and producing the materials that allow urban living, we know that for each rise in degree of temperature Celsius, there is a correlated decrease in agricultural production of approximately 10 percent. This presents a sobering conundrum: how to provide for the projected 9 billion humans who will be alive in 2050, on a planet that will be one to two degrees warmer and produce 10 to 20 percent less terrestrial-based food?[4] As authors and editors, and speaking for both our contributors and the University Press of Kentucky, we are alarmed by these figures, and believe that agricultural production is the key to understanding (and solving, as much as possible) these interconnected environmental and social issues. We aim in this volume to better understand the dominant industrial paradigm and, more important, to study and build a foundation for better understanding religiously inspired sustainable agriculture.

Industrial Agriculture and the Green Revolution

In this book, we will operationally define agriculture as a deliberate intervention into either naturally occurring or human-managed ecosystems, with the goal of breeding and producing a secure supply of calories for, ultimately, human consumption. Humans have practiced some form of agriculture for thousands of years, creating various technologies and management regimes geared toward producing plant- and animal-based calories.

In recent years, archaeologists have discovered more nuanced understandings of early gatherer-hunters and their food needs, including means of food storage, trade, and cultivation, so that our knowledge of early human diets is becoming more robust.[5] Furthermore, archaeological and anthropological evidence suggests that a key shift occurred in agriculture approximately ten thousand years ago. Beginning at this point in time, and occurring in various places around the planet over the next few thousand years, the onset of agricultural civilizations began, with their formation based upon the domestication of a few key plants and animals. Cristiano Grottanelli writes that these included a wheat zone, stretching from Europe to China, where barley, spelt, and oats were domesticated and grown with the aid of the plow; a rice zone, stretching from India to Indonesia to southern China, aided by the plow and hoe; a sub-Saharan Africa zone, where domesticated millet and sorghum were cultivated by using a hoe; a yam and taro zone in

Oceania, where coconut palm and sago palm were also domesticated via the digging stick and hoe; and a maize/corn zone in the Western Hemisphere, along with beans and pumpkins in the North and potato and quinoa in the South, with their domestication aided by the use of digging sticks and spades; and finally, a manioc/cassava zone in the Amazon and tropical Americas, with secondary domesticated calories derived from peanuts and sweet potatoes, spread and cultivated by digging sticks and spades.[6] After the cultivation of these various grains, tubers, and pulses came the domestication of animals, including sheep, goats, cows, chickens, llamas, rabbits, and pigs, so that taken together, these domesticated forms of calories created the context in which human numbers began to grow and societies became stratified, warlike, and hierarchical. All of the major world religions encountered today are also post-agricultural revolution constructs, which have spread around the world often hand in hand with agricultural seeds and technologies. These same domesticated plants and animals also form the backbone of today's industrial agriculture, with corn, soy, rice, potatoes, sugar, and wheat proving to be the key crops; cows and chickens the key animals; and also coffee and cotton being the key commodities produced and traded the world over.

Today's modern agricultural system, commonly known as "industrial agriculture," is the system of agriculture against which the contributors to this book are researching and writing. Industrial agriculture is based upon this ten-thousand-year-old history of both genetic stock and farming knowledge, but its advent has radically changed the trajectory of agricultural practice: today it is also a form of agriculture based upon the fossil fuel economy. The addition of fossil fuels and industrial technologies, like tractors and combines and refrigerated forms of transport, has turned agriculture into an industrial process carried out at a massive scale on farms that are often thousands of acres in size—but that grow only one or two commodity crops in the form of a "monoculture." Moreover, it is a process that is entirely dependent upon inputs of oil and gas in the forms of herbicides, biocides, insecticides, fertilizers, and fungicides, as well as fuel for transportation of industrial agriculture products. All of these inputs are added to farmlands so that the few key crops mentioned above can be grown and traded in global markets. Some scholars claim that agriculture, and especially domesticated agriculture based on a few key crops, can by definition never be sustainable.[7]

Some critics of industrial agriculture lament its perceived impacts on social health, focusing especially on the influence of both free trade agreements and proprietary agricultural patents and knowledge owned by international seed and agribusiness corporations. Such critics charge that international free trade agreements erode both democracy and the ability of small-scale peasant farmers to maintain their lifeways with dignity.[8] These critiques rest on a foundation of recognizing unequal global power relations that have existed since the beginning of the Columbian Exchange, which became the basis of modern industrial agriculture. This exchange was created when Columbus "reknit the seams of Pangaea" and brought both the potato and guano, "the world's first intensive fertilizer," to Europe.[9] This same process also began the transatlantic trade in human flesh—slaves from Africa and indentured servants from Europe—and led to the Dust Bowl in the United States in the 1930s, created over many decades when indigenous farming and grazing were systematically and violently replaced by monocultures of wheat, aided by the plow and mechanization of agriculture. The history of agriculturalists replacing small-scale herders and nomads and pastoralists and polycultural (in terms of a variety of plant and animal species being cultivated) farmers is a long one, and one that critics claim continues to this day. Readers will encounter many of the above-mentioned criticisms, and much of the history of unequal power relations, in the chapters that follow.

However, today industrial agriculture is buttressed by the World Trade Organization and other national and international political bodies that often rule in favor of streamlining world agricultural markets, and the North American Free Trade Agreement and other neoliberal trade agreements that contain farm subsidies and give knowledge and patent rights to agribusiness corporations, which so far has resulted in these agribusinesses gaining more and more shares of the global food market.[10] The impact of industrial agriculture on social and political structures the world over is another emergent theme in this book's chapters.

On one level, agriculture is built into the fabric of modern nation-state politics, first as a result of the Columbian Exchange; its inclusion was further solidified during the aftermath of World War II, when war technologies based on oil and natural gas (the latter is the basis of industrial fertilizers used to make modern farm crops grow, based on the Haber-Bosch process of producing ammonia-based nitrogen) were applied to agricultural fields. This use of technology precipitated a mass migration in developed countries off farmlands and into cities, where the development of an urban and suburban leisure class ensued. The calories to support this lifestyle were made possible by the "Green Revolution" technologies that form the basis of industrial agriculture and were coupled with post–World War II trade agreements that were made on the basis of American economic, political, and technological power and exceptionalism.

Furthermore, a unique American trajectory of farming and farm laws has created the context for this book, including the passage of the Morrill Land-Grant Act in 1862 and the Agricultural College Act in 1890, and then technological efforts to address growing global concern about widespread famines that dramatically impacted countries such as Pakistan, India, and the Philippines in the 1950s and 1960s. The result was that agricultural technologies developed before and after World War II in America were disseminated via American philanthropic and government entities to farmers in developing countries. This transfer of farming knowledge and technologies, based on developments of high-yield varieties of wheat, corn, and cotton, all of which require large amounts of chemicals and water to grow, meant that "the Green Revolution was not only an agricultural revolution but also a full-scale social revolution."[11] The impact of Green Revolution industrial farming technologies, and the politics and economics that underlay them, presents another instance in which "Western powers come to the economic, humanitarian, and occasionally military 'rescue' of 'developing' nations as predominantly 'white' civilization rides to the rescue of those needy 'others,'" a trajectory mirrored in the spread of first Christianity and then market capitalism out of Europe; and the results of which are "postcolonial ecologies," an emergent theme and critique that ties together many of the chapters in this book.[12]

The final interlinked issues of this revolution are its impacts on ecological, political, and social health. For some farmers and activists the world over, these issues have religious underpinnings, forming both the key drivers of the growing movement toward sustainable agriculture and the foundation of this book. Agriculture based on only a few key plants or animals is inherently detrimental to ecosystem health, especially over a long period of time. The historical record of this is what drives the critiques of agriculture broadly mentioned above. Contemporary environmental critiques of Green Revolution technologies and high-yielding varieties of seeds that form the basis of industrial agriculture are robust, with mounting evidence over recent years providing strong evidence that the critiques are well founded. One concern is the contribution of industrial farming, especially its dependence on animal agriculture, most notably cows, to overall human-induced climate change; some suggest that "the global system producing and distributing food—from seed to plate to landfill—likely accounts for 31 percent or more of the human-caused global warming effect."[13] Another critique is the impact of industrial agriculture food products on human health, as mounting evidence suggests that high-yielding varieties of crop foods are leading to higher rates of obesity, diabetes, and malnourishment.[14] Agriculture broadly, including especially industrial agriculture but also the expansion of peasant farms (the latter often as a result of global agribusiness economics), has an impact on global biodiversity, soil erosion, carbon storage, water tables, and habitat destruction, so that "conventional approaches to intensive agriculture, especially the unbridled use of irrigation and fertilizers, have been major causes of environmental degradation."[15] The problematic environmental impact of industrial agriculture is summarized by David Pimentel: "Fertilizers and animal manure-nutrient losses have been associated with deterioration of some large fisheries in North America," while "runoff of soil and nitrogen fertilizer from corn-belt corn production has contributed to the anaerobic 'dead-zone' in the Gulf of Mexico," which some studies suggest is now the size of Connecticut.[16]

Perhaps one of the most comprehensive and damning critiques of industrial agriculture, in both its ecological and social forms, is levied by the environmental lawyer Andrew Kimbrell, who writes that industrial food production yields a "fatal harvest." Fatal to consumers, as the massive amounts of chemicals used in growing crops dramatically increase the incidence of cancer and other diseases. Fatal to our rivers, lakes, and oceans that are polluted by pesticides and chemicals. Fatal to the genetic diversity of plants, as countless species are destroyed and replaced with universal high-yield crops and genetically engineered varieties. Fatal to our farm communities, which are wiped out by huge corporate farms. Fatal, potentially, to the entire biosphere as modern agriculture contributes to global environmental threats such as ozone depletion, the greenhouse effect, and mass species extinction.[17]

It is this concern with the fatality of industrial agriculture—fatal to the health of democracy, soils and landscapes, biodiversity, the physical health of both humans and domesticated farm animals, the well-being of all forms of future life on this planet, and most important, to the health of human-nature relations mediated by the context of religion—that motivates the research and writing found within this book. Given that humans must eat, then where might we find an antidote to the interlinked fatalities of industrial agriculture? This book argues the solution is in the technological practice of agricultural sustainability and, as important, in the uncovering of the interlinked values—emotional, affective, psychological, economic, and religio-spiritual—that underlie this method of cultivating calories.

Agricultural Sustainability

The agronomist Gary Fick reminds us to use the term *sustainable agriculture* cautiously because at the end of the day it can be interpreted to mean any form of agriculture. We give this point credence: Monsanto's use of the phrase, for example, is an empty gesture, devoid of meaning considering the company's farming and business practices. Rather, Fick discusses the need to cultivate *agricultural sustainability,* a form of farming that strives to find a balance of three interlinked aspects: (1) to "conserve or enhance" ecological relationships; (2) to nurture social relationships that generate fair wages and holistic knowledge for all involved with food production; and, key to this book, (3) "spiritual and ethical relationships that . . . engender stewardship or care of the natural and social foundations of the food system [and] address injustice, economic abuse, and other social or cultural failures that endanger the continued health of any part of the food system."[18] The farming practices outlined in this book present exactly this form of agricultural sustainability, and this book is the first to specifically bring together case studies that focus on the very key third element of agricultural sustainability.

Thus, agricultural sustainability is a method of producing calories from domesticated plants and animals in which the underlying view of human-nature relations is holistic and restorative, not mechanistic and reductionist. Farmers who attempt to grow food in a sustainable manner recognize that they intervene in natural systems, yet rather than basing this intervention on concepts of human superiority that typically lead to dominating these systems, as is the modus operandi of industrial agriculture based on its metric of increased yields, such farmers work with and mimic the systems found in nature.[19] Farms engaged in agricultural sustainability exhibit several common characteristics: they tend to be polycultural, smaller in scale, and less mechanized; they typically eschew the fossil fuel-based inputs upon which industrial agriculture depends; they preserve and save heirloom varieties of seeds and animals, and use cover cropping and crop rotation and rotational grazing; they treat their workers fairly; and, most important, they compost their "wastes" so that a closed-loop system is created on-site. The overall goal of these farming methods is to create, maintain, and enhance soil fertility, as soil health is the basis upon which agricultural sustainability rests. This insight was recognized by Sir Albert Howard, who wrote on the first page of his classic *An Agricultural Testament*, "The maintenance of the fertility of the soil is the first condition of any permanent system of agriculture. In the ordinary processes of crop production fertility is steadily lost: its continuous restoration by means of manuring and soil management is therefore imperative."[20]

Yet, there is one more link of even greater significance—the health of the soil is codependent upon the health of society, so that a sustainable society will by definition practice agricultural sustainability. This book enters into this dialectic by focusing on the variable of religion, highlighting how practitioners all over the globe are interpreting and putting into practice their religions in order to help generate resilient biosocial farming systems.[21] Each chapter takes a unique look at the interaction between religion and sustainable agriculture in a particular geographic region and as practiced by farmers influenced by varying religions, helping us to better understand the human, and especially religious, motivations behind practicing sustainable rather than industrial agriculture.

Brief Introduction to Chapter Discussions

In chapter 1, Leonor Hurtado Paz y Paz and Cristóbal Cojtí García take us to the heart of Maya political ecology in Guatemala and interrogate it vis-à-vis the sanctity of the *milpa* system and corresponding agroecological cosmovision. The chapter presents a dialogue between what is possible within the milpa system and the long history of discrimination against the Maya: from their dispossession from their ancestral lands to the excesses of the Green Revolution to recent massive investments by USAID aiming to incorporate peasant agriculture into the global market. The authors equate such externally induced erosion of peasant agriculture with an attack on Mayan spirituality. The recent revival of Mayan spirituality is discussed in the context of the science of agroecology and initiatives to create community-based food sovereignty.

Frédérique Apffel-Marglin (chapter 2) brings a promising story from the Peruvian high Amazon, where she is active in what she calls making and bioculturally regenerating *yana allpa*, a Kichwa-Lamista term for Amazonian Dark Earth (ADE), also known as *terra preta do Indio* in Portuguese. Through the Sachamama Center for Biocultural Regeneration, which she founded some five years ago, the Chacra-Huerto project is already working with four native communities and six local schools. The chapter weaves the conversations the author has had with Kichwa-Lamista indigenous communities and their elders with serious exploration of the properties of ADE, including the biochar and the broken pieces of ceramics (called *shaño* in Kichwa) found in those ancient soils. The core of this chapter is a captivating narrative of this anthropogenic soil, whose existence in the Amazon could have been as early as five thousand years ago, or even earlier. The author describes the significance of the Kichwa-Lamista culture of reciprocity/respect and how and why they present offerings to four deities associated with the *chacra* (the agricultural field): Mama Allpa, the earth; Yakumama, the rain/water; Mama Qilla, the moon; and Tayta Inti, the sun. For the author, "These reciprocal exchanges of gifts enact a *regenerative cycle,* which constitutes the opposite of the extractive actions characteristic of a capitalistic economy." She then articulates this act as an essential feature of the cosmocentric cosmologies of indigenous communities. In this cosmocentric universe, "the soil becomes Mama Allpa, a being to whom prayers and offerings are made, who is endowed with understanding, agency, and sentience, and who responds to the actions of humans." Mama Allpa is both a spirit and the soil/earth; Mama Qilla is both the moon and a spirit. The offering is not simply a gift exchange among humans; it is an exchange between humans and the nonhuman entities who also reside in the chacra. Additionally, the chapter discusses the significance of yana allpa in addressing deforestation in the Amazon, sequestering greenhouse gases, and contributing to community food sovereignty.

Chapter 3 on the Malawian permaculture initiative, written by the Reverend Eston Dickson Pembamoyo, offers a vivid example of how the values and skills embedded in permaculture are introduced and practiced in Malawi to regenerate *malimidwe a makolo*—ancestral farming methods. The ritualistic nature of such traditional techniques in Malawi are contrasted with the mechanistic methods of the Green Revolution. The author approaches permaculture as a corrective, a religiously infused path to bring authentic abundance back to Malawi, where the Green Revolution has severely impacted soil, seeds, and productivity. How can coordinated human initiatives and actions help to arrest food insecurity and create a sustainable supply without destroying the land? What can be done generally to develop designs people could use in order to combat hunger and poverty? The chapter also blends permaculture principles with theological traditions. The progress of this initiative since 1994, and lessons it yields from one of the younger African nations and one of the remaining agrarian economies of the world, is worth watching and celebrating.

In his chapter 4, "Nature Spirituality, Sustainable Agriculture, and the Nature/Culture Paradox," Michael Lemons pulls on years of participant-observation and permaculture expertise to weave a "dramaturgic" analysis of the permaculture subculture located in Puna, on the Big Island of Hawaii. Lemons's chapter weaves together summaries of lifestyles and the motivations behind them of various participants in Puna's unique permaculture subculture, exploring how these are indicative of a nature/culture duality that underlies the attempts of those on the Big Island to generate a localized understanding of permaculture and thus sustainable farming regimes. The author also adds a unique methodological lens to the book, in that his analysis is driven not only by qualitative research methods but by statistical, quantitative methods that allow us to gain a better understanding of the various motivations that inspire permaculture practitioners to engage in sustainable agriculture practices in Puna. Lemons expertly situates his findings in a variety of academic domains and literature, from anthropology to religious studies to postmaterialism and literature on human religio-spiritual views of nature. Lemons concludes that the urge to engage in a back-to-the land type of sustainable agriculture in Puna is very much motivated by connections to nature whereby nature is seen and felt to be sacred and holy, and these religious motivations and sensibilities thus inspire the practice of permaculture. Lemons's essay adds a nuanced perspective to the book as a whole, highlighting differences between the practice of permaculture in the Global North and the Global South, where different material, political, and religious backgrounds create localized understandings and practices of religiously inspired sustainable agriculture.

Jagannath Adhikari (chapter 5), a longtime researcher on the agroecology of Nepal and a frequent commentator on food and agricultural policies/trends in Nepal, assesses Hindu/Buddhist traditions and peasant farming in the Himalayan foothills of Nepal. The beauty of this chapter is the live dialogue Dr. Adhikari creates between the Vedic and Hindu textual wisdoms and the everyday practices of smallholding peasants and farmers. After outlining a Vedic system of an agricultural household and some remnants of everyday beliefs such as *dharma* (merit), *paap* (sin), and *karma* (deed), and live but declining practices such as *satjbeej charne*, the chapter concludes on a sobering note. The religious values of the indigenous farmers might not suffice to counteract the prevailing trends toward chemical and industrial agriculture propelled by Nepal's rapid pace of modernization and urbanization. The author's closer examination of the Pokhara Valley in Nepal testifies that the use of land for farming has become a losing proposition due to the region's recent real estate boom and allocation of land for other economic activities. While Adhikari acknowledges that peasant farming has been substantially displaced and eroded, regeneration of what he calls the *kisan dharma* holds deep promise in the future agrarian and food economy of Nepal.

In chapter 6, Pankaj Jain explores *dharmic* ecology and brings unique voices from the Gujrat and Maharashtra states in western India: practitioners of the new religious movement Swadhyaya, which literally means "self-study." This movement challenges both the utilitarian and protectionist motives behind environmental efforts. Instead, Swadhyayis attribute environmental concern to their devotional motive and a search for the "Indwelling God." Building on the statement of Pandurang Shastri Athavale, founder of the Swadhyaya movement, that dharma is "that which sustains both the personal order and the cosmic order," Jain argues, in the case of Swadhyaya, that the concept of dharma can be "successfully applied as an overarching term for the sustainability of ecology and agriculture." Jain argues that Swadhyayis strive to develop their dharmic teachings into practice by developing reverential relationships with the trees, water, cows, and other ecological resources of their community. Some examples of this are how Swadhyayis have collected rainwater on their roofs as well as recharged their water wells and rivers. What they call *Yogeśvara Krsi* (agriculture) includes four components: trust of neighbor, love for animals, faith in God, and respect for nature.

A. Whitney Sanford's chapter 7, "Gandhi's Agrarian Legacy," interprets official texts and the lived lives of members of Brahma Vidya Mandir (BVM) and associated farms such as Samvad and Nilayam Nivedita. The author describes everyday life on these farms, placing them within the context of the values of the *Bhagavad Gita*—a central Hindu text—and the legacy of Mahatma Gandhi. Established in 1959 by Vinoba Bhave (Gandhi's spiritual successor), BVM is a women's intentional community whose members consider the *ashram* their spiritual and social laboratory. As they attempt to address the failures of modern agriculture, their core principles are geared toward inculcating the values of self-reliance, self-sufficiency, nonviolence, bread labor, and self-discipline in a community setting. The story of BVM highlights an irony of modern India: while many Indians revere Gandhi as a national hero, they also proudly proclaim the country's large and growing middle class, who have rejected Gandhian austerity in favor of US-style consumerism.

Alexander Harrow Kaufman (chapter 8) digs deeper into what farmers and peasants in Thailand call the "Moral Rice" culture and network they have created over time. Within these networks there is a fertile debate between whether to accept the global organic rice certification or to replace it with a standard based on the teachings of the Buddhist Five Precepts. Since the goal of Moral Rice is to cultivate Buddhist values among organic farmers, one can become a member of the Moral Rice Network by joining the Dharma Garden Temple, learning to sacrifice oneself, and giving to others by joining the savings cooperative. Akin to what Frédérique Apffel-Marglin describes in the Peruvian high Amazon, Kaufman portrays how Thai farmer households pay reverence to Khwan Khao (Rice Soul), Mae Phosop (Rice Mother), Mae Thoranee (Earth Mother), and Mae Khongka (River Mother) through rituals and offerings. Rituals are performed on behalf of specific deities at each stage of the process: plowing, sowing, transplanting, harvesting, and threshing. The Asiatic water buffalo, which is used in farming, is also ritually thanked for its contribution to the rice culture. Overall, this chapter offers an excellent case study of the biosocial evolution of sustainable rice farming in Thailand.

Elaine Solowey's chapter 9, "The Seven Species and Their Relevance to Sustainable Agriculture in Israel Today," shares the insights of one of Israel's leading plant scientists, who is working on ways to cultivate sustainable forms of perennial agriculture in the region's arid climate. In this chapter, Dr. Solowey combines biblical exegesis, focusing on the sacred Seven Species of the Hebrew Bible, with her own experience as a Jewish farmer/plant researcher in Israel, crafting a compelling narrative about how the traditional ecological knowledge of the early farmers of the region has much to teach modern-day farmers in the quest for sustainability. She examines ecological, religious, cultural, and historical facets of traditional indigenous farming, applying the lessons of ancient documents and practices to the creation and maintenance of sustainable ways of growing food in Israeli agriculture today. Overall, this chapter provides a comprehensive history of how agriculture and religion have intermingled in Israel, and what this history might suggest for how to farm sustainably in an area of the planet that will have to navigate changing weather patterns brought about by anthropogenic climate change.

Yigal Deutscher, in chapter 10, "Tending the Garden of Eden: Sacred Jewish Agricultural Traditions," grapples with the implication of the biblical narrative in the Hebrew Bible in which the early Israelites were commanded to farm and steward the Promised Land. Inspired by his own experience practicing permaculture and other forms of sustainable agriculture in both Israel and North America, Deutscher articulates a modern Jewish agrarian hermeneutic that attempts to recapture the relevant teachings and insights about sustainable farming practices contained within the Torah. Deutscher argues that a clear blueprint is offered to the ancient Israelites, and this blueprint is one that modern-day farmers, especially Jewish farmers, can follow if they wish to restore soils and health to both farmland and cultural institutions. Unfortunately, through history, and especially with the onset of industrial agriculture, this blueprint has been neglected. Deutscher's hope is that the biblical teachings of rest and release; the concept of sacred land; the use of perennials; and other biblical insights can move Jews, and society at large, toward a more sustainable form of agriculture.

Anna Peterson's chapter 11, "Religion, Local Community, and Sustainable Agriculture," investigates sustainable agriculture at the scale of community, and specifically collectives of humans united under shared religious identities. One such community is the Old Order Amish, another displaced Roman Catholic peasants in postwar El Salvador. The Amish have a long tradition of low-tech farming in North America, but they face issues of rising property values and the need to navigate the modern economy while maintaining fidelity to religious views of work and technology. Peterson shares how close-knit Amish communities, held together by shared beliefs that inform their farming practices, are able to create an estimable track record of eco-friendly farming (when compared to metrics of industrial agriculture) that is more profitable and requires lower inputs of nonrenewable forms of energy—key attributes for any type of agriculture to be sustainable in a world of climate change and peak oil. Amish farmers are not "organic purists," as they do use small amounts of chemicals, but this use is part of a larger management regime that relies on more traditional practices, such as crop rotation and the planting of polycultures, as is evidenced in their healthy levels of topsoil. The key insight shared by Peterson is that "religious priorities and values shape both social and agricultural choices" for the Amish, demonstrating the clear role of religion in forming how some humans choose to produce food. Similarly, Peterson finds that small communities of Roman Catholic "repopulators" in El Salvador are also inspired to utilize practices associated with sustainable agriculture regimes, but their choices are shaped by postcolonialism, national policies (including the fallout from a civil war), and local religious identities, including liberation theology. Thus, for repopulators, sustainable agriculture goes hand in hand with democracy, social justice, and economic security. Their success in generating resilient farming practices is hampered by the lack of needed structural and democratic changes in El Salvador. Despite their geographical and historical differences, both case study groups can be classified as peasants, as they are low on the "energy food chain," and so are marginal groups within a larger capitalist society. Yet members in both groups use religious teachings and beliefs to organize their identity, and these also shape their farming practices; both groups also are critical of larger trends in society: the selfishness and wastefulness of American society, and the subjugation of peasant voices in El Salvador. Peterson expertly teases out implications of scale, levels of technology, economics, and politics to extrapolate possible sustainable agriculture solutions that are grounded in small religious communities, thereby adding a unique perspective to the book.

Chapter 12, "Heideggerian Reflections on Three Mennonite Cookbooks and a Mennonite Farm in Northwest Ohio" by Raymond F. Person Jr. and Mark H. Dixon, explores possible answers to the question "How do we, as human beings, interpret our connection to the earth as embodied and emplaced beings, and live our lives accordingly?" Person and Dixon writes as both scholars and insiders to the Anabaptist tradition; whereas Anna Peterson focuses on Old Order Amish from within this tradition, Person and Dixon explore theological insights about sustainable agriculture from the perspective of Midwestern (USA) Mennonites. They find answers in an "embodied theology" of simple living that underlies many Mennonite teachings and practices. Combining spiritual inspiration taken from within the Mennonite tradition with insights from the philosopher Martin Heidegger and his work on being, dwelling, and building, Person and Dixon tease out ethical values regarding how we should dwell and build as farmers, basing our activities on tenets of sustainable agriculture. The authors find evidence for sustainable dwelling and building in Mennonite cookbooks and the work of a small group of farmers tending a Mennonite homestead in Ohio. Person and Dixon's application of Heidegger's thought provides a needed engagement with the Western philosophical tradition, especially with phenomenology, and the implications of this for a religious sustainable agriculture; yet their deft treatment of Heidegger is grounded within their own work as farmers as well as within larger Mennonite views of creation and the human enterprise of food production and consumption.

Ragan Sutterfield's chapter 13, "Steward or Priest?" presents the "insider" work of a theologian who grapples with the fundamental question of what the human role in creation should be: that of "steward" or "priest." Taking the practice of saying grace before mealtime as his starting-off point, Sutterfield asks, "In a time when our food systems are not always good, when our meals represent sins against the creation, how can Christians bless a meal that represents a desecration as much as the blessings of God?" Sutterfield rightly points out that grace is really a second-level concern; the real issue is how the food that is being blessed was produced. Comparing the industrial stewardship paradigm of production (typified by megacorporation Tyson Foods) with the grace-giving priesthood model (exemplified by Joel Salatin of Polyface Farm), the chapter explores how cultivating a Christian sense of being priests presiding over creation can inspire an emergent Christian sustainable agriculture. Sutterfield expertly teases out the implications of these two models, on both religion and the planet, by focusing on the practice of raising chickens, rightly concluding, "What is at stake [in how we farm and the religious values behind this] isn't simply our moral status, but our very life. That is what 'sustainable agriculture' means—a way of farming that continues and draws forth the being of creation."

Chapter 14, by Maximilian Abouleish-Boes, explores the Muslim farming and business group SEKEM, based in Egypt. It briefly provides the context in which SEKEM originated, including a short history of the life of its founder, Dr. Ibrahim Abouleish. SEKEM's remarkable achievements in fostering sustainable agriculture and biodynamic farming in Egypt and abroad are detailed. For example, SEKEM has succeeded not only in reducing synthetic pesticides in its own operations but, in cooperation with the Egyptian Ministry of Agriculture, in cutting chemical use by more than 90 percent on Egyptian cotton farms. SEKEM's basic values are informed by a blend of European culture and philosophy with principles of Islam. The organization draws from the core Islamic values of *tawhid* (unity), *fitra* (natural state), *ilm* (knowledge), *ihsan* (beauty and excellence), *khalifa* (stewardship of the earth), and justice (*adl*) to embrace an agricultural ethic that is "future oriented and sustainably competitive."

Tristan Reader and Terrol Dew Johnson offer another perspective on indigenous farming in their chapter 15, "Tohono O'odham *Himdag* and Agri/Culture." They paint a vibrant ethnographic picture of the traditional farming methods and lifeways of the Tohono O'odham (the People of the Desert) of Arizona's Sonoran Desert. They discuss the traditional ecological knowledge of the Tohono O'odham, and how this knowledge is encapsulated within a functional cosmovision in regard to the harvesting of the fruit of the saguaro cacti as well as other important agricultural activities. As they explain, for the Tohono O'odham, "Food is health. Food is culture. Food is community. Food is the basis of a sacred economy of what you grow, what you harvest, and how you share." As is seen in other chapters of this book, the separation of agriculture, religion, culture, and nature is a false distinction—there is an explicit recognition that the health of all are entwined. This is evident for the Tohono O'odham in the traditional songs they sing, the rituals they undertake, the organization of their farming calendar, and the structure of their community, all of which are embedded within their food culture. As is the case with other indigenous farming agricultures covered in this book, the Tohono O'odham, too, are having to navigate postcolonial and post–Green Revolution lifeways.

The conclusion to this edited volume is written by coeditor Pramod Parajuli. He provokes readers to imagine an agricultural food system in which what he calls the "'soil to supper/sustenance and back to soil' continuum" is not only sustainable but also abundant and resilient—and he argues that its creation is still within our reach. Within the rubric of "searching for Annapurna," he develops a theoretical framework that looks at the intersection of biosphere, ethnosphere, and learningsphere in the context of humanity having entered an Anthropocene epoch in which we have exceeded planetary boundaries. He proposes that various religious traditions, including those covered in this volume, have a role to play in an emergent regenerative food system. However, the severity of our current situation demands that we reimagine and reconfigure what religion is and give room to the diversity of "earthbound" religiosities, worldviews, cosmovisions, and practices that millions of farmers, peasants, food growers, and gardeners on our planet are already engaged in. This emergent religiosity might nurture a mix of sacred and secular tenets and might even transcend the customary boundaries of religious traditions. Parajuli builds on conversations with his late mother in the Himalayan foothills of Nepal to illustrate the thread between the human household and the earth household. He focuses on the act of building soil and carbon, including Amazonian dark earth (ADE), as the foundation of such earthbound worldviews and cosmocentric visions. Parajuli urges us not to burn and send valuable carbon into the sky but to sequester it in the soil where it belongs, thus making, as William McDonough and Michael Braungart describe it, a "high-carbon diet for the planet." In an emergent abundant and resilient food economy, carbon should not be considered a waste to be minimized but an asset to be enhanced and multiplied. Finally, Parajuli ends the chapter with suggestions to the reader, and the larger scholarly community, about fruitful avenues of exploration that can be built upon the themes and methods of this book.

A final note to the reader before you proceed on your journey through religion and sustainable agriculture: we recognize that this book has its limitations. First, not every guiding research question listed at the beginning of this introduction was systematically addressed by each and every author, although each did her or his best to work through them using the data available. Nonetheless, we feel that the data and themes that emerge are compelling, and that their implications demand scholarly attention. Second, we recognize that although this book is global in scope, it is only a snapshot in time, and it lacks case studies and input from scholars or activists in key parts of the world. For example, we have no chapter on agriculture and religion in China, despite repeated attempts by the editors to track down submissions from scholars, journalists, or activists in East Asia. Another gap in this book is an exploration of what religious institutions and their teachings have to say about sustainable agriculture from a top-down, hierarchical perspective. We hope readers will search out information on religion-agriculture issues in their own communities and geographic areas as well as research that is being generated at institutional levels in religious traditions and published in academic journals or books to augment the rich material covered in ours.

We invite you now to read on to learn about the active, compelling interface between religion and sustainable agriculture around the globe. We hope that in your reading and study of these chapters, you too may find yourself engaged in the noble, needed practice of supporting sustainable agriculture, campaigning for food security and food justice issues in your local community, wherever you may reside.

Notes

1. "Religious Agrarianism," in *The Oxford Encyclopedia on Food and Beverages in America*, ed. Andrew Smith (New York: Oxford University Press, 2012).

2. For good introductory surveys that look at religion-nature interactions, including some mention of agricultural issues, see Martin Tyler and Victoria Finlay, *Faith in Conservation* (Washington, D.C.: World Bank, 2003); Mary Evelyn Tucker, *Worldly Wonder: Religions Enter Their Ecological Phase* (Chicago: Open Court, 2003); Darrell Posey, ed., *Cultural and Spiritual Values of Biodiversity* (London: Intermediate Technology, 1999).

3. Jan Douwe van der Ploeg, *The New Peasantries: Struggles for Autonomy and Sustainability in an Era of Empire and Globalization* (London: Earthscan, 2008).

4. S. Chakraborty and A. C. Newton, "Climate Change, Plant Diseases and Food Security: An Overview," *Plant Pathology* 60 (2011).

5. Alan Outram, "Hunter-Gatherers and the First Farmers: The Evolution of Taste in Prehistory," in *Food: The History of Taste*, ed. Paul Freedman (Berkeley: University of California Press, 2007).

6. Cristiano Grottanelli, "Agriculture," in *The Encyclopedia of Religion*, 2nd ed., ed. Lindsay Jones (Detroit: Macmillan, 2005), 185.

7. Daniel Quinn, *Ishmael* (New York: Turner/Bantam, 1992); Richard Manning, *Against the Grain: How Agriculture Has Hijacked Civilization* (New York: North Point, 2004); Wes Jackson, *New Roots for Agriculture* (Lincoln: University of Nebraska Press, 1985).

8. Vandana Shiva, *Stolen Harvest: The Hijacking of the Global Food Supply* (Cambridge, Mass.: South End, 2000).

9. Charles Mann, "How the Potato Changed the World," *Smithsonian Magazine*, November 2011; William Doolittle, "Agriculture in North America on the Eve of Contact: A Reassessment," *Annals of the Association of American Geographers* 82, no. 3 (1992).

10. Piero Bevilacqua et al., "Manifesto on the Future of Knowledge Systems: Knowledge Sovereignty for a Healthy Planet," International Commission on the Future of Food and Agriculture, http://www.weebly.com/uploads/1/3/6/7/1367341/manifesto_on_future_knowledge_1.pdf (accessed April 3, 2013); Raj Patel, *Stuffed and Starved: The Hidden Battle for the World Food System* (New York: Melville House, 2012).

11. Murray Leaf, "Green Revolution," in *Berkshire Encyclopedia of Sustainability*, vol. 4, *Natural Resources and Sustainability*, ed. Daniel Vasey et al. (Great Barrington, Mass.: Berkshire, 2012), 199.

12. Bonnie Roos and Alex Hunt, "Introduction: Narratives of Survival, Sustainability, and Justice," in *Postcolonial Green: Environmental Politics and World Narratives*, ed. Bonnie Roos and Alex Hunt (Charlottesville: University of Virginia Press, 2010), 2; Elizabeth DeLoughrey and George Handley, *Postcolonial Ecologies: Literatures of the Environment* (New York: Oxford University Press, 2011).

13. Anna Lappé, *Diet for a Hot Planet: The Climate Crisis at the End of Your Fork and What You Can Do about It* (New York: Bloomsbury, 2010), xvii.

14. Patel, *Stuffed and Starved*. Meanwhile, in a recent editorial, José Domingo called for much more robust assessments of the potential risks to both biophysical and ecological systems posed by genetically modified food organisms (GMOs; this is the insertion of a gene from one species into another species, and is not to be confused with selective breeding of a single species). "Human Health Effects of Genetically Modified (GM) Plants: Risk and Perception," *Human and Ecological Risk Assessment* 17, no. 3 (2011).

15. Jonathan Foley et al., "Solutions for a Cultivated Planet," *Nature*, October 20, 2011, 3.

16. David Pimentel et al., "Organic and Conventional Farming Systems: Environmental and Economic Issues," *Environmental Biology*, July 2005, 1, http://dspace.library.cornell.edu/bitstream/1813/2101/1/pimentel_report_05-1.pdf (accessed November 28, 2013).

17. Andrew Kimbrell, foreword to *Fatal Harvest: The Tragedy of Industrial Agriculture*, ed. Andrew Kimbrell (Sausalito: Foundation for Deep Ecology, 2002).

18. Gary Fick, *Food, Farming, and Faith* (Albany: State University of New York Press), 169.

19. Whitney Sanford, *Growing Stories from India: Religion and the Fate of Agriculture* (Lexington: University Press of Kentucky, 2011).

20. Sir Albert Howard, *An Agricultural Testament* (New York: Oxford University Press, 1972), 1.

21. This approach differs from those of other recent works aiming to help us better understand the emerging sustainable food movement, from investigations of the ethical dimension of eating local foods to nuanced ecofeminist critiques of place-based foods. See Gregory Peterson, "Is Eating Locally a Moral Obligation?" *Journal of Agricultural and Environmental Ethics* 26, no. 2 (2013): 421–37; Chaone Mallory, "Locating Ecofeminism in Encounters with Food and Place," *Journal of Agricultural and Environmental Ethics* 26, no. 1 (2013): 171–89.

1

Our Flesh Was Made from Corn

Leonor Hurtado Paz y Paz and Cristóbal Cojtí García

> Ixmukane, our grandmother of creation, ground the white and yellow corn kernels, and formed from the dough the four bodies of our grandfathers: Balam Kitzen, Balam Aq'ab', Majukutaj, and Iq B'alam and then she made nine drinks that became the blood of our first grandparents and parents.
> —*Popol Vuh: Sacred Book of the Quiché Maya People*

The Mayan civilization unites spirituality, science, and agriculture, thus creating an agriculture that is in harmony with nature, the individual, and society, weaving all the elements together as part of the cosmic fabric. Harmony is created by respecting principles that consider the human being as one integral element of the whole system rather than the main or dominant element of that system. In this context, the person respects the life of all other elements, understanding that all elements are alive and have their own mission. Another principle is "You are my other self"—this principle considers that one's life influences another's life and vice versa; thus, one must love the other as much as one loves oneself. And that "you" is not only other people; "you" is nature, water, soil, plants, animals—everything, because every thing is alive. That is why nature and agriculture go together: they look to reinforce each other.

This way of life and agriculture honors existing resources, biodiversity, and the preservation of future generations, and is integrally oriented by both the Mayan sacred calendar, Chol q'ij, and the 260-day lunar calendar, marked by nine lunar phases (the gestation time for human life), together with the solar calendar Ab', the 365-day agricultural calendar. The understanding of time is incredibly important because apart from governing agricultural work, each day offers advice for life, because each day has its own energy and numerical figure; each day has its own charm and secrets, its own name, its *Nagual,* a living being that encourages us; each day is inspired by its cardinal direction, containing the force of one of the essential elements: fire, air, water, or earth. This means that each day has its own life force, one that repeats with the same vitality in a cycle of fifty-two years, when Chol q'ij and Ab' coincide once more. This wealth of ideas, spiritual facts, and materials allows people and their community to be in unity with agriculture and to adjust their actions accordingly.

Guatemala's population is currently 14,713,763, and 60 percent of the inhabitants are indigenous: Mayan, Xinca, or Garifuna, who between them speak twenty-three languages.[1] These groups preserve their ethnic tradition, rituals, and social organization in various and different ways. Guatemala's ethnic, cultural, and linguistic diversity and the rights of indigenous people to live by their own cultural norms were legally recognized in 1995 with the signing of the indigenous rights accords. Since then, indigenous peoples in Guatemala have been reinforcing their cultural and ethnic organization, including in the realm of agriculture.

Historical Mayan agriculture required keeping track of time, which began to have a spiritual, social, and scientific function in which everything belonged to a whole, and agriculture itself wasn't differentiated from the human being, from society, or from the cosmos. In this integration, the union of the cosmos, nature, people, and all living things is valued and respected, as both the interdependency and the complementarity of everything is recognized. For in the end, people, like all beings, are part of the cosmic fabric. The planet is known as Mother Earth, who offers everything that makes life possible: she produces the trees and all of the plants, she provides water and calls down the rain, she houses and feeds the animals who then create music and dance—all of which allows us to feed our communities. Thus, people belong to Mother Earth, not the other way around, for she is not our property; she cannot be sold or bought.[2]

The products of agriculture are granted to us with love from Mother Earth, Father Sky, Sister Water, and Brother Sun to nurture the social ideal of living well, meaning that the whole community has enough to share in harmony. The "good life" doesn't allow for some people to have more than they need while others suffer without, because it encompasses the fundamental value that "you are my other self." The Sky's Heart and the Earth's Heart make up all people, giving them their heart, mind, and body as well as a full capacity to live with dignity while loving and respecting all that allows them to exist.[3]

The Maya are an agroecological civilization. Over the course of thousands of years, they developed a deep, profound knowledge and agrarian practice that afforded well-being and allowed them to achieve complete economic, political, social, and cultural development. Mayan life and agriculture are driven by a holistic vision of physical and spiritual interdependence, and by a worldview rooted in spirituality. These characteristics are vital and persist as an inspiration and guide to creating integrated development that is complementary to and in harmony and equilibrium with nature, family, and community.[4] The Mayan culture believes all beings have their light and dark side, hot and cold, masculine and feminine; moreover, the Maya view these supposed opposites as not contradictory or competing with one another, but rather as complementary, diverse, and necessary for existence in a cycle of constant development. Diversity, then, is recognized as an essential ingredient of life; to be different is not to be opposite but complementary. This principle also applies to agriculture and has enhanced agricultural diversity.[5]

The Spanish invasion in 1524 and the subsequent conquest destroyed the economic, political, and social structures of the Maya, and deeply attacked their spirituality. During the colonial period and even up to today, the Mayan people have been evicted from their lands, discriminated against, segregated, exploited, and murdered. Their spirituality and its manifestations are condemned and attacked.[6] Spanish conquerors imposed the Catholic religion on the indigenous populations as an ideological instrument of domination. True to the ideology of the Maya, attacks against their agrarian economy and attacks against their spirituality are one and the same. Dispossession, discrimination, and exploitation have historically been the principal aggressors against Mayan spirituality, because without land or dignity there cannot be the kind of communion that feeds spirituality.[7]

Originally, the Maya occupied extremely productive lands and developed a technology that is known today as agroecology, based on the application of concepts and principles of ecological design, development, and sustainable agrarian system management.[8] Pre-Columbian people lived in valleys with abundant water resources and a six-month rainy season; in this environment agriculture imitated the natural life of plants and animals in a harmonious interdependence. Intercropping was used during planting: seeding corn, beans, and different types of squash all in the same space. Corn was the main source of sustenance, beans both complemented the diet and fertilized the soil, and squash added to the diet, retained soil moisture, and suppressed weeds. Thus, the population had a rich and balanced diet, while sustainable agricultural practices preserved the fertility of the soil. There was abundant production that allowed people to live well and maintain surpluses without exploiting the land. The main driver of agricultural production was community life, not the generation of commodities. Producing in order to sustain life was and remains linked to spirituality, and so one must ask permission and forgiveness to use the earth; bless the seeds, water, sun, air, and labor; and finally, give thanks and share the fruits of Mother Earth.[9]

During colonization, the Maya were suppressed and condemned to live in the mountains on rugged, fragile soil, so that they were forced to cut down trees and jungle to create the agricultural land necessary to stay alive. In contrast, the colonists appropriated the highly productive lowlands, and in 1525 this act of expropriation was legitimized by the pope, creating the Latifundio-minifundio system, now the backbone of Guatemala's agricultural production system.[10] Practices of expropriation, marginalization, exploitation, discrimination, and murder of the indigenous population have continued into the twenty-first century. As late as 2011, fourteen communities in the Polochic Valley (Baja Veraz, Guatemala) experienced cases of eviction, destruction of homes and crops, violence toward peasants, and assassinations.[11]

Through colonization, the Europeans classified the indigenous populations as an inferior race, treating them as subhuman. The concept of race is a political system defined through inheritance and social categorization, symbolically constructed from skin color. Race categorization was invented by the Europeans to justify attacking and usurping the land and labor of indigenous peoples and African slaves. In this way, the colonizers acquitted themselves of being characterized as criminal, illegal, and terrorist for their violent acts of conquest. "Race" criminalizes nonwhites and decriminalizes whites, and so history has been built on a basis of racism.[12] Such structural racism created by the imposed system has destroyed much of sustainable agriculture and the spirituality of indigenous people all throughout the Americas.

In 1871, liberal reform in Guatemala established a new system of exploitation of land and the labor force, as dominant Creoles protected by the law expropriated communal lands and imposed forced labor on the indigenous community for export production. The objectives of the reform were to subjugate the indigenous and poor mestizo populations, forcing them to sell their labor cheaply and survive from what they produced in their *minifundios,* or smallholdings. Historically, this phenomenon reinforced the impoverishment of Guatemala's indigenous peoples.[13] With the revolution of 1944 there were significant changes, but coerced labor, though forbidden, continues and persists to this day under disguised forms.[14]

Under these conditions of exploitation and domination, the Maya engaged in religious syncretism, combining the imposed Catholicism and their own spirituality, a practice that still exists today. Some writers explain this by attributing it to the "devil," but this characterization fails to take into account the indigenous consciousness and spirit of rebellion. Hidden within the mystical shadows of Catholicism, the Maya maintain their own expression and spirituality, a reflection of their own firm, powerful mentality.[15]

After World War II, in order to make use of the surplus nitrates (for explosives) and poisons (for gas), the United States turned these chemicals into fertilizers and pesticides. To create international markets for these products, it was necessary to replace traditional farming with the industrial model of production. In an effort to spread industrial farming and at the same time provide a technological alternative to the agrarian reform being demanded by Guatemala's peasant movements, the United States, through its Agency for International Development (USAID) and international institutions such as the FAO (Food and Agriculture Organization) and the CGIAR (Consultative Group for International Agricultural Research), promoted the Green Revolution. This

Map of Guatemala. (Based on *Problèmes d'Amérique latine* 43 [1976].)

consisted of increasing the production of basic grains with high-yielding varieties and improved hybrid seeds. With these came the demand for and use of external chemical inputs produced with petroleum that combined fertilizers, pesticides, herbicides, irrigation, and mechanized agriculture. Proponents of the Green Revolution, such as William Gaud, publicly prided themselves on their achievements, noting that the external assistance they financed increased agricultural production.[16] What they did not mention was that this "assistance" was made possible by loans, and that the new system required agricultural expansion, destroying forest and jungle for the sake of production.

In Latin America, the Green Revolution was promoted between 1960 and 1980 and even persisted into the 1990s.[17] Initially, Latin American states had to invest more in agriculture during the Green Revolution, providing subsidies, price incentives, infrastructure, and research. These programs were established as conditions for loans for agricultural development. In Guatemala, such loans were quickly plagued by corruption, racism, and structural inequalities. The Green Revolution changed the mode of production and the agricultural markets, decreasing the access of the poor population—mostly indigenous Mayans—to land and staple foods. On top of the harmful conditions created by the Green Revolution, in Guatemala discrimination and oppression against indigenous peoples further deteriorated their way of life.[18] During this period, missionary action linked to the Green Revolution continued as a useful tool for colonization. As the United States became a stronger presence in Guatemala, missionaries arrived en masse, with pastors and investments from Christian sects promoting their own concept of development and organization, further dividing communities.[19]

After Jorge Ubico's overthrow in 1944 by the "October Revolutionaries," a group of left-leaning students and professionals, along with liberal-democratic coalitions led by Juan José Arévalo (1945–1951) and Jacobo Arbenz Guzmán (1951–1954), instituted social and political reforms that strengthened the peasantry and urban workers at the expense of the military and big landowners, like the US-owned United Fruit Company. With covert US backing, Colonel Carlos Castillo Armas led a coup in 1954, and Arbenz took refuge in Mexico. A series of repressive regimes followed, and by 1960 Guatemala was plunged into a civil war between military governments, right-wing vigilante groups, and leftist rebels that would last thirty-six years—the longest civil war in Latin American history. During the conflict, control, discrimination, segregation, and assassination of the Maya increased. During the 1980s and 1990s, model villages, development centers, and civil defense patrols were forms of segregation controlled by the army. The permanent presence of the army re-created colonial conditions and maintained a regime of terror. Mayan spiritual leaders were directly repressed and killed, and their sacred sites were destroyed or used by the army in carrying out massacres.[20] The kidnappings, torture, and murders were carried out not only to obtain information but also to terrorize the people. The counterinsurgency developed in phases: (1) violent eradication of the guerrilla outbreak, (2) civic action, (3) devastation of land, (4) hunting down those who fled, and (5) civil patrols.

During this time, the Maya resisted. The love of relatives and neighbors was the spark of life that reunited the scattered and lost, giving rise to new forms of social organization. What emerged was a versatile human collectivism that retained the feel of the family home while respecting personal and cultural needs. This served as a testimony of hope, affirming that the weak and poor can defeat strategic violence and resist techniques used to divide the innermost depths of a person, that which defines identity and loyalty.[21] This resistance was aided by Catholic liberation theology: progressive Catholic forces mobilized in favor of the people's struggle, though they were unable to transform the reactionary and conservative church hierarchy.

The role of the Green Revolution as a counterinsurgency measure impoverished and radicalized the peasantry. The Green Revolution finished the four-hundred-year process of destroying peasant agriculture and degrading Mayan spirituality. Agriculture itself was destroyed because of the Green Revolution due to these realities: (1) it increased yields only for the first few crops and harvests, (2) it imposed monocultures, (3) it destroyed the soil's organic matter, (4) it brought indebtedness and destroyed peasant livelihoods, and (5) many peasants went bankrupt and were forced to migrate to the agricultural frontier, to cities, or to the United States.[22] The Green Revolution divorced peasant farmers from the spiritual practices they had performed for centuries, from preparing the field to be sown to blessing the seed and thanking it to sharing and celebrating the harvest.

Although the alleged triumphs and production miracles of the Green Revolution were heralded internationally, from its inception the new system was criticized by both scientists and activists for several reasons: the excessive cost of seeds and their required technology, the subsequent dependency on that same technology, the danger of eliminating traditional crops that were better suited to their environment, and the emergence of new pests. All of this was evidence that the Green Revolution was not ecologically sound. Although it increased crop yields for a few years, the crop products did not nourish people.[23] In Latin America, the Green Revolution mainly favored the large- and medium-sized producers developing industrial agriculture totally dependent on inputs from the United States. Small producers were persuaded and/or pressured to use the "improved seeds" and chemical inputs, which they obtained on credit. Since all peasants were producing the same products, at harvest time there was an excess—and prices dropped off. Peasants, unable to pay their debt, lost their land.[24]

The attack continued and still continues on other fronts. In 1980 the International Monetary Fund (IMF) and the World Bank implemented structural adjustment programs (SAPs) to reduce the fiscal imbalance between lending and borrowing countries and supposedly reduce poverty. The SAPs promoted the privatization of basic services and resources (education, health, electricity, water, and the like) as well as currency devaluation, deregulation, and the reduction of trade barriers. Other conditions of various SAPs included the reduction of social programs, opening domestic stock markets to direct foreign investment, and special rights for foreign investors.[25] Severe international fiscal discipline is applied to countries that don't adopt these programs, effectively marginalizing them. Ultimately, such measures have undermined both the economy and sovereignty of poor countries, converting basic needs into commodities to which the majority of the population has no access.[26] The imposed neoliberal model excludes rural producers of staple foods for the domestic market, favoring financial investment in agricultural exports for the liberalized global market. For this reason, low-priced and unstable production doesn't receive subsidies, usually leading small producers to bankruptcy and contributing to the concentration of landownership by larger corporate farms, most of which export their products.

Another neoliberal measure is the establishment of free trade agreements (FTAs), such as the Dominican Republic-Central America Free Trade Agreement (CAFTA-DR), which expand and diversify the region's trade, remove trade barriers, and facilitate cross-border circulation of goods and services to increase investment opportunities and enforce intellectual property rights. Before the Guatemalan Congress voted to allow FTAs in 2006, there was massive opposition from farmer organizations, workers, trade unionists, and students. Letters of protest were written to Congress from the Republic (the legislative chamber), the Higher University Council of Universidad San Carlos, and the Episcopal Conference, all expressing revulsion for the agreement and demanding a popular consultation prior to its adoption. Ignoring popular sentiment, Congress approved the Central American Free Trade Agreement (CAFTA).[27] Since its implementation, CAFTA has caused small farmers to go bankrupt and migrate to cities or abroad because the domestic market is saturated with subsidized imported products, driving domestic producers out of the market. Likewise, it allows foreign companies to violate labor treaties created through popular struggles. CAFTA "cements" the SAP programs into international treaties such that national congresses and international parliaments have little to no authority. CAFTA represents the loss of national sovereignty.[28]

The Green Revolution and the subsequent FTAs only added to Guatemala's history of the denial of its indigenous people, which has led in certain periods to policies of extermination and to massacres. Massive elimination of Indians took place in the sixteenth, eighteenth, nineteenth, and twentieth centuries via *tierra arrasada,* or scorched-earth, policies and the immense displacement of more than a million indigenous peoples. The acts were committed with an intent to destroy, in whole or in part, numerous indigenous groups: they were not isolated actions, nor were they the inadvertent result of out-of-control troops or the handiwork of rogue military commanders.[29] The idea that indigenous peoples might become the subject of their own history and join large-scale political life through revolutionary organizations unleashed a wave of purposeful extermination that brought about the death of thousands. The fear of an Indian rebellion and the desire to exterminate them overlapped in a historic political juncture that resulted in true ethnocide. According to a report prepared by the Commission for

Historical Clarification, CEH, called "Memory of Silence," of the 42,275 investigated and verified cases of murder during the years 1960–1996, 83 percent of the victims were full Mayan.

Recently, linked to the Efraín Ríos Montt genocide trail, US journalist Allan Nairn, who was in Guatemala in 1982, reported: "Ríos Montt was the dictator of Guatemala during 1982, '83. He seized power in a military coup. He was trained in the U.S. He had served in Washington as head of the Inter-American Defense College. And while he was president, he was embraced by Ronald Reagan as a man of great integrity; and someone totally devoted to democracy. And he killed many tens of thousands of civilians, particularly in the Mayan northwest highlands. In this particular trial, he is being charged with 1,771 specific murders in the area of the Ixil Mayans." The testimony of a general named "Tito"—actually Otto Pérez Molina, Guatemalan president—Nairn stated, suggests that "the use of U.S. and Israeli aircraft and U.S. munitions against the civilians in the Ixil highlands was actually much more extensive than we understood at the time." Moreover, "beyond that, beyond the material U.S. support, there's the question of doctrine. . . . It is often difficult for soldiers to accept the fact that they may be required to execute repressive actions against civilian women, children and sick people." When asked about this, Ríos Montt responded, "Well, that training document which we use is an almost literal translation of a U.S. training document." Nairn affirms that the army "didn't view [Mayans] as individuals, but they saw them collectively as a group as sold out to subversion. And this was a doctrine that the U.S. supported."[30]

Currently, the US government's plan in Guatemala with the Rural Value Chains project, Cadenas de Valor Rural, as part of USAID's Feed the Future initiative, has the following goals: (1) to develop a competitive market-driven agriculture, (2) to prevent and treat malnutrition, and (3) to improve humanitarian aid. Today, most investment areas are in horticulture and coffee, with products grown for export. This investment and export will be implemented in five areas: Huehuetenango, Quiché, San Marcos, Quetzaltenango, and Totonicapán. These areas have the highest degree and concentration of both national poverty and indigenous peoples. The horticultural investment regimes will also work with private authorities using transgenic and improved seeds, chemical inputs, and irrigation. All required products and services must be purchased from US companies. Projects will be funded for five years with US$40 million.[31] This investment is intended to stimulate agriculture by inserting peasants into the global market—without consideration of the fact that it was the global market that provoked the current crisis in the first place. It also seeks to improve humanitarian assistance, contradicting sovereign rights and the country's capacity for self-sufficiency. This kind of foreign investment disguises market expansion as humanitarian and development aid. As Eric Holt-Giménez of Food First has commented, these million-dollar investments annihilate peasant farmers and their spirituality, sustainable agriculture, and the country's possibilities to build food sovereignty.

Initiatives such as those sponsored by USAID aim to destroy peasant farming despite its advantages: peasant farming is more efficient and productive, it produces at least half of the global food supply and most of the employment in rural areas, and it can cool the planet. The pattern of the Green Revolution is repeated here, only this time using genetically modified seeds. However, this is not a counterinsurgency measure because there is no insurgency. Instead, this is purely an investment to advance capitalist agriculture by removing the peasantry—which in Guatemala happens to be primarily indigenous. This kind of capitalist "development" emanating from the Global North is unjust and unattainable. Unjust, because the Global South is treated like a servant territory. Unattainable, because for all its devastating nature, the development does not offer solutions to poverty or to hunger.[32]

The Mayan people survived colonial and later government repression by valuing their identity, culture, and spirituality. However, it was not until 1992, during the commemoration of the five-hundred-year anniversary of Spanish arrival in the Americas, that talk turned to indigenous rights, and Mayan spiritual practices were demonstrated in public celebrations. Guatemala has ratified international conventions and declarations recognizing the collective rights of indigenous peoples, and the country condemns all forms of racism, discrimination, and violence against indigenous and tribal peoples. The International Labor Organization (ILO) agreement no. 169 was ratified by Guatemala in 1996. The Declaration of the Rights of Indigenous Peoples, crafted by the United Nations, was adopted in 2007, although in Guatemala these collective rights have been recognized only in a very limited way, due in part to truncated constitutional reform in 2001. At the national level, the peace accords signed in 1996 after thirty-six years of internal war included the Agreement on the Identity and Rights of Indigenous Peoples, opening a space for the recuperation and revitalization of Mayan spirituality. The Ajq'ijab', the counters of days and Mayan priests, appear publicly in Mayan communities and nationally, expressing their right to be different under equal conditions.[33]

In this space, spirituality linked to the practice of some agricultural Mayan Kaqchikel communities and other ethnic groups has begun to strengthen. In present studies, work is being done with Kaqchikel communities which, through practicing agroecology, are returning to sustainable agriculture, knowing that it is their only way to survive. As these communities revive their spiritual traditions, they privilege the sanctity of agriculture and the primacy of the *milpa,* or "the three sisters" of polyculture—corn, beans, and squash. Kaqchikel women in multiple communities plant and breed seeds to preserve agrobiodiversity, to share their knowledge, and to be the guardians of fertility, as they were taught by their grandmothers. Spirituality and sustainable agriculture are seen as harbingers of hope, a positive testimony to the possibility of recovering a full life.[34]

Let us present a case study of an indigenous initiative to address some of these issues. Cooperation for Rural Development in the Western Region of Guatemala (CDRO, Cooperación para el Desarrollo Rural de Occidente) was created in 1983 in Nimasac, Totonicapán, Guatemala. The founders' main idea is to promote social, economic, and political development in the rural area according to indigenous spirituality. CDRO considers that development is an integral process of organization and production that allows a better life for all the people in the community. This process must be controlled by the community and must respond to the people's own vision, interests, and needs. CDRO integrates several activities: agriculture, education, health, use and trade, and development of infrastructure. The organization and work of each activity is a process that begins with consulting the elder people to understand how the undertaking was done traditionally, why, and who are the best persons to guide the endeavor. (Elder people are older than fifty-two, those who have lived all the energies of the days, since the solar calendar and the sacred calendar match every fifty-two years.) On issues related to health, nutrition, hygiene, children, and education, women's advice is more powerful; on those related to agriculture, water, transportation, and infrastructure, men's advice is preferred. This knowledge is shared with all the people in the community and together they decide how they are going to organize and work on a project. CDRO confirms that indigenous communities always relate their daily activities to spiritual beliefs and rituals.

CDRO reinforces agroecology as a way of living and producing, because it is the way people want to plant, work their fields, and produce their food. CDRO encourages the elders, the sacred people, Ajq'ij, to teach their community the original ideas, values, and practices. There are many commonalities, but each community has distinctive practices, rituals, and prayers. As expressed by Carmelo Chan, CDRO *facilitador* (field worker), "In Xecul, Totonicapán, for example, seeds are submerged in water the day before planting them. This entails several rituals: asking permission to gather the water from a certain water place; expressing gratitude to the water, soil, sun, air, and seeds; expressing their commitment to be loyal to their field and to respond to its needs; and finally asking for the needed strength to work. The night before sowing women and men don't sleep together, because they need to save all their energy and inspiration. We respect these practices, but we don't ask any other community to follow it."[35]

At the same time that CDRO reinforces agroecology, the community reinforces spiritual activities linked to the practice. This has begun a process of rebuilding the confidence and organization in communities that were deeply affected by the civil war. CDRO is promoting development reinforcing the community's own identity.

Additionally, in 2004 twenty organizations working on agriculture, food, nutrition, and fair food markets created the Network for Security and Sovereignty for Food in Guatemala (REDSSG). The association is promoting practical activities for participatory community food assessment, food production, school participation, coordination between organizations working with the community, the creation of community food markets, festivals for sharing their experiences, and many other activities.

Elder people began to organize spiritual activities, inviting people to recognize the strength and love of Mother Earth, because she was encouraging them to join, share, and build a better life. Thus a spiritual group was organized inside REDSSG, one willing to walk with communities that want to rescue their spiritual tradition linked to agriculture and food. And participants were not only people interested in spiritual practices linked to agriculture; they included as well technical people teaching agroecology—and both groups know that their work is better done working together.

People realize that working together and learning better agricultural practices are very important, but at the same time they recognize that spirituality is the inspiration that makes it possible. "Spirituality brings our hearts and minds together, and that is why we're able to work together and share."[36]

It is essential that any framework used to analyze the state of agriculture and Mayan spirituality be political and structural, because it is only in this way that one can assess the severity and complexity of the problem as well as the urgent need for change in the conditions of agricultural production and life in Guatemala.

Mayan peasant agriculture has suffered several profound systematic blows: (1) the dispossession and land-use change by colonists and the institutionalized discrimination against indigenous people, (2) the expropriation of communal lands and forced labor legislated by the liberal reforms of 1871–1944, (3) the imposition of the Green Revolution, which catered to outside capitalist interests, (4) the imposition of free trade agreements, and (5) the massive US investments used to incorporate peasant agriculture into the global market and to maintain dependency. All of these attacks on peasant agriculture are also attacks on Mayan spirituality.

In spite of this, peasant agriculture still persists in Guatemala. There is now a new political space created by the peace accords. Many Mayan peasant farmers are returning to sustainable agriculture as a strategy for survival. At the same time, they are reviving their spiritual traditions, like the sanctity of the milpa. Mayan spirituality renews a way of life that respects the cosmos, nature, and all living beings. This way not only allows one to be maintained as a person but also serves as a vital link to one's community and to all indigenous peoples. The process of revitalizing Mayan spirituality allows for the reconstruction of those damaged by racism, violence, and exploitation; it favors each person's capacity for expression and decision, maintaining identity and sense of self; and it enables the reconstruction of destroyed social fabrics and the establishment of a respectful relationship with nature. This relationship creates better production, adapting and evolving and learning from nature's cycles and the interdependence of the elements.

Agroecology is an agrarian-environmentalist facet of food sovereignty, which is the right of people to take democratic control of their food system; the right to access healthy, culturally appropriate food produced in an ecologically sound manner and with sustainable methods; and the right to define their food and agricultural systems for themselves. These concepts reflect Mayan spirituality by treating life as an interdependent whole, whereby the goal is to produce enough while also respecting the environment and equitably distributing resources so that all people live in dignity. Land, water, air, seeds, the workforce, and the animals are sacred elements, and all should benefit in their relationship with production. Mayan spirituality is being revived and is strengthening as an expression of resistance and healing. This strength is supported by the adoption of a political position that defends indigenous rights and food sovereignty. Food sovereignty is considered just as important as Mayan language and identity, and is as irreducible as their spirit. Therefore, spiritual affirmation is also political. It recognizes that the struggle to defend food sovereignty is also a way to defend Mother Earth, demonstrating the integration of the material and spiritual life. This position challenges all of Guatemalan society, asserting that there can be no talk of national sovereignty if there is no food sovereignty.

Spirituality is a vital component of resistance and healing, and it is a catalyst and a resource that allows people to strengthen their sense of self and further their well-being. It allows survival as a person, family, and community, but not as a whole people. The Mayan people need another structure: they need land, access to resources and services, and a decent and fair system that allows them to exist with self-determination. The Mayan narrative as we presented it above attests that spirituality is threatened and damaged not only by exploitation, discrimination, and repression, but also by the lack of land and by structural marginalization. If the Mayan people have no land on which to produce and subsist, then they lack what is necessary to inspire and give life to their society and their own spirituality.

What is required is a structural change to develop agroecology while at the same time recuperating Mayan spirituality. Structural change implies a fair distribution of land and resources, and the implementation of national programs that foster agroecology both economically and technologically so that the country can achieve food sovereignty. The transformation must be freely celebrated and accompanied by Mayan spiritual practices. Despite the new political space created by the peace accords and the reaffirmation and empowerment of Mayan identity, agricultural change and spiritual renewal are not possible without systematic change within the country to completely overcome the discrimination that exists against indigenous people. We are left with some vital questions: Can Mayan spirituality be integrated back into modern Mayan culture? Could this be a path toward food sovereignty? Toward autonomy?

Study Questions

1. What were your assumptions about the relation between indigenous Mayan spirituality and agricultures before reading this chapter? Have those views and assumptions changed, and if so, how?
2. The Green Revolution has destroyed traditional agriculture in Latin America and Asia, but now the Gates Foundation finances it in Africa. Why is the foundation pushing a failed process?
3. Why is USAID promoting the program Feed the Future, knowing that it benefits only US investors, not the poor people?
4. The US government has played a determinant role in destroying the agriculture and the economic and political life of many countries around the world. What might be the role of the new US generation? What should be your own role?

Notes

1. http://www.ine.gob.gt/np/generoypueblos/documentos/Perfil%20Estadiscio%20final.pdf (accessed May 20, 2013).
2. The traditional concept that land cannot be privately owned was overturned with the invasion of the Spaniards. Currently, 80 percent of the arable land is owned by 8 percent of the population. Economic, political, and military powers determine the ownership of land, so that indigenous and community lands are currently violated even though Guatemala's Constitution, article 67,

established legal recognition of Native community land.

3. Cristobal Cojtí García, *El valor sagrado del maíz* (Guatemala City: Cholsamaj, 2012); Leonor Hurtado, *La vida complementaria entre las personas requiere la soberanía alimentaria* (Guatemala City: Cholsamaj, 2010).

4. Saqb'ichil-COPMAGUA, *Más allá de la costumbre: Cosmos, orden y equilibrio*, 2nd ed. (Guatemala City: Cholsamaj, 2000).

5. Hurtado, *La vida complementaria*.

6. Severo Martínez Pelaez, *La patria del criollo* (Guatemala City: Editorial Universitaria USAC, 1998). Despite the accords signed in 1995, respect for indigenous spirituality has improved very little. Indigenous organizations and communities have included spiritual ceremonies as a regular practice in public activities, but it is mainly a show. There are still extended preaching activities taking place in indigenous communities, mainly done by US evangelical churches.

7. José Luis Chaicoj Sian, personal communication, 2012.

8. Miguel Altieri, *Agroecología: Bases científicas para una agricultura sustentable* (Norda Comunidad, 1997).

9. Saqb'ichil-COPMAGUA, *Más allá de la costumbre*.

10. Carlos Guzmán Böckler and Jean-Loup Herbert, *Guatemala: Una interpretación histórico-social* (Mexico City: Siglo XXI SA, 1970).

11. Convergencia por los Derechos Humanos Guatemala, *Comunicado sobre los despojos en polochic* (Guatemala City: FLACSO, 2011).

12. Steve Martinot, *The Machinery of Whiteness* (Philadelphia: Temple University Press, 2010).

13. Martínez Pelaez, *La patria del criollo*.

14. Flavio Rojas Lima, *Los indios de Guatemala* (Hegoa, 1992).

15. Martínez Pelaez, *La patria del criollo*.

16. William Gaud, "The Green Revolution: Accomplishments and Apprehensions" (speech given at the Universidad Cuernavaca, 1968).

17. Karlos Pérez, *La revolución verde* (Hegoa, 2000).

18. Felipe Gómez et al., *Racismo y genocidio en Guatemala* (Guatemala City: Cholsamaj, 2004).

19. Guzmán Böckler and Herbert, *Guatemala*.

20. Comisión para el Esclarecimiento Histórico, *Guatemala memoria del silencio* (Guatemala City: Comisión para el Esclarecimiento Histórico, 1999).

21. Ricardo Falla, *Masacres de la Selva, Ixcán-Guatemala, 1975–1982* (Guatemala City: Editorial Universitaria USAC, 1992).

22. A critique of the Green Revolution and documentation of its side effects (intended and unintended) has been done by the Oakland, California-based Food First; see www.foodfirst.org. For Latin American case studies, see Eric Holt-Giménez, *Campesino a Campesino: Voices from Latin America's Farmer to Farmer Movement for Sustainable Agriculture* (Oakland, Calif.: Food First, 2006).

23. Pérez, *La revolución verde*.

24. Armando Bartra, *Fin de fiesta: El fantasma del hambre recorre el mundo* (Mexico City: Endira, 2008).

25. James Greenberg, *A Political Ecology of Structural-Adjustment Policies: The Case of the Dominican Republic* (Culture and Agriculture, 1997).

26. Eliana Cardoso and Ann Helwege, *Latin America's Economy: Diversity, Trends and Conflicts* (Cambridge, Mass.: MIT Press, 1992).

27. Violetta Yagenova, *Cronología de la lucha contra el TLC en Guatemala durante marzo 2005* (Guatemala City: FLACSO, 2005).

28. Holt-Giménez, *Campesino a Campesino*; Andrés Cabanas, *TLC en Guatemala: Represión contra diálogo* (Guatemala City: Cholsamaj, 2005).

29. Comisión para el Esclarecimiento Histórico, *Guatemala memoria del silencio*.

30. "Exclusive: Allan Nairn Exposes Role of U.S. and New Guatemalan President in Indigenous Massacres," *Democracy Now!* April 19, 2012, http://www.democracynow.org/2013/4/19/exclusive_allan_nairn_exposes_role_of (accessed April 25, 2016).

31. USAID, Request for Applications (RFA) Number: RFA-520-11-000003, in *Rural Value Chains Project* (Guatemala City: USAID, 2011).

32. Isabel Rauber, *Dos pasos adelante, uno atrás: Lógicas de ruptura y superación del dominio del capital* (Vadell, 2010).

33. Saqb'ichil-COPMAGUA, *Más allá de la costumbre*.

34. Cojtí García, *El valor sagrado del maíz*.

35. Personal communication.

36. Colectivo Social por el Derecho a la Alimentación, *Informe alternativo por el derecho a la alimentación* (Guatemala City, 2008).

2

Soils, Spirits, and the Cosmocentric Economy

Re-creating Amazonian Dark Earth in Peru

Frédérique Apffel-Marglin

> Agriculture ... can provide "balance for well-being" through relationships not only among people, but also nature and deities. In this concept, the blessing of a new field represents not mere spectacle, but an inseparable part of life where the highest value is harmony with the earth.
>
> —Darrell Addison Posey

Girvan Tuanama, our indigenous farmer at Sachamama Center for Biocultural Regeneration in the Peruvian high Amazon, said to me: "My grandmother and the elders used to offer pieces of ceramics in their *chacras* to Mama Allpa.[1] Now most people no longer do that." So when we went to his Native community at the time of the *mikuna*—communal meal—we went to his grandmother's house to eat. The little old lady sat on the threshold welcoming everyone in. After everyone had eaten and left, Girvan, his grandmother, and I stayed in the now-quiet room. Carmen Tapullima Salas looked to me to be in her nineties, frail but lucid. Girvan had told me that she did not have a birth certificate since in those days such niceties were not always observed. After some chatter to make us comfortable with each other, I asked Carmen about the offerings in the chacra. "Yes, I used to always bring with me my *shaño* and give it to Mama Allpa.[2] That is what we used to do then." Girvan explained: "It is the *ingenieros* who made us ashamed of this

Ceremony and broken ceramics. (Photo by Frédérique Apffel-Marglin.)

practice by telling us it was a silly superstition. So now only a few of the old ones still do it, but in many communities this *pago* [offering] is no longer done.[3] But you yourself heard the *comuneros* [members of a Native community] of Shukshuyaku yesterday tell you that they keep finding pieces of ceramics in the old chacras. That tells them that their ancestors used to offer them to Mama Allpa."

Indeed, the day before we had gone to Shukshuyaku, a Native community with which we have created a permanent communal chacra, for the first planting of the June agricultural campaign.[4] Since archaeologists have discovered in the last few decades that the *yana allpa* (black earth) we are re-creating in our center is full of ceramic shards, the comuneros were enthusiastic about remembering their ancestors' practice and offered broken ceramics to the soil before planting.

Combing the literature on Amazonian dark earth (ADE), popularly known in Brazil as *terra preta do Indio* (black earth of the Indians), reveals that most authors consider that the many pieces of broken ceramics found in ADE come from middens. A few are found in conjunction with burials, and one anthropologist studying the ceramics of the contemporary Asurini do Xingu peoples in Brazil reports that at the death of a woman, her ceramics are broken.[5] (Among these peoples, as among the Kichwa-Lamistas, women are the ceramicists.)

The only scholar to report that the broken ceramics found in terra preta come from offerings to the deities/spirits is Alfredo Narváez Vargas, a Peruvian archaeologist and anthropologist who has excavated one of the major archaeological sites of Chachapoyas, Kuelap. Dr. Narváez Vargas told me that when he excavated in a few archaeological agricultural terraces near the site of Kuelap, he typically found a dense layer of black organic soil that was easily differentiated from the yellow color of the geological layers. In the black layer he found fragments of ceramics he considered to be remnants of offerings.[6] His findings correspond closely to archaeological descriptions of terra preta sites in the low Amazon. This information is particularly relevant here since Sachamama Center in Lamas, department of San Martin, is situated at the edge of the Chachapoyas pre-Columbian culture area.

However, to my knowledge, no one has reported on a practice of the Kichwa-Lamistas of offering pieces of broken ceramics to the various spirits of the chacra. The four most important spirits of the chacra are Mama Allpa, the earth; Yakumama, the rain/water; Mama Qilla, the moon; and Tayta Inti, the sun. Many of the Kichwa women elders keep some pieces of ceramics for offerings. They also make the traditional pre-Columbian offerings of corn beer (*chicha*).

Offerings of pieces of ceramics are made before planting and at harvest in each of the two major agricultural campaigns; they are also made at solstices. This means there are quite a few occasions during the year when offerings are made by people involved in agriculture, and since agriculture is the principal economic activity of the Kichwa-Lamistas, almost everyone used to do this. However, the seductions of modernity, combined with the scornful attitude of the dominant mestizos toward such practices, have deeply eroded it. Although poor mestizo farmers practice the same agricultural techniques as the Kichwa-Lamistas— swidden agriculture, planting according to the phases of the moon, polyculture, and so on—they are much less likely to make offerings to the spirits of the chacra. The Kichwa-Lamistas keep their ritual practices to themselves, careful to protect them from the contemptuous gaze

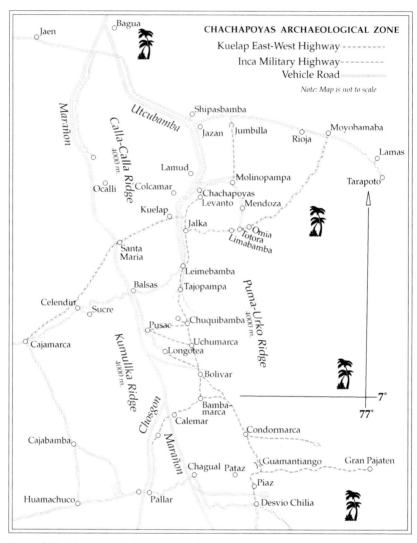

Map of Chachapoyas cultural area. (Based on Morgan Davis, "Chachapoyas, the Cloud People: An Anthropological Survey," 1985, Tozzer Library, Harvard College, Cambridge, Mass.)

of others. The colonial laws of "extirpation of idolatry" may no longer be enforced (although they have never been removed from the books), but their spirit is very much alive.[7] The memory of the death-dealing ferocity with which they were enforced has not completely vanished among the indigenous people. It must also be pointed out that Kichwas belonging to evangelical movements, which have made deep inroads in the Peruvian high Amazon, have ceased making such offerings because their faith considers them to be the work of the devil.

I believe that the ancient practice of offering ceramics could well account for the amount of broken ceramics found in the ADE all over the Amazon basin. It is well known that at least in the highland Wari pre-Columbian culture, elaborate ritual ceramics were dashed to pieces after an offering. I have admired such reconstituted vessels in the archaeological museum in Ayacucho as well as the one in Pueblo Libre in Lima. According to Harvard professor Thomas Cummins, renowned specialist in Peruvian pre-Columbian ceramics (among other topics), the practice of dashing the ceramics used in offerings could very well have been a widespread one, extending as far as the low Amazon.[8] The same was communicated to me by Dr. Alfredo Narváez Vargas from

Traditional pre-Columbian terraced fields in Chachapoyas territory at high elevation. (Photo by Frédérique Apffel-Marglin.)

his findings in another archaeological site, one on the northwest coast of Peru in the town of Tucume. There Dr. Narváez Vargas has found broken ceramics in altarlike sites associated with sacred objects that were clearly put there as offerings.[9] Since ceramics in the low Amazon antedate ceramics in the rest of the Americas (see below), it might even be possible that the practice originated there. Vargas's discovery of ceramics in the Amazonian dark earth in the Chachapoyas culture area (ceramics he claims are the remnants of offerings to the deities/spirits) is the strongest evidence so far for a continual practice among the indigenous population of making such offerings in their food fields. The Chachapoyas center of Kuelap is at an altitude of some three thousand meters and is full of evidence of stone terracing as well as of terracing made with perishable material since only the abrupt change of height is left, indicating such terracing was constructed with the use of perishable materials.

Offerings to the Spirits and the Cosmocentric Economy

Although most of the literature on ADE is overwhelmingly quantitative and scientific, silent on its religious aspect, Gerry Gillespie comes close to referencing spirituality. Pointing out that successful communities the world over have always returned organic matter to the soil, as this is the very basis of those communities' sustainability, he maintains that such acts were "an intentional and conscious attempt to maintain a link between the individual and their food producer. As much as any other action, it was an act of respect."[10] In fact, such respect for the soil, as well as respect for all the other beings/spirits involved in growing food, is typical in peasant and indigenous societies worldwide. It was also typical in premodern European agriculture. As I have argued elsewhere, it was not until the advent of the enclosure movement, the Reformation, and the scientific revolution that gift exchanges between humans and nonhuman entities involved in growing food (the soil, the water, the sun, and the moon, among others) began to be seen as magical acts devoid of efficacy.[11]

Offerings made to the soil, locally embodied in its spirit Mama Allpa (Earth Mother), emphatically enact a conscious and intentional bond of respect between humans and the soil, simultaneously recognizing the agency of the earth. There is a concomitant awareness that without a return gift to the soil/earth, the success of agriculture is in danger. The earth/soil gives of its produce, enabling humans to live, and humans enact their awareness of their dependence on those gifts by offering something to the soil. These reciprocal exchanges of gifts enact a *regenerative cycle,* which constitutes the opposite of the extractive actions characteristic of a capitalistic economy. In the latter, the soil is considered a "natural resource" rather than a sacred mother. I have borrowed the term *cosmocentric* from Stefano Varese, who contrasts Euro-American anthropocentrism with the cosmocentric cosmologies of indigenous peoples.[12]

Agriculture is a series of concerted actions carried out by various actors: humans, who prepare the fields, choose the seeds, prepare the soil, and so on, and nonhuman actors. Without the actions of the soil, the water, the sun, and the moon, whose phases affect the rising and falling of all liquids, from the sea with its tides to the sap in plants, agriculture would not be possible.[13] As I have documented at some length in my book *Subversive Spiritualities,* the agency of the nonhuman world is a feature that has become accepted, particularly in the field of critical science studies and among certain quantum physicists. The French anthropologist/philosopher Bruno Latour, whose work has contributed so much to this recognition, has titled one of his books *We Have Never Been Modern.* The classical Cartesian/Boylian/Newtonian paradigm of a mechanical, agency-less, insentient world has been effectively displaced, although this paradigm shift has barely begun to percolate in modern culture at large.

As the discoveries concerning ADE and other anthropogenic phenomena in Amazonia have revealed, the Amazonian rain forest is anything but a virgin forest. It is the result of millennia of concerted actions carried out by pre-Columbian humans and all the nonhumans in that locality. I have suggested that concerted actions between humans and certain nonhumans crucial for human welfare have been carried out over long periods of time, giving rise to entities—or rather, beings—who embody those concerted actions. For example, the soil becomes Mama Allpa, a being to whom prayers and offerings are made, who is endowed with understanding, agency, and sentience, and who responds to the actions of humans. In modernity the soil has become a "natural resource" bereft of agency, sentience, and understanding. A natural resource is simply something there for us humans to make use of, to exploit, without feeling that by doing so we are showing a lack of respect. In other words, a natural resource does not require us to reciprocate for what we take since it has no agency. As is well known, where the preservation of natural resources has been a concern, the logic of modernity has led to the calculation of their monetary values or, in economic parlance, making the externalities visible as well as quantified: what today is called green economics. However, green economics is no more able than neoclassical economics to help us humans to enact regenerative rituals of reciprocity with the nonhuman world since giving a monetary value to aspects of that world in no sense recognizes its agency.

I have written so far only in terms of the soil; however, both in Amazonian indigenous cosmology and in premodern European peasant cosmologies, the world was enlivened by a multiplicity of presences, all belonging to an overarching "world soul," or anima mundi that did not exclude us, the humans. In other words, nature and culture were not conceived of as mutually exclusive categories.[14] Acts that make us aware of the need to return to the soil are fundamental to sustained civilization and need to be recovered. Such acts are simply taken for granted among peasant and indigenous societies and in general among what anthropologist Pramod Parajuli calls "ecological ethnicities."[15] Some others have discussed such an economy under the rubric of "the gift economy."

The Gift Economy or the Cosmocentric Economy?

The term *gift economy* was coined early in what became a classic in the field of anthropology, French anthropologist Marcel Mauss's *The Gift* (*Essai sur le don*), originally published in 1925. Mauss captured with the words "the spirit of the gift" elaborate rituals of exchange in nonmodern small-scale societies. Mauss states that in Maori custom "this bond created by things is in fact a bond between persons."[16] This view that gift exchanges are really about creating bonds between human persons, thus making the presence of the spirits invisible or ineffective, has dominated the anthropological literature on this topic. Gift exchanges always pass through gifts to invisible beings, namely, the deities or spirits. It is only then that these so-called gift exchanges become truly regenerative. Viewing gift exchanges as being "really" between human beings is in keeping with the strong tendency in anthropology to consider spirits, deities, and the like as representing a reality belonging to some other aspect of the human domain, such as the semiotic, social, political, or economic, or some other aspect of human reality. In general, scholars in that field have not taken the route of giving agency to those aspects of the natural world responsible for our (us humans') sustenance and recognizing the need for humans to reciprocate to those through offerings in order to regenerate the gifts.

The bounties of the nonhuman world are regenerated by constant ritualized offerings between people and the spirits, such as in our case Mama Allpa, the earth mother/spirit, and all the other spirits of the chacra. Those constantly reiterated actions of making offerings and thus reciprocating ensure the continual possibility of receiving sustenance and of not depleting the source of sustenance. In other words, it is the constant reiteration of rituals of reciprocating offerings that ensures the regeneration of the source of the gifts. The gift, according to Mauss and many after him, increases sociality: namely, bonds between humans. I would argue that it also increases spirituality if we understand by this term a continual flow beyond oneself and beyond one's human community toward the nonhuman world and its spiritual embodiment. "If this were not so, if the donor calculated his return, the gift would be pulled out of the whole into the personal ego, where it loses its power."[17] Such enactments deserve to be labeled *cosmocentric economy*. The cosmocentric economy is one in which the regeneration of the sources of sustenance is not taken as an automatic natural or biological process but is recognized to be dependent on the proper ritual reciprocating action of humans. It is a radically nonanthropocentric type of economy—in other words, one profoundly different from the capitalist economy.

The spirits of the chacra for the Kichwa-Lamistas of the Peruvian high Amazon are not projections of the human world onto the nonhuman world, metaphors used by humans to *represent* the nonhuman world. Rather, they literally embody that which enables the humans living in a particular place to draw from the food-producing aspects of the nonhuman, or natural, world without depleting it. Humans reciprocate with their nonhuman sources of sustenance with offerings to these sources, whose agency is recognized and whose need to receive respect and gratitude is equally recognized by endowing them with a humanlike identity: Mama Allpa is both a spirit and the soil/earth; Mama Qilla is both the moon and a spirit, and so on. Such actions constitute a severe constraint on the exploitation of the nonhuman, or natural, world since exploitative behavior would open one (and one's group) to the reprisal of the spirits. Such reprisal consists in the degradation of the soil, the sustainer of civilization. Ritual offerings to the sources of sustenance ensure the regeneration of those sources and embody an implicit recognition that such regeneration depends on proper human actions and is not automatic.

The spirits or deities make certain types of action possible and others difficult, reprehensible, or sinful. The consequences of such actions are undeniably material or physical and simultaneously also discursive and spiritual.[18] The difference between a natural resource and the spirit of the forest or of the soil/earth spells the difference between exploitation and the regeneration of those sources of sustenance. The consequences also serve to place severe limits on the human uses of these gifts. These rituals *enact* or *perform* continual biocultural regeneration through a variety of actions and utterances that embody the entanglement of humans, nonhumans, and the spirits. It is such entanglements that I want to capture by the term *cosmocentric economy*. The cosmocentric economy is a regenerative, nonexploitative economy in which reciprocity and redistribution among humans, nonhumans, and the spirit realm tend to produce equity rather than sharp differences in wealth among the humans, as well as ensure the regeneration of the sources of livelihood for humans.

The Reformers in sixteenth-century Europe called such rituals "magic" due to their insistence on the total separation between humans, nonhumans, and the religious—namely, God. For the Reformers, agency, voice, and meaning became exclusively human attributes as well as the prerogatives of a God removed from the material world. Ever since the Reformers' separation of matter and spirit, such rituals of regeneration could be understood only as humans *representing* symbolically or metaphorically the nonhumans, who thus became passive and silent. Just as the Puritans in New England, such as John Winthrop, could feel justified in expelling the Indians from the land, the conquistadors and the colonial authorities in New Spain felt justified in making laws mandating the "extirpation of idolatry," aimed at preventing the Amerindian natives from performing their regenerative rituals, which were seen as inspired by the devil, since for the Spaniards the spirits did not really exist.[19]

Today, perhaps, we are sufficiently removed from the hold of the Reformation on everyone, including a secularized anthropology, to reassess our assumptions.[20] The key enactments needed are with those entities of the natural world that have agency and interact with humans to ensure their continued sustenance. This is to my mind at once a material and spiritual endeavor.

The Ecological Situation Today in the Peruvian High Amazon

When I first discovered the existence of terra preta a few years ago through reading Charles Mann's book *1491*, it immediately seemed worthy of further study and possible application in the Peruvian high Amazon.[21] I had been going to the town of Lamas in the Peruvian high Amazon for some fifteen years and was well aware of the urgency of the situation regarding the local practice of slash-and-burn agriculture by both poor mestizo and indigenous farmers. Lamas is situated in the department of San Martin to the northeast of Lima on the eastern tropical foothills of the Andes, and that department has the dubious distinction of having the highest rate of deforestation in the whole of Peru.[22] Every year I return to Lamas there is less forest.

The deforestation is the result of several factors. In the 1980s and 1990s, the region saw an enormous increase in the production of coca for the cocaine trade, locally identified as the "pocketbook of the guerrilla movements." This illicit trade prompted an enormous amount of deforestation to make way for coca fields.[23] In addition, ever since the region was opened up with the construction of a road in the 1960s by the government of Belaunde, there has been a dizzying amount of immigration by land-hungry peasants from the highlands and/or the coastal areas of the country. The migrants' ignorance of the special nature of Amazonian soils, combined with the government's official stimulus for industrial monoculture agriculture, has resulted in large tracts of forest being clear-cut to make way for either cattle ranching or cash cropping, principally with rice, corn, sugarcane, or cotton.

Slash-and-burn agriculture was introduced when the Spaniards brought steel tools in the sixteenth century (see below for details). Experts calculate that for this method of agriculture to be sustainable, a family must own a minimum of fifty hectares of land, whereas the mean amount of land of indigenous farmers today is between three and ten hectares.[24] This has led to shorter fallow periods between clearings, with the result that the secondary forest that regenerates in abandoned food fields or chacras is progressively less vigorous. It has also led to the rapid growth of degraded lands where nothing grows except weeds.

The situation of the Quechua-speaking Kichwa-Lamistas in this region is the result of their conquest in 1656 by Martin de la Riva Herrera. As was the Spaniards' practice in the sixteenth and seventeenth centuries, the indigenous population was resettled and concentrated—the Spanish term is *reducidos*—into new settlements for ease of control and evangelization. The descendants of these indigenous people can still be found in the part of Lamas called Wayku, where they were first resettled, as well as in more than two hundred small rural settlements in the region where they subsequently fanned out in the course of the last four hundred years. The few Spaniards who settled the town of Lamas at that time interbred with the local women, creating a mestizo population, a process that has continued to this day. When the children of such unions are recognized by the father, they take on Spanish surnames and identify with Spanish language, culture, and mores. Ever since the conquest, the more powerful members of the mestizo population have dominated and exploited the indigenous inhabitants, who have been relegated to the poorest lands on the steepest slopes and were kept from access to education until as recently as the 1980s.[25]

The government of Velasco Alvarado in the late 1960s and early 1970s passed legislation that enabled indigenous communities to become legally demarcated and recognized as *comunidad nativas:* that is, Native communities that ran their own internal affairs and whose territories were off limits to non-Natives.²⁶ In the Lamas region, it is only since the 1980s that indigenous communities have begun the onerous, costly, and bureaucratic process of becoming legally recognized as Native communities. To date, there are some seventy such communities in the region among the more than two hundred potentially eligible. Furthermore, as Calderón Pacheco points out, this law, although at the time of its enactment was a revolutionary move, for the first time giving indigenous people protection from outsiders wanting their lands, has the unintended effect of preventing them from freely resettling in more forested areas as they need to do, given their form of swidden agriculture.²⁷ This has led to the present situation in which the vanishing forest in their territories is making their customary swidden shifting agriculture less and less feasible. Such sedentarism is putting a great deal of stress on the environment of these Native communities, and thereby slowly but surely increasing the amount of degraded lands.

The leadership of the Kichwa-Lamistas is well aware of the problem. The former president of the largest of the four Kichwa-Lamista organizations, the Ethnic Council of the Kichwa People of Amazonia (CEPKA is the Spanish acronym), Misael Salas Amasifuen, stated this in his talk to US undergraduates at the Sachamama Center on July 31, 2011:

> The vast majority of the Kichwa-Lamistas depend on their *chacras* for survival. Due to communal labor, called *maki-maki* or *choba-choba* in Quechua, where kin, neighbors and friends form work parties to work on each other's fields or projects, the need for cash is minimal. Nowadays, ever since the federal government requires all children to attend school, which means uniforms and other supplies have to be purchased by the parents, the need for cash has steeply increased. Kichwas typically sell the surplus produce they raise in their *chacras* in the local markets and also sell their labor at times of peak demands during the coffee harvest for example or other peak agricultural labor demands in order to acquire cash. However, their ability to grow their own food is not only an economic necessity but also the guaranty of their autonomy. It is that autonomy that in turns enables them in no small part to keep their distinct identity. Thus food sovereignty for the Kichwa-Lamistas not only spells their physical survival but their political and cultural sustainability as well.

The Kichwa-Lamistas are very actively engaged in negotiations with the regional government for stewardship of the local conservation reserve forest situated in the nearby mountain range of the Cordillera Escalera, which is part of their ancestral territory. Unexpectedly, however, the Kichwa-Lamista leadership is not envisaging carrying out their form of slash-and-burn agriculture in those forest reserves. What surprised me in this attitude of the Kichwa-Lamistas is that it seems to accept the logic of the separation between utilitarian activities on the one hand and conservation or preservation on the other. Even though they envision activities of hunting and gathering in the forest, they reject the making of chacra and thus food sovereignty in it through their traditional form of swidden agriculture.

Forest conservation areas were created in the United States in the second half of the nineteenth century as nature parks. As William Cronon's history of such parks has made us all aware, these "wilderness preserves" were recast as "God's Temples." As Cronon shows, the enclosure of the Indians on reservations at the end of the Indian Wars in the second half of the nineteenth century as well as the disappearance of the frontier were key ingredients in the invention of such preserves. A nature park or preserve was constituted as a nonutilitarian space of sacred national wilderness. Cronon points out, "To this day the Blackfeet continue to be accused of 'poaching' on the lands of the Glacier National Park that originally belonged to them and that were ceded by treaty only with the proviso that they be permitted to hunt there."[28] As is well known, the idea of nature parks and later of biodiversity preserves was exported worldwide with the attendant requirement of excluding the indigenous peoples who lived in and from these lands. However, more recently, several governments have changed course and allowed the people who used to live in such preserves the right to remain in them.[29] Clearly, for the Kichwa-Lamistas, as for indigenous peoples everywhere, the sacred and the utilitarian are not separate.[30]

The Kichwa-Lamista leadership explains that their people have agreed not to practice agriculture in the area of conservation forest because there is so little forest left that were they to practice slash-and-burn techniques there it would lead to the same situation now existing in the Native communities. The leadership emphasizes the urgency of preserving forests to protect the very identity of future generations.[31] In his talk to US undergraduates in 2011, Misael Salas Amasifuen stressed the need for projects under the direction of experts to introduce new forms of permanent agriculture in Native communities. As he confirmed to me at the end of his talk, he had in mind our own Chacra-Huerto project at Sachamama Center.

It was precisely recognition of this need that led me to create Chacra-Huerto, a permanent agriculture project conducted in collaboration with the Kichwa-Lamistas and launched in the spring of 2010. The project's name is meant to immediately communicate that the food chacra need not be in the forest, away from the settlements, but can be created near them, like the house gardens, or *huertos*. Such huertos contain fruit trees, herbs, and medicinal plants as well as the local tomato and chilis, cultivars originating from the high Amazon region. They are tended mostly by the women of the house and are permanent.

Amazonian Dark Earth

The recent discovery of Amazonian dark earth in the Amazon basin has generated a great deal of excitement and already produced many changes in disciplinary boundaries and the role of soil scientists such as Johannes Lehmann of Cornell University and William Woods of the University of Kansas. Soil scientists and others in related fields are becoming innovators intent on solving issues of food security and climate change, increasing biodiversity, and producing clean energy.[32] The Food and Agriculture Organization of the UN (FAO) has taken a keen interest in these discoveries and the possibility they hold for solving such urgent global issues. As W. I. Woods and W. M. Denevan write: "The topic is now of major scientific interest, of relevance both to prehistory and to agriculture development and global climate change today."[33]

Although these discoveries have generated a great deal of research, writing, and revisions of established knowledge as well as efforts to recreate these amazingly fertile soils, some of them several millennia old, there has been a remarkable absence of focus on the religious or spiritual aspect of this phenomena.

ADE is a pre-Columbian black anthropogenic soil full of broken ceramics. Such patches of dark earth have been found in many places in the whole Amazon basin, not only in Brazil but also in Ecuador, Peru, and

French Guyana.[34] These patches of ADE vary greatly in size, from two hectares to ninety hectares.[35] The dating of those soils also varies with the sites. What seems established is that agriculture in Amazonia may have begun four thousand to five thousand years ago, or even earlier. Manioc, an Amazonian domesticate, has been found in archaeological sites in coastal Peru dating some 4000 years BP, thus establishing the existence of its cultivation in Amazonia well before that time since it is a tropical cultivar not native to the Pacific coast of Peru.[36] The remarkable thing about ADE is that those soils are still fertile today. In several places contemporary farmers are using those patches for plantation of various fruit trees as well as for marketing it as potting soil.[37]

As Charles Mann's lively narrative in his book about the Americas before Columbus, *1491*, testifies, older archaeological/anthropological theories about the pre-Columbian Amazon have been turned on their heads with the more recent discoveries. This region was not sparsely populated by bands or tribes of horticulturalists practicing shifting cultivation through slash-and-burn techniques but rather was the home of large, permanent settlements with ceremonial centers and permanent agriculture. The older theories of Betty Meggers about the Amazonian environment putting serious constraints on the development of permanent agriculture have been proven wrong by these new publications.[38]

What made this level of complex civilization possible were the ADE and permanent agriculture and not swidden—or slash-and-burn—shifting agriculture. In swidden agriculture, a clearing is made in the forest: trees are cut and burned to give the poor yellow Amazonian soil some fertilizing phosphate in the ashes. These clearings can be cultivated for a few years—from one to four, depending on soil fertility—and then the forest is left to regenerate. Families stagger their various food fields and are thus constantly clearing new chacras by burning a patch of forest. Swidden agriculture is not able to support a large enough population to enable permanent settlements to emerge.

As the anthropologist Robert Carneiro of the American Museum of Natural History established in his experiments in the 1970s, the amount of time required to clear a field of an acre and a half (the average size of a food field) with a stone ax is over five months. Using steel axes, workers he hired cleared a field the same size with the same number of workers in eight workdays.[39] According to the geographer William Denevan, metal tools brought in by the Spaniards in the sixteenth century largely created slash-and-burn agriculture. This is the form of agriculture used by *campesinos*, both indigenous and mestizo, in the Amazon region today. Slash-and-burn or swidden agriculture has become one of the important forces behind the loss of tropical forest. Although it permits the forest to regenerate, it is very inefficient and environmentally unsound under the present land-holding patterns. The burning sends most of the nutrients up in smoke and releases carbon dioxide into the atmosphere. Swidden agriculture is the third-largest cause of carbon dioxide emissions in the Amazon region, and in general forest conversion is estimated to contribute globally some 25 percent of the net carbon dioxide emissions to the atmosphere.[40]

The demographics research of Henry Dobyns has shown that in this region a major demographic collapse occurred shortly after the Spanish invasion.[41] Due to a combination of a lack of immunity to such European diseases as smallpox and influenza, enslavement, and war, the Native population declined with amazing rapidity within a few decades after the conquistador Francisco de Orellana and his friar, Gaspar de Carvajal (who wrote of this voyage), first traveled down the Amazon in 1541, only nine years after the first conquistador, Fancisco Pizarro, reached what is now Peru. The arrival of the Spaniards spelled disaster for the Amerindian population: in a few decades nine out of ten Amerindians died. New scholarship now is taking Carvajal's report of his trip down the Amazon much more seriously. There he describes large settlements and thousands of people lining the banks of the Amazon to greet the passage of his ship. Since subsequent voyages down the Amazon reported only forest and very few people, earlier researchers had concluded Carvajal had made it all up. However, Carvajal's report and descriptions from Captain Altamirano, who was with Aguirre in 1561, mention villages stretching for many kilometers along high bluffs, with houses touching one another.[42] Archaeological digs have confirmed that at the time of Carvajal's voyage, large permanent settlements and imposing ceremonial centers, complete with pyramids, existed. Beautiful, complex ceramics have also been found.[43]

Ceramic production in Amazonia may have begun before 7000 BP, according to evidence from a site called Pedra Pintada, as reported by Anna

Roosevelt. This date places ceramics in Amazonia among the oldest not only in the Americas, but indeed in the world. It would appear that with the disappearance of the vast majority of the Amazonian peoples, the memory as well as the ability to create ADE, and with it the complex civilization it made possible, was lost. Slash-and-burn agriculture became the norm for the Native population as well as for the poor mestizo population in the region, generating a dispersed demographic pattern of small, semi-permanent settlements. Slash-and-burn agriculture was supplemented with hunting, fishing, and gathering, a pattern that still exists among the Native population. This population has no memory of what their ancestors knew and of creating a permanent agriculture.

Amazonian Dark Earth's Miracle Ingredient: Biochar

The black color of ADE is due to charcoal carbonized with little or no oxygen. Scholars are not able to pinpoint the technology used by these pre-Columbian farmers, but they have analyzed this charcoal and determined that it is produced by a slow, oxygen-less method called *pyrolysis* from a variety of biomass. Biochar is porous and can retain organic nutrients over extremely long periods, even under conditions of the typical torrential downpours of the region. This particular characteristic is credited for the remarkably long-term fertility of those ADE soils. In addition to this key characteristic, biochar has the capacity to sequester carbon dioxide from the atmosphere once put in the soil. Furthermore, the process of creating biochar, pyrolysis, is one that does not release carbon dioxide into the atmosphere while burning the biomass but rather sequesters CO_2.[44] In other words, as Albert Bates notes, the process is a clean one.[45]

These characteristics inspired a desire to re-create ADE not only for food security in the Amazon basin but also for atmospheric carbon dioxide sequestration, addressing issues of climate change not only in that region but worldwide. One of the pioneer soil scientists who initiated research on ADE in the 1960s, the Dutch scientist Wim Sombroek, worked incessantly to safeguard the threatened Amazonian rain forest and develop economically viable systems of land use. He called this effort the Terra Preta Nova project. Charles Mann writes in a foreword to *Amazonian Dark Earths: Wim Sombroek's Vision*: "The *Terra Preta Nova* project, with its vision of the soil as a key element in our common future, has attracted enormous public attention to soil science. On a professional level, the scientific collaboration links soil scientists, archaeologists, geographers, microbiologists, engineers, ecologists, economists, and atmospheric scientists around the world in a common project that promises to reveal much about the workings of soil, may have an enormous impact on agriculture and could even play a role in climate change."[46]

Chapters 18 and 28 in *Amazonian Dark Earths* are devoted to Sombroek's project. In the early 2000s, the Brazilian state of Pará granted the use of one of its municipalities, Tailandia, for this project. In the last chapter of the book, soil scientist Johannes Lehmann, who has contributed enormously to the study of ADE in his soil science laboratory at Cornell University, concludes, "Biochar is able to directly address actions 3, 4, and 7 of the UN Hunger Task Force. . . . It is a promising approach for the suggested entry point by the Task Force to invest in soils as a battle against world hunger. . . . The conversion of biomass into biochar can either be primarily a net withdrawal of carbon dioxide from the atmosphere or a net emission reduction or both."[47]

If the biomass utilized for the creation of biochar is chosen responsibly, that is, mostly from agricultural by-products rather than through cutting trees, this would constitute a renewable source of clean energy. Since many of those agricultural by-products are burned in the open air, releasing large quantities of both methane (from their rotting) and carbon dioxide into the atmosphere, the systematic conversion of such agricultural by-products into biochar would reduce CO_2 emissions twice over: once by sequestering carbon dioxide from the atmosphere and again by preventing the release of carbon dioxide and/or methane into the atmosphere through the burning or open-air rotting of such biomass waste.

Working with Native Communities: Sachamama Center's Chacra-Huerto Project

Our demonstrative chacra-huerto at Sachamama Center has been visited not only by the Kichwa-Lamista leadership but by many other indigenous farmers. This has led to several communities requesting us to help them create similar fields in their settlements. The amount of degraded lands in Native communities is growing and the forest is receding. Many Kichwas are looking for an alternative. To date we are working with four Native communities, all within a half-hour ride from Lamas.[48] The comuneros, with the help of our chacra-huerto team, have prepared terraced fields in their community on degraded lands.[49] The communities decided that these fields would be communal ones, the produce to be used for village feasts, and everyone works on them in the traditional form of communal labor known as *choba-choba*.[50]

In the three communities where they have already had a harvest, the chacra-huertos have produced well. We are of course well aware that only time will tell whether this experiment can replace the chacra in the forest and swidden agriculture. It requires that ADE be applied to the fields after each harvest and mixed with the existing soil in a 50 percent proportion in order to build up a depth of black soil sufficient for nourishing plant roots.

Thus the Chacra-Huerto project of the Sachamama Center nonprofit organization in Lamas, carried out in collaboration with several Kichwa-Lamista communities, is indeed a Terra Preta Nova project corresponding to Wim Sombroek's vision of re-creating these amazingly fertile and sustainable pre-Columbian Amazonian dark earths for the purposes of food sovereignty for the smallest farmers and of ameliorating the carbon dioxide level of emission in the region, thus contributing to ameliorating the emission of greenhouse gases as well as strengthening indigenous spiritual agricultural practices. The terrain in the Lamas district is mountainous, the flat river valley floors having been taken by more powerful mestizos and devoted in large part to irrigated rice cultivation. Lamas and its mountainous environment is the region where the Kichwa-Lamistas are settled. As mentioned above, there is evidence of ADE in terraced fields at the higher altitude in this region, the archaeological sites of Chachapoyas.[51]

Concluding Remarks

At Sachamama Center we follow the lead of our Kichwa-Lamista collaborators and give offerings of broken ceramics to our reconstituted ADE, or yana allpa. We involve US undergraduates in these enactments, suggesting that they speak in their own words, from their hearts, to the soil, the water, the forest, the moon, or the sun. We do this as a regenerative action, one that embodies the recognition that the soil and all the other elements contributing to sustainable permanent agriculture have agency and are not part of a mechanical world. We need not subscribe to the Kichwa-

Lamista theology, "belief system," or "cosmovision" in order to do so. For us, the spirit of the earth, of the moon, of the water, of the sun, and of the forest embody the agency of those nonhuman entities. The agency of those aspects of the natural world implies the rejection of the modernist anthropocentric worldview of an inert, insentient, mechanical universe, full of "natural resources" there for us humans to exploit according to only our own exclusively human will. Making offerings to the spirits is our way of recognizing that without reciprocating with our nonhuman food source, we are endangering the regenerativity and therefore the sustainability of the source of our sustenance. In other words, we engage in cosmocentric economic activities. At the same time, the presence of ceramics in ADE also increases its fertility, according to soil science.[52] Whether one considers this a happy coincidence or the gift of the spirits depends on one's spiritual inclinations, but either way the nonhuman sources of our sustenance are regenerated through such reciprocating actions and our dependence on them is enacted with offerings, a behavior I would argue deserves the label "cosmocentric economy."

If we are to ameliorate the climate crisis and avoid the apocalyptic conflagration of the planetary atmosphere that a pronounced disequilibrium between oxygen and CO_2 would lead to in less than ninety years if we do not radically alter our course, then we need to start behaving differently, nonanthropocentrically, and recognize the agency of the nonhuman world and our dependence on it.[53] It is not sufficient to change our thinking; we need to change our behavior as well. We need to start enacting a different reality through actions that are at once technical and spiritual, and we need to involve our youth in such enactments. Studying the Kichwa-Lamistas in the classical anthropological fashion is not enough; we need as well to learn from them so as to enact together a healthier and sustainable world in which regeneration becomes an inescapable human responsibility.

Study Questions

1. What are the key terms, roles, and deities associated with indigenous and traditional Peruvian high Amazon farming? How and why are these important, and how do they function in helping to create a sustainable farming culture?

2. What is a cosmocentric economy, and how does this relate to sustainable agriculture in Peru?
3. What is a regenerative action, and how is this important to cultivating a sustainable agriculture? How might religions articulate and encourage regenerative actions, and how might they articulate and discourage actions that are degenerative?

Notes

1. *Chacra* is a Quechua word (which has entered the local Spanish language) for a food swidden garden. Mama Allpa is the earth spirit; *allpa* also means "soil."

2. *Shaño* is the local Quechua word that refers to pieces of broken ceramics that the women who are the ceramicists also grind to a powder and add to the new clay to make new pots.

3. *Ingenieros* refers to agronomists who overwhelmingly are mestizos (descendants of indigenous and Europeans persons) and whose scientific education generates this kind of attitude. However, as Fernando Baez writes in *El saqueo cultural de América Latina: De la conquista a la globalización* (Mexico City: Random House, 2008), the act of making the indigenous people ashamed started as soon as the Spaniards invaded South America.

4. There are two major agricultural campaigns in these tropical eastern foothills of the Peruvian Andes, June and December, when indigenous farmers practice swidden agriculture, using the slash-and-burn technique to open food fields in the forest that produce from one to four years until the soil is exhausted and another chacra has to be opened in the forest.

5. Fabiola A Silva, "Cultural Behaviors of Indigenous Populations and the Formation of Archeological Record in Amazonian Dark Earths: The Asurini do Xingu Case Study," in *Amazonian Dark Earths: Origin, Properties, Management,* ed. Johannes Lehmann, Dirse C. Kern, Bruno Glaser, and William I. Woods (Dordrecht: Kluwer Academic, 2003), 373–85.

6. Alfredo Narváez Vargas, personal communication, August 11, 2012. Dr. Vargas bases his identification of broken ceramics as offerings on his findings in a site in Tucume on the northwest Peruvian coast, where he found a great quantity of such fragments in a small altar of the Sacred Stone of Tucume, a site contemporaneous with Kuelap.

7. The literature on the brutal eradication of indigenous spirituality during the colonial period is extensive. I would refer the reader to an excellent study ranging from colonial times to the current era of globalization: Baez, *El saqueo cultural de América Latina*.

8. Thomas Cummins, personal communication, December 2011.

9. Alfredo Narváez Vargas, personal communication, August 31, 2012.

10. Gerry Gillespie, "City to Soil: Returning Organics to Agriculture; A Circle

of Sustainability," in *Amazonian Dark Earths: Wim Sombroek's Vision*, ed. William I. Woods, Wenceslau G. Teixeira, Johannes Lehmann, Christoph Steiner, Antoinette WinklerPrins, and Lilian Rebellato (Springer Digital Book, 2009), 465.

11. Frédérique Apffel-Marglin, *Subversive Spiritualities: How Rituals Enact the World* (New York: Oxford University Press, 2011).

12. Stefano Varese, "El dilema antropocéntrico: Pueblos indígenas, naturaleza y economía política" (paper delivered at the PUCP conference, Lima, Peru, September 17, 2009).

13. R. Alavarado and Jaime Walter, "Fases de la luna y su influencia en los cultivos agrícolas" (PowerPoint published by the Faculty of Agronomy of the Universidad Nacional de San Martin, Tarapoto, 2005).

14. Apffel-Marglin, *Subversive Spiritualities*, 21.

15. Pramod Parajuli, "Learning from Ecological Ethnicities: Toward a Plural Political Ecology of Knowledge," in *Indigenous Traditions and Ecology: The Interbeing of Cosmology and Community*, ed. John A. Grim (Cambridge, Mass.: Harvard University Press, 2001), 559–89.

16. Marcel Mauss, *The Gift: Forms and Functions of Exchange in Archaic Societies*, trans. Ian Cunnison (New York: Norton, 1967), 10.

17. Lewis Hyde, *The Gift: Imagination and the Erotic Life of Property* (New York: Random House, 1983), 128.

18. For an argument on the entanglement of the material and the discursive using Niels Bohr's understanding of quantum mechanics, see chapter 4 in my *Subversive Spiritualities*.

19. On this point, see especially chapters 5 and 6 of ibid.

20. As Leora Batnitzky shows in her book *How Judaism Became a Religion: An Introduction to Modern Jewish Thought* (Princeton: Princeton University Press, 2011), the category of "religion" is a Protestant one, since for Judaism, as for Hinduism and probably many other spiritual traditions, one's spirituality is not a private affair of "beliefs" held in the privacy of one's mind and heart. But she also shows how this new category became inescapable for everyone through the establishment by the nation-state of a separation between a political public sphere and a private sphere where "religion" belonged. The secularization of anthropology—and for that matter of the modern education system as a whole—is part of this phenomenon. For anthropology this development is both ironic and problematic since typically the people it studies do not recognize something like a totally secular sphere.

21. Charles Mann, *1491: New Revelations of the Americas before Columbus* (New York: Alfred Knopf, 2005).

22. Rider Panduro and Grimaldo Rengifo, *Montes y montaraces* (Lima: PRATEC, 2001).

23. Coca is a traditional plant originating from the high Amazon region and used in the highlands for both medicinal and ritual purposes since time imme-

morial. The increase in coca production during the 1980s and 1990s was due to the enormous upsurge in US demand for cocaine. The traditional uses require the unprocessed dried coca leaf with an alkaloid content of no more than 2 percent, whereas the production of cocaine chemically treats the plant and transforms it into a drug with an alkaloid content of some 80 percent and requires a much higher quantity of the leaf. Since the government of Fujimori imprisoned the leader of the Shining Path guerrilla movement (in 1994) and arrested the leadership of the other guerrilla movement, the MRTA, the growing of coca in the region has dramatically decreased.

24. Lecture in my study abroad program given by agronomist Cesar Enrique Chappa of the Faculty of Agronomy at the University of San Martin, Tarapoto, January 2007; personal communication, Oro Verde Fair Trade Coffee co-op, Lamas, San Martin, Peru.

25. Luis Calderón Pacheco, "Relaciones entre mestizos y Kechwa en Lamas en el contexto de la globalización," in *Comunidades locales y transnacionales: Cinco estudios de casos en el Perú*, ed. Ivan de Gregori (Lima: Instituto de Estudios Peruanos, 2005).

26. The person Velasco chose to determine what to do with Amazonian indigenous people was the Peruvian anthropologist Stefano Varese, who had just (1968) published his justly famous book on the Asháninkas of the central Amazon in Peru, *La sal de los cerros: Resistencia y utopía en la Amazonía Peruana*, 6th ed. (Lima: Fondo Editorial del Congreso del Perú, 2006). Varese created both the law of Native communities and the law of bilingual education in indigenous areas, which were put in effect around 1975.

27. Calderón Pacheco, "Relaciones entre mestizos y Kechwa en Lamas."

28. William Cronon, "The Trouble with Wilderness; or, Getting Back to the Wrong Nature," in *Uncommon Ground: Rethinking the Human Place in Nature*, ed. William Cronon (New York: Norton, 1995), 79.

29. Luis Vivanco, *Green Encounters: Shaping and Contesting Environmentalims in Rural Costa Rica* (New York: Berghahn Books, 2006).

30. For the peasant and *adivasi* cosmovisions as well as practices of sacred groves in South Asia, see Frédérique Apffel-Marglin and Pramod Parajuli, "Sacred Groves and Ecology: Ritual and Science," in *Hinduism and Ecology: The Intersection of Earth, Sky, and Water*, ed. Christopher Key Chapple and Mary Evelyn Tucker (Cambridge, Mass.: Harvard University Press, 2000), 291–316.

31. Sachamama Center fully supports the Kichwa-Lamistas' efforts to have control over the biological reserve Cordillera Escalera.

32. For a sample of publications on this topic, see Mann, *1491*; Lehmann et al., *Amazonian Dark Earths*.

33. W. I. Woods and W. M. Denevan, "Amazonian Dark Earths: The First Century of Reports," in Woods et al., *Amazonian Dark Earths*, 2.

34. www.clas.ufl.edu/users/caycedo/iquitos (accessed January 6, 2012); www.en.wikipedia.org/wiki/Terra_preta (accessed January 6, 2012); Alfredo Narváez Vargas, personal communication about ADE in the Peruvian high Amazon.

35. See E. G. Neves, J. B. Petersen, R. N. Bartone, and C. A. da Silva, "Historical and Socio-cultural Origins of Amazonian Dark Earths," in Lehmann et al., *Amazonian Dark Earths*, 29–50. A hectare is the equivalent of slightly over two acres.

36. Ibid, 34.

37. For an informative journalistic portrayal of the story of ADE, see the BBC film *The Secret of El Dorado* (2002), available through Google videos.

38. Among these are Anna Roosevelt, *Moundbuilders of the Amazon: Geophysical Archaeology on Marajó Island, Brazil* (San Diego: Academic Press, 1991); Mann, *1491*; Lehmann et al., *Amazonian Dark Earths*; Woods et al., *Amazonian Dark Earths*.

39. Robert Carneiro, "Tree Felling with the Stone Axe: An Experiment Carried out among the Yanomamö Indians of Southern Venezuela," in *Ethnoarchaeology: Implications of Ethnography for Archaeology*, ed. C. Kramer (New York: Columbia University Press, 1979), 21–58.

40. S. N. Swami et al., "Charcoal Making in the Brazilian Amazon: Economic Aspects of Production and Carbon Conversion Efficiencies of Kilns," in Woods et al., *Amazonian Dark Earths*, 419.

41. Henry Dobyns, *Their Number Become Thinned: Native American Population Dynamics in Eastern North America* (Knoxville: University of Tennessee Press, 1983).

42. See Gaspar de Carvajal, "Discovery of the Orellana River [1542]," in *The Discovery of the Amazon According to the Account of Friar Gaspar de Carvajal and Other Documents*, ed. H. C. Heaton, trans. Bertram B. Lee (New York: American Geographical Society, 1934), 167–235; Mann, *1491*.

43. C. Steiner, W. G. Teixeira, W. I. Woods, and W. Zech, "Indigenous Knowledge about *Terra Preta* Formation," in Woods et al., *Amazonian Dark Earths*, 193–204.

44. Lehmann et al., *Amazonian Dark Earths*; Woods et al., *Amazonian Dark Earths*.

45. Albert Bates, *The Biochar Solution: Carbon Farming and Climate Change* (Gabriola Island, B.C.: New Society, 2010).

46. See Charles Mann's foreword to Lehmann et al., *Amazonian Dark Earths*, xiii.

47. Lehmann et al., *Amazonian Dark Earths*, 482.

48. Ideally, we hope that this Terra Preta Nova experiment will spread to as many Native communities as possible. In the summer of 2012, in collaboration with Professor Pramod Parajuli (also one of the editors of this volume), we started a pilot project with the local education board to introduce the teaching of ecological literacy in schools in the Lamas district. Through this project we hope this form of indigenous permaculture will spread in the region.

49. There are three types of degraded lands recognized locally according to the weeds growing in them: *shapumbales, yaraguales,* and *cashukshales*. The forest no longer can regenerate on these unproductive lands.

50. The practice in choba-choba is that the family on whose land the work party labors that day prepares a large meal for everyone. Through the Chacra-Huerto project, Sachamama Center has been providing the food, which is cooked by the women of the communities.

51. This does not mean that there were no terra preta sites in the Lamas region, only that no archaeological excavations have taken place in the area, since there are no spectacular ruins there.

52. This is confirmed in John Vandermeer and Yvette Perfecto, *Breakfast of Biodiversity: The Truth about Rain Forest Destruction* (Oakland, Calif.: Institute for Food and Development Policy, 1995), 34. I am grateful to Dr. Peter Sherman of Prescott College for this reference.

53. According to scientists, the upper acceptable limit of the amount of CO_2 in the atmosphere, namely 350 parts per million, has already been passed. As of the end of May 2013, we are now at 400 parts per million. At 600 parts per million, the planetary atmosphere is predicted to self-combust, ending all life on earth (Bates, *The Biochar Solution*, 76). If we continue with the same industrial, economic, and cognitive paradigm dominant today and do not change course, this is supposed to happen around 2100.

3

Renewal of Non-Western Methods for Sustainable Living

Eston Dickson Pembamoyo

For several decades, largely due to colonial influences, many Malawians have turned toward a more "Western" culture of speaking, dressing, and eating. As a result, they have become ashamed of traditional foods, refusing to eat or serve them to guests, even at the expense of their own health and lives.[1] Importantly, for the purposes of this book, many have also adopted "industrial" farming.[2] However, not all of this change is necessarily bad. We in Malawi have much to learn from other cultures' successes, but we can also learn from their mistakes and should avoid repeating them. Such a dialogue about what methods of farming and thus eating—traditional, Western, or a mix of both—and which are the most helpful and sustainable for Malawi, both culturally and environmentally, is the focus of this chapter.

Of key concern is that if this replica of Western farming and lifestyle continues, Malawi might never go back to its historical, precolonial, community-based food sovereignty. This is a valid concern, and we in Malawi should adhere to practices and principles of "socio-economic cultural systems that can be carried out in perpetuity."[3] From my perspective, in Malawi there must be a deliberate sociopolitical and religious will to foster change and to develop policies that are eco-friendly, familiar to symbols and signs already present in Malawi's historical practices.[4] Given Malawi's historical antecedents of community-based, sustainable agriculture, my experience teaching permaculture in Malawi is that those Malawians who attend permaculture courses find themselves walking on familiar ground toward *mwanaalirenji*, or the "Growing of the Kingdom of Abundance."[5] Permaculture presents a mix of Western and traditional biocultural farming practices that can help farmers and citizens of Malawi move back toward a domestic sus-

Malindi permaculture design course, December 2009—two weeks' participation results in a certificate. (Photo by Eston Dickson Pembamoyo.)

tainable agriculture that is in line with their cultural and religious values of sharing and environmental protection.

This chapter explores the impact that church-sponsored permaculture is beginning to have on a return to sustainable farming in Malawi, particularly in certain parts of the country. One key area where this is occurring is Malindi, which is in the Mangochi District. Malawi is a representative republic in southeastern Africa; formerly the British protectorate of Nyasaland, it gained complete independence in 1964. It is bordered on the north by Tanzania, on the east by Lake Malawi, on the southeast and south by Mozambique, and on the west by Zambia, with its capital being Lilongwe.

The following concepts, defined from an African perspective, will help guide our discussion about religion and sustainable farming in Malawi.

Sustainability. This requires that biotic and abiotic elements of an ecosystem interact in a way that does not compromise the ability of future generations of organisms in that ecosystem to live and produce necessary things for themselves. People who are living sustainably within an ecosystem keep all their natural resources healthy so that they remain useful for their own and subsequent generations while ensuring that other species have adequate access to ecosystem services so that they can continue to flourish.

Permaculture. This is an agricultural design system that aims to create stable productive systems that wholesomely provide for human needs and harmoniously integrate various land uses.

Food sovereignty. This means total control over diversified agricultural resources, and includes the way people access resources, policies

Map of Malawi. (Based on http://www.maphill.com/malawi/location-maps/flag-map/blank-outside/free/.)

concerning equitable distribution, and the socioeconomic factors attached to the resources in a free society.

Food and nutrition security. This means that a community has enough varied foods available year round for the physical, mental, and spiritual health and growth for everyone.

Main Agricultural Challenges in Malawi

A key motivating concern in this discussion is the "environmental destructiveness" of human society, which in Malawi is becoming evident as the country adopts certain Western thought systems and lifestyles.[6] For example, the adoption of monoculture farming has led to forests being cleared away for the creation of huge single-variety fields. Anything else that grows in such fields is a weed and is either uprooted or sprayed by chemicals. Historical forestry products such as wild fruits, mushrooms, and wild animals are no longer available to supplement grown foods, and yet the food grown is still not enough for everyone. As the result of Western-influenced Green Revolution farming practices, there is too much loss of topsoil, which is washed away by just a little amount of rainfall, and this is exacerbated during Malawi's rainy season. Siltation therefore takes place in almost all lakes and rivers, and water levels keep dropping yearly. Fish catches in various lakes are affected to the point that it is difficult to supply fish products to the growing population of people; this is also a cultural issue, as fish products are a part of Malawi's culinary identity. Furthermore, industrial farming uses large amounts of freshwater, so some water tables are rapidly dropping compared to historical levels, and boreholes in some areas now dry up yearly. Lastly, human-induced climate change is impacting rainfall patterns, which affects maize production—and maize is a key crop for almost all farmers in Malawi.

Over the years, the Malawian people as well as the settlers, missionaries, and colonists who came to the country have debated agricultural issues intensely and developed the methods they thought were helpful to meet the needs of their own time. Neglected in these efforts are rich seams of intellectual and cultural history: the indigenous agricultural practices, crops, fables, and local ecoreligious knowledge. Since the onset of the Green Revolution, and more so since the independence of Malawi and our slow entry into the global marketplace, a lot of resources and time have been invested on cash and staple crops that use synthetic fertilizers and chemicals that have affected soil alkalinity, farmer livelihoods, and the quality of local waters—similar to processes that have occurred in South America and South Asia, as seen in other chapters of this book.

But a new story is emerging now in Malawi, and it is a story of equality, based on deep historical antecedents. In this story, we in Malawi are not duplicates, destined to repeat the mistakes and follow the lifestyles of the West, but we are not separate, either, given that we live in a globalized world. We are profoundly connected on this planet. Our interconnectedness is mutual, not parasitic, at both the macro and micro levels, between humans, and between humans and nonhumans, and this connection can be acted out and realized via sustainable agriculture operating under a paradigm different than that held by Western industrial agriculture. In this emergent story we will reenter a land of mountains and valleys, watered by rain, where everyone responsibly takes care of and watches over all things together. This is the vision for religiously inspired sustainable agriculture in Malawi.

The story is preparing a new Malawi to be connected with Mother Earth, with our Malawian ancestors, and with the creator.[7] Together we in Malawi shall fix issues of soil erosion, deforestation, inadequate potable water, high population growth, the catastrophic increase of HIV/AIDS, and the decline in fish catch in our lakes. This new Malawi will have "rain fall [that] is adequate and the earth [will] bear 'things' and fish [will] multiply in all lakes."[8] Our solutions are in part based on permaculture, "a practical and effective way [that] uses natural patterns and natural principles for increasing sustainability and quality of life."[9]

Colonial Methods of Agriculture

A critical analysis of recent Malawian political-environmental history is important for understanding the relationship between colonizer and colonized, white European and black African—between Western farming knowledge celebrated as superior and African practices presented as environmentally wasteful, in need of Western technology for improvement. The truth is, however, that colonial and postcolonial farming methods use extensive synthetic fertilizers and chemicals that contribute to the food security and food sustainability problems that Malawi and other African countries now face. Within this history, the predatory character of some white settlers and missionaries, and their imperial hunting and farming methods, greatly reduced wildlife and depleted large tracts of forests. They were not only responsible for the extermination of indigenous animal and plant species, but they also introduced new species that affected climate and brought about new difficulties—diseases and droughts.[10]

Fortunately, there is still a vast store of ancient farming knowledge in communities in Malawi. It is good that sustainable farming researchers have discovered that indigenous farmers, who have sometimes been portrayed as ignorant and not adaptive, have actually been utilizing very sophisticated methods of agriculture that could help deal with the current challenges. These methods can help both developing and developed worlds to grow more and better food with fewer chemical inputs, slow erosion, control pests, decrease dependence on fossil fuels, and feed an expanding global population.

The harmful effects of pesticides, inefficient fossil fuel usage, chemical fertilizer inputs, genetic monocropping, and laboratory-enhanced farming of livestock and fish have become increasingly hazardous and must be checked (see Todd LeVasseur's introduction to this book). We in Malawi should build upon the traditional methods that evolved during the "primitive" years of agriculture. Unfortunately, despite all the disadvantages of industrial farming methods, the current and recent past generations of Malawians have neglected much of the knowledge of traditional systems in pursuit of unsustainable modern industrial methods geared toward economic profit as the sole determinant of success.

My experience as a lifelong citizen of Malawi is that Western methods of agriculture have failed us; hundreds of thousands of people in Malawi are undernourished and still go hungry, and the efforts of the developed nations to arrest hunger have failed, too. It should teach us that not all that is "West" is "best"—to be black is not to be blank, and to be white is not to be right. Together, black or white, yellow or red, we should be focusing increasingly on sustainable methods of farming that are environmental friendly, just, and secure, bringing about the well-being of all communities the world over.

Alternatives to Intensive Chemical Farming

If current agricultural conditions are to be corrected, we need a science of design based upon biological principles that can provide a sustainable way of living. It should be able to blend harmoniously the way humans live with nature and use the diversity, stability, and resilience of natural ecosystems to provide a framework and guidance for people to develop their own solutions to the problems facing their world, including the ability to sustainably produce food. Malawi needs a system that is based upon the use of organic farming in order to produce good-quality food without using synthetic chemical fertilizers and pesticides, a system that will help reduce diseases and lower the current negative environmental impact of industrial farming.

One large-scale alternative that can help move Malawi toward sustainable food security and food justice is permaculture. This neologism, a combination of "permanent" and "culture," denotes integrated land use and lifelong habitual processes and practices designed for permanent sustainable living. This permanent culture works toward a sustainable lifestyle that creates abundance on earth by designing our lives and livelihoods to be in harmony with nature. It is about applying a few strategic inputs to agricultural production systems and achieving a high output with the help of nature. And, key to the concept, permaculture is an interlocked cultural farming system that endeavors to bring about an environmentally sustainable, spiritually fulfilling, and socially just human presence on the planet.

Creating Abundant and Better Food in Society

There are numerous environmental challenges in many parts of Malawi, including Malindi, and they synergistically contribute to the growing food insecurity of the region and country. Many households in Malindi do not have enough food toward the end and at the beginning of each year. It was never like that four decades ago: Malawi was a green land of "plenty."[11] Our climate and rainfall patterns were stable and predictable. A term that captures this is *mwanaalirenji*, which means being in possession of all necessary materials in life. It is not only human beings who need enough and better food, but also the land that needs to be continually well fed for it to support plant and animal life sustainably. But how can we speak about abundance when in many parts of Malawi some families have no food during the hungry season? When local farmers cannot afford to buy farm inputs? How can Malawians become food secure when we grow maize only?

Permaculture Testimonies

People use the term *food security* differently throughout the world. In Malawi, some political leaders think that enough maize production to last the country all year means food security and health. They measure food security by the amount of calories each household can obtain from an acre of maize.[12] This is why *sasakawa*, or planting the maize at only ten centimeters apart, has become standard in Malawi. Perhaps what people need is not food security but rather *food sovereignty*—the creation of abundant varieties of foods and good soils.[13]

One example of Malawian food sovereignty is seen in the work of Hugh Brown in Mzimba in the north of Malawi. He writes, "Since 2002 we have been doing careful trials with the agroforestry legume *tephrosia vogelii*. In December 2005 we planted maize in some plots on which *tephrosia* had been growing for the previous three years."[14] Brown and his colleagues claim that yields have exceeded their expectations, and yet the soil in which they grow their crops is no better than the average Malawi soil, and their test plots include some stony and gravely patches. No chemical whatsoever is being used in this farm ecosystem. Interestingly, the highest yields they got were from the zero-tillage farming system they adopted from permaculture.

Another permaculture-inspired example of increasing food sustainability is seen in the work of Harry Phiri, a primary school teacher in the Zomba-Eastern region. He writes, "In the past 40 years Bill Mollison and David Holmgren discovered permaculture in Australia and began to develop their own eyesight to see what happened to their land through foreign farming methods. Using their [permaculture] methods, it has taken 12 years of design work to change our home at Matawale Township and we have brought nature back to our door step."[15]

Mac Justice Betha, a specialist in approved natural medicines (ANAMED) from Chikhwawa in southern Malawi, reports on efforts in his area: "Now Permaculture information is becoming widely available nationally, advocating the 'Growing and Eating of More Good Food Using Less' and the importance of natural medicines, with these messages spreading. Shortly the 14 posters of Permaculture Design Course will be printed, so that the information on 'how to grow your own better food and heal yourself' will be available to many pupils in primary schools in our area where the posters will be hung."[16] Relatedly, Kamchira Kampandira, a permaculture coordinator in central Malawi, notes, "It is always amazing to see that Permaculture practitioners and groups are increasing. Permaculture is real, we are happy we can grow better food around our home if we only go nature's way. We do not sweep our designed grounds any more. Permaculture is now paying us back more abundantly than sweeping that only made our home[s] grow taller—as they lost inches to erosion."[17]

We all need food produced without "chemical fertilizers, pesticides and use of hybrid seeds."[18] Permaculture does not demand huge sums of money to buy agrochemical inputs, and it is emerging as a viable strategy to solve the dilemma of food insecurity. It applies techniques and principles from ecology, cooperative economics, appropriate technology, sustainable agriculture, and the wisdom of indigenous people to create sustainable human environments. As such, the promise this practice has for both society and soil extends far beyond abundant food production systems to explore new horizons for creating a diversely sustainable life on earth.

Permaculture is a helpful strategy for all—rich and poor alike—facilitating the growth of crops and aiding communities to equitably distribute and share healthier foods. Richard Matondo, a farmer from Malindi in Malawi writes, "Permaculture is real, and can help change life for [the] better. . . . I have food and medicinal forests around my house, and today people are able to differentiate . . . the house of a permaculture member from the rest."[19] Those in Malindi who are adopting permaculture are being rewarded by nature, which provides more bountifully, and the houses of permaculture members, surrounded by varied crops, enjoying cool conditions provided by trees, and evidencing improved soil texture and fertility, strongly suggest that extensive single-crop farming under the Green Revolution has "undernourished" us.[20]

Historical and Sociological Attempts to Achieve Sustainable Agriculture

Interlinked issues concerning the nurturing, keeping, and growing of food are central in any society. In fact, without this, there could be no society, as we need food to live. People worldwide have different beliefs, values, and traditions surrounding such nurturing and growing, including the way food is kept and used. Families with enough and varied foods all year round in a community are held with high esteem and accorded dignity, while those with less or no food are often seen as being no better than "slaves," or they may be considered "lazy freeloaders." Before the coming of Western lifestyles, those without enough food were helped by fellow local Malawians, who helped the sick, elderly, and even the lazy prepare their gardens through *dima*.[21]

Traditionally, Malawians ensured that they had a continuous year-round food supply through safe storage of yields, and they grew a wide variety of crops. This is evident in the songs, stories, and rituals of the Yao of Malindi about the goodness of working hard in a crop field, the need to grow more than one type of crop in a garden, and the importance of early and timely planting and weeding, as seen in the lyric "tachilapa sala kogoya" (lest you shall espouse famine). Permaculture, like the "primitive" and traditional farming of Malawi seen in such traditional songs, encourages intercropping. Given good rainfall, an agrochemical farmer will harvest a single crop on the same piece of land where a permaculture farmer could continuously grow corn, cassava, millet, soybeans, sorghum, pumpkins, vegetables, cucumbers, beans, pigeon peas, groundnuts, cow peas, bananas, and some fruits.[22]

The Green Revolution—Scientific Discoveries, Inputs, and the Quest for Sustainable Living

The pursuit of abundance is a perennial human undertaking. At the dawn of the Enlightenment in the eighteenth century, a search for minerals took the place of the fifteenth-century pursuit of spices. New technologies spread rapidly, leading to the scramble for the earth's minerals, oils, and human resources—mostly for the benefit of political and economic elites, many based in Europe and later North America. An example is the "triangle trade" of slavery, which included detestable treaties made with greedy chiefs in Malawi, all of which greatly benefited Europeans.

The Green Revolution continues this reprehensible trajectory. After World War II, efforts shifted from manufacturing weapons to developing agricultural products and processes, including chemical inputs and hybrid crops, which became the precursor of today's genetically modified organisms. Throughout Africa, including in Malawi, huge hectares of forests were depleted to establish monocropping estates. These require enormous sums of monies for high-tech equipment and pesticides, and not many Malawians can afford this kind of farming. Hence, since the 1950s and 1960s, food insecurity has remained an unresolved issue despite numerous governmental initiatives and interventions. New agrochemical policies and increased rules and regulations do nothing to cultivate food security, as huge swathes of forest land continue to be wantonly cut down for farming. The locals not only end up being colonized, but they also lose their natural green woodlands. "Innocent" people who do not follow the new agricultural policies, agricultural systems, and new colonial farming methods face charges and penalties.[23] The promise of the European Green Belt, with all its chemical inputs, has spectacularly failed in Malawi.

Modern Malawi and Permaculture: A Little Light Glowing in the Darkness

Although they were suppressed, old thought systems and agricultural lifestyles remained in the blood of some Malawians. For decades these old thoughts have conflicted with those of post-Enlightenment modernity, but permaculture seems to be hearkening back to some old systems and practices. Even though we are faced with so many environmental challenges now, this is the most scientific age Malawi has ever experienced. It is an age that knows industrial chemical elements in the atmosphere can be reduced by planting many fruit trees.[24] Furthermore, the average Malawian knows the benefits of using renewable sources of energy and of recycling, although not many seem to bother. It appears that the mindset of many Malawians rests on industrial technologies and industrial farming. Yet, when "people look back to 40 or 50 years ago . . . crops were more abundant and easier to grow without use of agrochemicals and . . . trees still covered the hillsides."[25] In light of our current undesirable circumstances, we must ask ourselves why conditions were better fifty years ago and yet now "hunger still looms about in most homes."[26]

Ancestral farming views and practices probably hold the answers. If only we saw the little light in this "age of darkness" and used it now, we should be all right with ourselves and the earth. The ancestral views of Malawians on food include strong ties with the spiritual world as well as close links with the forests, deep water, and Mother Earth. Thick forests and individual huge trees are sacred, not to be cut down without permission from the elders, who have direct acquaintance with the living dead and the spiritual world through prayer and worship. In industrial farming, trees in a garden are merely weeds to be cleared away and deep waters are pumped for irrigation.

Traditionally, just before the first rains come, sorted seeds are brought to the elders for prayers. The life of the seeds has to be dedicated to God the creator, the one who raises life in the seemingly dying seed before it is put in the womb of Mother Earth to germinate. The elders among the Nyanja, Chewa, Yao, Lomwe, Ngoni, and many other tribes believe that when seeds are buried in an intercropped garden, each crop fixes its semen to boost fertility in the soil, helping to get a well-balanced garden ready for new healthy life so the young crops can multiply. The religio-spiritual views of this ancestral system sees soil, "earth crust," as a living sacred component and part of society that should be "respected." For example, Elizabeth Mzamu, sixty-one, from Pepmphero Orphan Care Centre near Lulanga Health Centre, stated during a six-day permaculture awareness course in July 2010, "We have always grown up believing the soil keeps communities of both plants and animals, and it is a final home for the living dead, yet we must revere it: that is where we shall all finally go and will be answerable for any wrong we did to it."

In regard to respect for nature, "primitive and spiritualist" farmers in Malawi's past looked after their crop fields to avoid loss of soil life.[27] They selected strong seeds that would ejaculate strong semen into the soil in the subsequent planting seasons. Seeds did not travel long distances, and they were preserved in clay pots or treated with smoke. In most cases they were free or bartered because it was believed that life could not be sold but must be supported to continue. In traditional agriculture, skills and religious beliefs concerning the growing, preservation, and storage of foods have been passed on from one generation to the next in homes, in annual public festivals, and in the initiation rite of passage at puberty. For a sustainable agriculture to arise in Malawi, such traditional skills and beliefs must be recognized and respected as valid, so they can be supported and put into practice.

Among the Yao of Malindi, a young person is allowed to marry if the uncles are certain that he or she can manage his or her own crop field, build a hut, make some basic tools, cook, and clean the hut. A young chicken is given to each inductee during the rite of passage as a necessary part of his or her new homestead. By the end of a month or two, when most rituals end, the inductees will have seen their chickens grow to maturity. Such community rites of passages within a defined kinship structure reinforce traditional ecological information, passing on needed farming knowledge to the next generation of subsistence farmers.

Permaculture and the Traditional Rite of Passage

A bridge is developed when permaculture models are compared to and fit within the traditional agricultural beliefs, values, and traditions of Malawi's various ethnic and tribal groups. There is still evidence to this agricultural past in songs, proverbs, and the teachings passed on during rites of passage among many tribes across the country that instruct participants to plant various crops that mature at different times; to plant maize and other crops together immediately after the first rains; to tend regularly their crop fields; and to support elders who are attending crop fields. During these ceremonies, girls are encouraged to feed babies with good foods and are taught to prepare enough good foods for the family at the right time.

One famous traditional initiation song among the Yao of Malindi goes, "Kulindima kwa wula kwende tutile-e, atati na mawo talime-e" (At the rolling of the thunder, let's run away; the father and mother will cultivate).[28] One proverb is "Jwamlume amloleje palutumbo" (Look at a man on the stomach).[29] If the initiation camp is near a river, as is often desired, inductees grow vegetables and other quick-maturing crops and also do some manual work such as mowing thatching grass, which they later sell.

Free-range goats and chickens provide fertilizer and help mulch the grounds at Chapanga in Chikwawa. (Photo by Eston Dickson Pembamoyo.)

The inductees communally share food in order to increase and strengthen their relationships. At the end of the period of the initiation, the inductees learn how to use manure from the chickens or goats reared during their stay in the bush. They are told that plants and animals are friends of humanity and they should be well looked after, as they provide timber, reeds for making coffins and mats, hides, meat, and manure.

During initiation, parents and guardians undertake certain responsibilities toward their inductees: they provide a sizeable clay pot used as a bucket for washing, and they give some cash and in-kind goods to the elders of the initiation in appreciation for their duties and efforts in instructing the boys or girls. Parents or guardians do not require any external donors, support partners, or subsidies to send their young into the bush to "eat honey."[30] On the final day they gather together for a big feast, thus increasing their sense of unity, belonging, and support for one another. These "principles and ethics in the rite of passage" among the Yao in Malindi are what perhaps make permaculture fit well into their system.[31]

Permaculture and the Renewal of Non-Western Methods of Sustainable Farming

Permaculture, through its interconnectedness with traditional Malawian cultural farming values and practices; its system of intercropping; its ethics of giving reverence to Mother Earth and valuing plants and animal life; and its admonishment to freely share the resources and wisdom in society helped people in Malindi to feel that traditional agriculture has come back to the peoples of "antiquities," to use the theologian Walbert Bühlmann's expression.[32] As Malawians learn the principles and ethics of permaculture, they find themselves walking on familiar ground, interacting with permaculture through their traditional lifestyles and ancestral teachings. Therefore, permaculture supports them in their pursuit of sustainable living and also helps them appreciate the goodness, love, humility, simplicity, and hospitality in "primitive" agricultural methods.

A Brief History of Permaculture in Malawi

In the mid-1970s, two Australian ecologists, Bill Mollison and David Holmgren, started to develop ideas for creating stable agriculture systems. "They called this approach 'permaculture' and published it in *Permaculture One* in 1978. Since then it has spread all over the world ... and many people have adopted [a] permaculture lifestyle."[33] Permaculture slowly spread into parts of Africa, and in 1994 Malawians, in the wake of Mozambican refugees depleting native Malawian forests, invited permaculture practitioners from South Africa to Chilema in Zomba District for the country's first official permaculture gathering.[34]

As a result of the training at Chilema, a Malawi network of permaculture practitioners was formed, which has subsequently helped train many other Malawians. Permaculture training posters were developed in both English and Chichewa, and a training guide for a permaculture awareness course (PAC) and a permaculture design course (PDC), all in line with set international standards, was developed and controlled by the Permaculture Network in Malawi (PNM). The network worked together with the government and with nongovernmental organizations such as GTZ (German Technical Programs Support), the European Union, the World Food Programme, and later the Ministry of Education to disseminate permaculture practices.[35] In 2005 the Regional School Permaculture Program for Southern Africa Region (ReScope) opened an office in Blantyre and joined hands with PNM members to introduce ILUD (integrated land use design) in primary schools. Furthermore, the participation of the PNM in the eighth International Permaculture Convergence (IPC8) in Brazil,

Iponga Primary School in Kalonga with permaculture garden. (Photo by Eston Dickson Pembamoyo.)

the hosting of IPC9 in 2009, and Malawi's presence at the 2011 IPC10 in Jordan strengthened the resolve of the Malawi government and line ministries to collaborate with the permaculture movement. Many people in Malawi now are able to view the future more hopefully as they work not only for food security but also toward the goals of generating food abundance, food sovereignty, and above all, agricultural sustainability as a means to "block" further land degradation.[36]

Permaculture and the Ancestral Knowledge of Agriculture

Permaculture as a name was born in the 1970s with Bill Mollison and David Holmgren, but as a practice, permaculture is as old as the earth. Permaculture "is a way of living that uses organic principles. It is about blending the way humans live with nature. Permaculture uses the diversity, stability and resilience of natural ecosystems to provide a framework and guidance for people to develop their own solutions to the problems facing their world, on a local, national or global scale. It is based on a philosophy of cooperation with nature and caring for the earth and its people."[37] It is a system of sustainable living based on the observation and imitation of natural systems, and it builds on the knowledge of our ancestors and applies this to modern life. If you ask old people what they used to eat in Malawi, the answer is almost always *nsima* (thick corn flour porridge). But the nsima was made from many different crops—millet, sorghum, yams, green bananas, and other traditional staples—all intercropped in one field.

Almost all the foods now common in Malawi—maize, cassava, sweet potatoes, onions, tomatoes, Chinese cabbage, rape/canola, mustard, and many others—were introduced as foreign crops. There are over six hundred local foods that we could grow and eat, yet we are still relying on a small handful of foreign, difficult-to-grow crops to meet our food needs. Instead of celebrating the resources we have to keep us healthy, we are ignoring them and have become ashamed of the very part of our culture that allowed our ancestors to survive for thousands of years.

Permaculture helps people to realize that all life is dependent upon a balanced ecosystem, and that dependence is unlikely to decrease in the foreseeable future, given the rise of modern industrial civilization based on the use of nonrenewable resources that are generating greenhouse gasses. As Meyer et al. state, "With the rapid population increases now confronting the peoples of the world the demands for plant products are at an all-time high and can be expected to rise continuously. . . . Whatever the future may hold for the development in atomic energy and direct utilization of solar energy, we can safely anticipate that green plants will long remain the most important source of food energy."[38]

In light of Meyer's comments, efforts should be directed toward nurturing the environment, creating a sustainable ecosystem in which green plants can continue to thrive. To reverse our current global warming and food insecurity requires our realization that when land is exhausted, without having manure applied or without being allowed to "recover its goodness during fallow periods, soil erosion occurs."[39] All these insights are "primitive" in the sense that our ancestors followed them, and it wasn't until relatively recently that we began following a different, destructive path: until about the 1970s in Malawi, forests were still intact and soil was generally healthy.

With the introduction of permaculture in the mid-1990s, Malawians found themselves singing a familiar song. They became strongly connected again not only to their roots but also to the desires of their Mother Earth. They are connected back to the farming traditions and the cultural values and practices that they knew before the coming of external powers and colonial methods of agriculture. Before foreign agricultural policies were introduced, Malawians planted crops with various functions in a single plant station, and this not only balanced the soil fertility and created healthy, diverse meals that came right from the field, but it also increased the amount of native crop diversity for a growing population. "Primitive" agricultural practices increased the relationship of organisms in what in permaculture is called a guild, and thus ancient people applied the truism of permaculture ethics in their own practices, which scaled up their food diversity and sovereignty. It is obvious that the indigenous farming methods enabled farmers to reap a wide choice of foods maturing at different times during the year, leading to a more resilient food security than is seen in Malawi today.

Little Light Refracts in the "Mirror"

The introduction of colonial methods of farming discouraged many Malawians from planting varied crops on the same plant station and in the same field. It is alleged from oral sources that penalties were given to those who defied European farming methods.[40] Farmers were rewarded for weeding clean all the ridges in their entire crop gardens and for uprooting all trees and other plants that grew as "weeds" inside their monocrop fields. These crop fields are now infertile and fail to feed their owners. Permaculture holds out the hope for rebuilding the soil once again, in that it encourages the making and use of manure and/or mulching, prescribes minimal or zero tillage, advocates for agroforestry practices and intercropping, and utilizes many other land, crop, and animal husbandry practices that improve soil fertility in gardens. Overall, it attempts to weave together climate, plants, animals, building designs, soil, water, and energy management into cohesive and sustainable socioeconomic systems. Therefore, the promise permaculture makes to communities is abundant life. This extends far beyond food production systems and sovereignty to explore new potential horizons for the sustainable life that Jesus claims in John 10:10b.[41]

Permaculture Awareness Introduced in Rural Communities

Some local farmers who pioneered permaculture in 1994 have proved to society and line government ministries that permaculture is a tool for sustainable living.[42] The practice of permaculture is growing and spreading into many parts of the country, and it recently found its way into various levels of governmental leadership and into the national education system. If permaculture succeeds, it will in part be thanks to the African continent's permaculture education. Malawi, endowed with all its natural resources, has the potential to show the world the path to sustainability if it fully adopts the principles of permaculture. It is a place where some communities are making efforts, including through education initiatives in the primary schools, to retain their bond with their natural heritage through permaculture. It is perhaps now being accepted that permaculture may provide a practical and commonsense approach to identifying and maximizing the country's available natural and social capital in order to empower people for a new sustainable and self-sufficient Malawi. There is even a recent desire seen in some sections of the government to adopt permaculture in order to improve food security and the social and economic well-being of people through the use of organic, low/no cost techniques, including participatory approaches to community development.[43]

The Mothers' Union Model

Women in Malawi increasingly hold positions of power in their communities and in the government. Their influence is especially evident in churches and local gatherings. Moreover, it is probably true that women and children provide the bulk of Malawi's workforce in both subsistence and industrial farming. The Anglican Church in Malawi has embraced permaculture to help women easily grow more and better food.[44] A book called *More and Better Food,* launched in Malawi in July 2011, intends to support "those who already grow some food but want to grow more, with less effort and fewer inputs."[45] The book aims to assist those affected and infected with HIV and AIDS and to reduce the workload of poor women involved in food production. Although the Anglican Church is not sanctioning child and women labor, it is trying to help women increase food yields through permaculture, at the same time allowing nature to do some of the work.

The Interfaith Permaculture Approach at Malindi

Surprisingly, when permaculture in its "new forms" was first introduced in Malawi in 1994, a good number of the participants invited to Chilema were Muslim and Christian clerics.[46] Unlike the case in other countries in Africa, members of the same family in Malawi often belong to different religions—without conflicts. According to the Qur'an, catastrophes like natural disasters do not happen only on account of humanity's violations of environmental composure; they also are caused by widespread human moral decay. The use of the following verses in conversation with Islamic groups in Malindi has helped to spread permaculture ethics and principles for sustainable living: "And the earth We have like a carpet; set thereon mountains firm and immovable; and produced therein all kinds of things in due balance" (15:19–21). Moreover, the following verses from the Qur'an surprised many first-time listeners in Mangochi: "And We send down water from the sky according to due measure, and We cause it to soak in the soil and We certainly are able to drain it off with ease. With it We grow for gardens of date-palms and vines, in them have ye abundant fruits . . . produce oil, and relish . . . for food. And cattle . . . We produce milk for you."

The Qur'an defines the key players in the ecological balance as follows: animals (birds, insects, mammals, fishes, and invisible living things), plants, air, mountains, rivers, oceans, soil, land, forests, and fields. Importantly, the care of all rests in people's hands. From a Christian perspective, the earth is not fully created if there is no water, no flora and fauna growth. After creating human beings God takes a break to rest. Thus, the creation story (Genesis 1:13–31) reaches its climax after man and woman are created and given responsibility to look after all God has made. Besides the creation story, most of the stories and parables Jesus shared, as in Luke 15, as well as the concepts of the prodigal son, the loving father, and many others are agrarian and centered around issues of crops and animals.[47]

Malindi Parish Permaculture

In December 2009, thirty-five members of the local community resolved to be trained in permaculture in order to turn their surroundings into better places to live. They realized that most houses had lost inches of soil as the surface was washed away by rain runoff and sweeping. They had noticed that the fish catch in the lake dwindled each year, that the human population kept increasing, that there was a scarcity of clean water, that forests were disappearing, and that the weather had become hotter and hotter.

They realized that although they strove in vain to add days to their lives, they failed to strive to add life to the days of the essential natural resources in their surroundings. As a result of these insights and for the sake of their children, these residents established a demonstration plot near the rectory. After only two rain seasons, it became a learning center admired and emulated by people in the area. The herbs, both medicinal and culinary, and other food crops grown were ready for use. The moringa trees provided relish for some guardians at St Martin's Rural Hospital. Aloe vera and neem in the plot were helpful to the community.

Overall, the project was sponsored by a UK-based charity, Malawi Association for Christian Support (MACS). The collaboration points toward more fruitful relations between Africa and Europe not based on the imposition of Western technologies and views.

Permaculture, the Real Solution for Sustainable Living

Permaculture is an appropriate solution for sustainable living in that it connects rules, values, religious beliefs, and the customary traditions of society. Furthermore, following the laws of nature helps people to fashion their conduct according to its model. Unless there is mutual interaction between the planet and all that live on it, care, love and peace shall never prevail on earth, and only if we strongly teach our children to honor all of nature's gifts—water, soil, air, trees, and animals—will the joys and beauties we desire exist. We are slowly rediscovering the principles that inspired ancestral farming but were lost with the onset of colonialism, principles central to sustainable living. Permaculture, when merged with the goals of justice and earth stewardship espoused by religious values, has the capacity to generate sustainable food security for Malawi.

Study Questions

1. How is the history of Malawi in relation to colonialism and the Green Revolution similar to that of other case studies in this book?
2. What are some of Malawi's traditional farming practices and concepts that are similar to the teachings and farming methods of permaculture, and why might these provide a more just foundation for sustainable agriculture in the country compared to Green Revolution technologies and colonial worldviews?

Notes

1. In an interview after a governance and leadership training organized by AID-Star Two Project for Non-governmental Organizations at Lilongwe Hotel, Malawi, from July 31 to August 4, 2013, senior traditional authority Chief Kalonga stated, "The country's dependence on maize [corn] alone has introduced the 'hungry season.' Malnutrition and hunger-related diseases are all relatively recent events in Malawi's history. They are probably caused by the way politicians have handled issues of food and also because people have stopped growing and eating local foods. They have opted for tinned and factory-made foods from Europe, [and these] are expensive, besides having lost essential food values in them as they travel miles and miles."

2. Robert Prosser laments, "The countryside [of Malawi] is still being destroyed at an alarming rate by the subsidy system of Common Agricultural Policy, which encourages ripping up of hedgerows, destroying woods and using too much fertilizer and pesticides. . . . [These] agricultural policies damaged the countryside." *Human Systems and the Environment* (Edinburgh: Thomas Nelson and Sons, 1992), 146–54.

3. Bill Mollison and Mike Holmgren, quoted in Terry Leahy, *Permaculture Strategy for the South African Villages* (Callaghan: University of Newcastle, 2009), 11.

4. Comment by Hogson Liposa, a participant in a two-week permaculture design course held at Malindi in December 2009. As a result of this design course, some Malawians at Malindi and many other places in the country have adopted permaculture as a bridge to cross back toward abundance.

5. *Mwanaalirenji* literally means "plentiful" and refers to the availability of all necessary resources in a household year round, the Growing of the Kingdom of Abundance, whereby a child does not cry because there is plenty to eat in the house. This relates to Malawi's agricultural history and the disasters of the country's recent participation in the Green Revolution. Malawi has three seasons: a rainy season from December to April, when crops are historically planted; what is now called the "hungry season" in April and May, when the previous year's food supply is running low, stretching into the cool/dry season of July and August; and the hot and dry season of October and November, when crops are harvested. This is followed by the next rain season, and the planting once again of crops. Thus, Malawi has recently suffered recurrent periods of food scarcity, so that having plenty of food in the house is seen as a blessing.

6. Michael Raw, *Manufacturing Industry: The Impact of Change* (London: Collins Educational, 1993), 11.

7. As stated in Deuteronomy 11:8–13 by the creator in his covenant with Israel: we shall remember to "love the Lord our God and serve him with all our heart. If [we] do, he will send rain on [our] land when it is needed . . . so that there will be corn, wine, and olive-oil for [us], and grass for [our] cattle. . . . [We] will have all the food [we] want."

8. Anglican Council in Malawi, *Mapemphero ndi Nyimbo za Eklezia* (Masiku Opemba: Likuni, 1996), 98.

9. Yayasan Idep, *A Facilitator's Handbook for Permaculture Solutions for Sustainable Lifestyles* (Bali: Permatil and IDEP, 2006), 1.

10. According to Brigldal Pachai, after colonial partition of the continent, Africa already had an emerging comparative advantage in the export of certain agricultural products such as maize, groundnuts, tobacco, and sugarcane, some of which demanded high chemical input and vast amount of lands to be cleared so they could be produced. In many parts of Malawi it was in the joint interests of some local chiefs, European merchants, and the colonial administration that created and enforced agriculture policies and bylaws to replace the existing "primitive" farming methods. The new policies and crops were intended only to further white settlers' access to foreign markets for export and thus profits. If the hunger season is to be reversed, Malawians need to look back to their precolonial farming history and adopt some of the good methods that kept our ancestors in harmony with their natural surroundings. Brigldal Pachai, ed., *The Early History of Malawi* (London: Longman Group, 1972), 42–47.

11. There used to be plenty of crop varieties that matured at different times of the year and that made our country bountiful. We can easily go back and see in the historical record that we supplied our agricultural excess to nearby countries.

12. Much of the corn grown in Malawi is for human consumption, whereas in the United States most of its production is subsidized and it ends up being used to feed livestock or to generate high fructose corn syrup, which is then used as a food additive/sweetener.

13. In terms of food security, agricultural research in Malawi is primarily focused on creating hybrid seeds and animals. The government subsidizes synthetic farm inputs to grow more, but this applies to maize only. Tobacco farming claims vast lands and heavy use of chemicals as well, so that in the end the profits are not enough to enable farmers to provide their families with essentials. When maize harvests fail, the government and many organizations respond by importing maize to distribute. This is opposed to the practices and beliefs of precolonial farming ideologies, which encouraged mwanaalirenji: abundance and balanced food right from the field, derived through intensive intercropping to improve both the soils and the quality of human meals.

14. Quote and data taken from an unpublished article sent to the editor of the *Permaculture Network Newsletter* compiled at the Secretariat in Zomba in Malawi for the November 2007 issue.

15. Unpublished article sent to editor of the *Permaculture Network Newsletter* for October 2009.

16. Unpublished article sent to editor of the *Permaculture Network Newsletter* for June 2008.

17. Unpublished article sent to editor of the *Permaculture Network Newsletter* for October 2009.

18. Anne Bayley and Walter Nyika Mugove, *More and Better Food: Farming, Climate Change, Health and AIDS Epidemic* (Oxford: Strategies for Hope Trust, 2011), 7.

19. *Permaculture Network Newsletter*, January 2012.

20. According to the Ministry of Health's nutritional rehabilitation unit posters distributed throughout the country in 2006, 50 percent of our Malawian children are developmentally stunted, 80 percent have iron deficiencies, and 59 percent have vitamin A deficiencies. Nutritional rehabilitation units are taking in new patients every day, and many adults are susceptible to disease, especially people living with HIV.

21. In this practice, different family members gathered together to help in the garden of one community member, and at the end of the labor they shared food and drinks. The following day, they would share labor in another garden, repeating the process.

22. The various crops grown by the permaculture practitioner are harvested at different times of the year, and therefore the farmer has food year round. The varied crops in the same garden offer one another natural protection, support, ground coverage, natural digging, soil improvement, and natural disease and pest control. Although the farmer gets steadily smaller quantities of varied yields, there is enough better and healthier food for the family.

23. John Weller and Jane Linden, *The Mainstream Christianity to 1980 in Malawi, Zambia and Zimbabwe* (Gweru: Mambo, 1984), 9. In "Precolonial Africa, everyday life had a mysterious dimension. A woman digging in her garden believed that good harvests depended on the help of the spirits as well as on her own efforts." Sadly, with the onset of colonialism and the introduction of new farming technologies after World War II, all such good and innocent people were forced to change their farming ideologies. The *thangata* system of tenants and lords always favored white farmers over the indigenous owners of the land.

24. Constance Mungall and Digby J. McLaren, eds., *The Challenge of Global Change: Planet under Stress* (Toronto: Oxford University Press, 1991), especially 19.

25. Bayley and Mugove, *More and Better Food*, 16.

26. In a permaculture design course in December 2009 at Malindi, it was discovered that families fail to produce enough food these days because the soils are exhausted. The same amount of land is used by subsequent generations of extended families, and agrochemical inputs, hybrid seeds, and animals are expensive. Government agrochemical policies encourage farmers to abandon *malimidwe a makolo*, ancestral farming methods.

27. Weller and Linden, *Mainstream Christianity*, 11. The authors also share that when the British missionary and governmental employee David Livingstone "visited the Shire in 1859, he was not allowed into the gardens for fear he would spoil the crops by disturbing the spirits" (9).

28. Although the song is sung positively, it is in fact meant to discourage

young people from running away during the farming period. The singer, typically a counselor, explains the meaning at the end, advising young people not to go away from their parents when the rains are about to fall because such behavior causes hunger in society.

29. This proverb teaches women to treat their husbands well, giving them enough good food so they will be energetic and healthy.

30. Among the Yao in Mangochi, children are taken into the bush to "eat honey"—a euphemism for circumcision used to avoid frightening the new initiates.

31. The ethics of permaculture are: care of the earth, care of the people, and fair share (or equitable distribution of resources and setting limits on consumption and populations). The permaculture principles are: each element in nature performs many functions; each important function is supported by many elements; use everything to its maximum; always see solutions and not problems; look for sustainable solutions; make things pay—recycle everything, work where it counts—use two-thirds of your time to plan and the other third for working; like nature, do not compete but rather cooperate; maximize the edge effects; place elements where they can assist other elements; intercrop; nature never monocrops; produce food where you live, even in cities; and remember that nature does not demand systematic and strategic planting in lines. http://permacultureprinciples.com/ethics/ (accessed December 17, 2013).

32. Walbert Bühlmann, *The Coming of the Third Church* (Maryknoll, N.Y.: Orbis Books, 1976), 19.

33. Neil Russell-Jones, *Your Own Allotment: How to Find and Manage One and Enjoy Growing Your Own Food* (Oxford: Spring Hill Books, 2008), 55.

34. Among the first participants in Chilema were June Walker, the current matron of the Permaculture Network, resident in Mangochi District; Mac Justice Betha, a board member from Chikwawa; and the Reverend Ndomondo of the Baptist Evangelical church in Liwonde, to mention just a few. The homes of these members are good permaculture sites in the country. Today the Permaculture Network has created other permaculture sites in almost all the twenty-seven districts and many primary schools throughout Malawi.

35. Through such collaboration, permaculture is presently incorporated into college curricula in some universities and colleges throughout Malawi. Many organizations adopted permaculture, and some selected particular areas in permaculture to target specific groups in society.

36. Martin Duddin, ed., *Aspects of Applied Geography, Rural Land Degradation* (London: Hodder and Stoughton, 1993), 3.

37. Russell-Jones, *Your Own Allotment*, 55.

38. Bernard Meyer, Donal Anderson, and Richard Bohning, eds., *Introduction to Plant Physiology* (London: D. Van Nostrand, 1960), 3.

39. Peter Zinditha, "Population and the Environment in Mt. Mulanje," *Mulanje Mountain Conservation Trust Newsletter*, December 2009, 7.

40. A colonial agriculture minister once even jailed some local farmers for not making ridges and continuing to use local methods. While the making of ridges and purchasing of chemical fertilizers drained the farmers' energy and financial resources, it also displaced manure and shifting cultivation and the making of dunes, all of which were considered primitive in the eyes of missionaries espousing Green Revolution industrial farming practices. The Reverend Canon Chisuwi, the pioneer indigenous/Malawian priest, while adhering to the missionary's mindset, could not see the wisdom in local farming worldviews. This is seen in his book, *Kalilole Wa Ana* ("Mirror for the Children"). He calls the primal worldviews of the Nyanja peoples "Nthawi ya mdima" (the time of darkness) and hails the coming of the missionaries and new, Western methods of agriculture, health, and education, claiming these will help bring Malawi into the modern age. If Chisuwi had only lived to the present day, he would see what the "Mirror" has done for the children of the darkness, leaving them with depleted forests, lost soil, and compromised food security.

41. "I came that they may have life and have it abundantly."

42. Kenneth Mwakasungula is leading permaculture by example and has turned his home into a veritable Garden of Eden in Kalonga, in the northern region. The Nordins family and the Chawawa family at Mchezi operate the best permaculture sites in Lilongwe. June Walker's site at Tanthwe in Monkey Bay, MOET's (Mangochi Orphan Education and Training) near Club Makokola, and Malindi are all shining examples of permaculture in Mangochi. Lastly, Panthunzi and Changoima in Blantyre and Chikwawa respectively are also viable permaculture sites in southern Malawi. At all of these locations, permaculture methods are building soil fertility, and the sites serve as learning centers for neighboring farmers so they can learn to do the same.

43. Government interest is indicated by discussions that took place outside a July 2012 parliamentary debate concerning GMOs, and the subsequent comments some members of Parliament made in support of permaculture ideology when the philosophy was shared with them by permaculture lobbyists. See also unpublished extracts from the 2005 design meetings reports between the Ministry of Education and members of the Permaculture Network in Malawi.

44. Permaculture has found its way into the theological training institution. For example, in 2008 the Reverend Canon Alinafe Kalemba, who is the college dean, and Canon Martin Mgeni, both of Leonard Kamungu Theological Seminary, related the principles and ethics of permaculture to the teaching of the church concerning sustainable growing of crops and rearing of animals and decided to request PAC training. The permaculture training and practice at the seminary now prepares lecturers and students to face diligently the real challenges of sustainable agriculture in the community.

45. Bayley and Mugove, *More and Better Food*, 9.

46. It is, of course, somewhat ironic to refer to permaculture's "new forms." When introduced to permaculture principles and ethics, many elderly Mala-

wians nod their heads, signaling that this is what they knew and practiced all along before the coming of Western agricultural methods.

47. The church at Malindi teaches Sunday school pupils permaculture principles with stories from both indigenous traditions and the gospel.

Nature Spirituality, Sustainable Agriculture, and the Nature/Culture Paradox

The Permaculture Scene in Lower Puna, Big Island of Hawaii

Michael Lemons

Nature Girls

Ellie Harris loves climbing trees.[1] She's from England, did her undergraduate work at Oxford, and obtained two master's degrees—one from Stanford and one from Berkeley—before ditching her Berkeley doctoral dissertation for a back-to-the-land lifestyle in the rainforests of Hawaii. She is a raw vegan who drinks concoctions of ginger and aloe mixed with bananas and coconut cream each morning. "Organic is good, local is better, and raw is essential"—that is her credo. She drinks the neighbor's homemade cherry wine but refuses beer from the local grocery store. She puts honey on her staph infections but turns down my antibiotic ointment. She likes drum circles but not rock concerts. She likes weeding but hates the mower. She also avoids commercial soaps and shampoos; instead, Ellie likes to stand naked under the spout of the water catchment and scrub her hair and body with the fruit of the noni tree.

Ellie and a handful of others lived for a while on a piece of property belonging to a middle-aged woman named Ana. Ana is from California and has a Ph.D. in biology. Cheerful and fit, she bought her property years ago when land was still cheap in Hawaii, and she is now dedicated to low-impact living. She no longer owns a car. She poops in a pit and covers it with soil and compost. Once a month, the pit gets buried and a new one is dug nearby. In a few months, the buried pit will become the home of a nut or fruit sapling. Ana has been doing this for years and now gets nearly all of her food from the trees, vines, and vegetables she tends on her property. The detailed list she gives me of the one-hundred-plus food items she grows or gathers on a regular basis shows coconut, eggs, breadfruit, avocado, banana, and papaya to be her top six food sources. Her property borders a native forest reserve, and she frequently heads into the reserve to kill nonnative species that have become intermixed with the indigenous plants and trees. Pigs and mongoose—nonnative to the island, as she points out—are forever wandering out of the forest into her backyard cornucopia, where they become fertilizer for her for food crops. She used to bury them when they died, but not anymore; she says it is unnecessary effort compared to just leaving them in the open where they've fallen and letting nature do the work. When I ask how the animals die, Ana just smiles and says, "I facilitate that process."

In many ways, Ellie and Ana are model "nature girls" whose back-to-the-land lifestyles might serve as an inspiration for many in an era increasingly concerned with deliberately fashioning a culture of sustainability. They are two of a handful of individuals who served as key informants at my research sites in the lower Puna District on the Big Island of Hawaii. Permaculture communities were defined as groups of two or more persons living on one or more acres of land who either (1) used the permaculture trope as part of community advertisement or community identification, or (2) were described by members of the first group as being examples of permaculture. I have been trying to understand exactly how Ellie's and Ana's worldviews—the pattern of perceptual interpretation and categorization by which they define and find meaning in themselves, their environment, and society in general—are different from those held by people who live more typical fast-food, air-conditioned, 9-to-5 lives in nearby neighborhoods. Ellie's and Ana's beliefs and actions invoke an image of deep concern, connectedness, and integration with the natural world that symbolizes the ideals of "sustainable living" and "harmony with nature" currently emerging in local, regional, national, and transnational contexts around the world, including other case studies highlighted in this book. Their apparent embodiment of these ideals lead them to be highly admired by

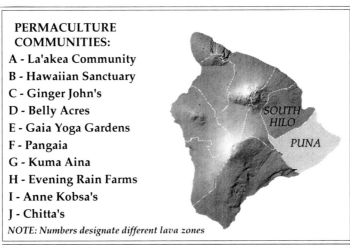
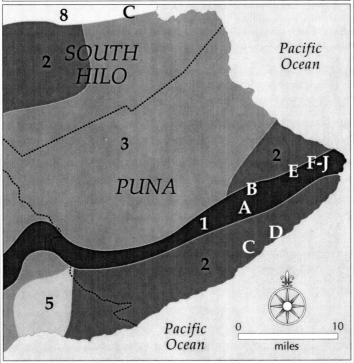

PERMACULTURE
COMMUNITIES:
A - La'akea Community
B - Hawaiian Sanctuary
C - Ginger John's
D - Belly Acres
E - Gaia Yoga Gardens
F - Pangaia
G - Kuma Aina
H - Evening Rain Farms
I - Anne Kobsa's
J - Chitta's
NOTE: Numbers designate different lava zones

Map of permaculture communities in lower Puna. Community C had properties in both lower Puna and the south Hilo District. Lower Puna unofficially comprises areas of Puna located in lava zones 1 and 2. Zone 1 constitutes the ridge of the Kilauea East Rift Zone. (Based on information provided by Michael Lemons.)

the throngs of young adults who arrive here from the mainland each year to experience "natural" living for a few months at the various permaculture communities in the area. But what emerges as most interesting to me about Ellie and Ana is that they seem to have achieved these ideals in a way that contradicts the message currently being espoused by academic spiritual ecologists and environmental ethicists and historians—for they have done so by accentuating the profundity of a perceived nature/culture dichotomy.

Current academic theorizing in Western thought seeks ways to frame a universal discourse of the environment. This trend denounces a perceived nature/culture dualism and seeks instead a coherent nondual approach intended to resemble non-Western thought.[2] But Ellie and Ana haven't transcended or erased the nature/culture distinction—they've emphasized it, and made a run toward the "natural" end of things. "Nature" has been conceptualized as a realm of beliefs, practices, processes, and things that are generally benevolent, healthy, desirable, sustainable, spiritual, and in opposition to beliefs, practices, processes, and things associated with the malevolent elements of a dominant mainstream industrial techno-"culture." In turning away from this realm of industrial techno-"culture" and running toward this realm of "nature," Ellie and Ana are demonstrating a faithful embrace of nature/culture duality in a manner that, ironically, ends up guiding them toward lifestyle choices that symbolize the evasive ideals of nature/culture harmony.

As I will argue, the concerns that give rise to Ellie and Ana's ideals of nature/culture harmony are more than solely environmental in origin; they include social and metaphysical concerns that, upon deeper examination, prove as important as environmental issues for many of those who pursue community forms of permaculture in Puna. Within the permaculture setting, this perceived dichotomy between the moral, pure world of the natural and the immoral, impure world of techno-industrial culture serves a role that Roy Rappaport termed the Ultimate Sacred Postulate.[3] Seeing the world through such a lens helps to catalyze initial interest and participation in permaculture communities. Furthermore, it serves as a latent conceptual point around which consensus is built within these communities during the ongoing process of deciding which beliefs, behaviors, objects—and sometimes which people—belong on the bus and which don't. But, like any conceptual platform that serves to define a movement's liturgical order, the same apparent formulaic simplicity that allows permaculture to become such an attractive no-brainer solution to many of the world's current pressing problems simultaneously leaves the meaning of permaculture itself open to an increasingly wide range of interpretations as the movement attracts growing numbers of adherents.

Nature/Culture and Nature Spirituality: Scripts Guiding Permaculture Drama

The term *permaculture* first appeared on the sustainable agriculture scene in Australia during the late 1970s; since then it has gained increasing acceptance as one of the leading tropes for implying a back-to-the-land, grow-your-own approach to sustainable living. Founding permaculture manuals focus on details of sustainable homestead design guided by three principles: care for the earth, care for people, and a conscious reduction of resource use and population. Permaculture's growth has paralleled growing concerns for sustainability and desires for alternatives to economic and cultural paradigms associated with industrial capitalism. The years following publication of its founding texts have witnessed a steady proliferation of books, magazines, societies, academic articles, and the like, nearly all in praise of the promises of permaculture and presenting permaculture practice as a solution to many impending social and environmental problems.

However, the realities of actual on-the-ground pursuit of permaculture can depart substantially from the meanings and aspirations attached to it by academics, literati, and the progressive urban collective. My goal in this chapter is to paint a picture of the permaculture scene as I experienced it from 2006 to 2010 during participant-observation sessions, interviews, and follow-up visits at locations throughout the lower Puna District of the Big Island of Hawaii. Using a dramaturgical perspective, I present permaculture as a theatrical performance, with participants as actors whose scripts are driven by an underlying theme of nature/culture dualism.[4] Staging the action in the volcanic tropical jungles of Hawaii's lower Puna District magnifies and spiritualizes this script. The end result is Puna's permaculture drama. I present here vignettes from various scenes in this drama—scenes that ultimately tie into a millenarian storyline that forever wends its way toward an intended climax of social and environmental salvation made possible through communitarian "back-to-the-land" living.[5]

It is important to note that I present this story not just as a scholar but also as a past client of and participant in the sustainable agriculture and community permaculture scene. My personal history outside academics is that of a postmaterialist nature-loving spiritual seeker. Not surprisingly, I found the community life that arose around sustainable agriculture one of the few places I could find myself rubbing elbows with those of like mind. It was through this elbow-rubbing process as a client and participant during periods throughout my early twenties to midthirties, and later through participant-observation and in-depth interviews as a scholar during my graduate career from my midthirties up to the present, that I learned I was not the only one drawn to the sustainable agriculture scene for metaphysical reasons that went beyond a pragmatic concern with food security. Nor was I the only one who found the on-the-ground pursuit of back-to-the-land sustainability somewhat less utopian than imagined—and often even downright disillusioning. I tell a small part of that story here with a reflexive conviction that comes from recognizing that the narrative is as much a critique of my own path as anyone else's, while at the same time recognizing that, although this may be a path less traveled by the majority culture, I have not walked this path alone.

It is also important to note that, although the story I tell may reflect many aspects of other sustainable agriculture projects, it is not intended to be representative of all forms of sustainable agriculture, back-to-the-land community lifestyles, or even permaculture lifestyles in particular. My intention here is to tell the tale of a specific brand of sustainable agriculture (communitarian forms of back-to-the-land living that use the "permaculture" trope for purposes of self-identification or are identified as examples of "permaculture" by residents of other back-to-the-land communities) as it occurs within a specific geographic location.[6] For these communities, identification with the permaculture trope is a means by which to declare a particular locus of shape and flavor within the broader spectrum of available sustainable agriculture forms and alternative identities in general.

For purposes of brevity, I am not providing a description or typology of the various identifiable forms of back-to-the-land projects that are to be found in lower Puna and elsewhere, the various flavors and expressions of nature spirituality that occur here and elsewhere within the Western world, or specific descriptions of permaculture and its meaning as it is interpreted and experienced in other settings.[7] What is worth noting here before continuing with the story is that back-to-the-land projects in lower Puna have gained a reputation—on the Big Island itself as well as within the greater Western permaculture scene and the alternative identity scene in general—for being particularly "woo-woo."[8] I argue that this reputation stems from the degree to which postmaterialism, egalitarianism, and mysticism influence lower Puna's permaculture participants' overall beliefs and attitudes toward nature and spirituality. Furthermore, I argue that the tropical Hawaiian volcanic island setting itself helps shape the cultural character of permaculture projects here and the degree to which spiritualized forms of nature/culture dualism tend to prevail among lower Puna's permaculture participants.

The Stage: A Short History of the Lower Puna District

"No other place in America combines a deep spiritual connection to the land with such a hot real estate market."[9] "The district of Puna is Hawaii's wildest outpost with a reputation for free spirits, hippies and other nonconformists."[10] Home to "pakalolo (marijuana) farms, FBI fugitives and the un-bathed."[11] Thus go descriptions of the lower Puna District, the lower half of one of nine districts of the Big Island of Hawaii. At 499 square miles, the Puna District as a whole is nearly the size of Oahu, Hawaii's main island and home to the capital, Honolulu. Yet the contrast between the Puna District and Oahu could hardly be more dramatic: Oahu is a wealthy cosmopolitan island of white sandy beaches holding over 1 million permanent residents, while Puna is a black-sand jungle frontier community of twenty thousand living on the side of an active volcano.

One portion of this volcano, the Kilauea East Rift Zone, is caused by a track of underground lava dikes running east of the Kilauea volcano summit through the lower half of the Puna District. The land area covered by the lava flows from this active dike system separate the Puna

District geologically into two informal sections: upper Puna and lower Puna. The lava flows of lower Puna result in black-sand beaches, while Puna's windward location on the island leads to rain-cloud skies, rainforest precipitation patterns, and thick growths of ferns, mosses, and lichens as well as various unplanned tangles of weedy exotics and invasives that find habitat along the roadsides. The overall effect is a raw, prehistoric vignette of nature that confounds the mainstream's white-sand visions of paradise and compels the typical tourist to seek sunnier skies on the other side of the island. These same traits have helped shape Puna's reputation as a hippie haven, which began with a steadily increasing trickle of alternative-minded individuals throughout the 1970s.[12] Lower Puna's relatively nonarable soil types and constant lava hazards result in dirt-cheap land in the tropics with little policing, zoning laws, or other interference from mainstream techno-industrial interests.

Overall, lower Puna embodies the necessary prerequisites of availability of open land, lack of rigid social limitations, and other economic and geographic peculiarities noted by historians such as Hicks and Edgerton as typical attractions for communitarian-minded imaginations.[13] In this setting, lower Puna has become a flower power paradise for a fair

Kehena Beach's black sand, volcanic soil, and rocky shoreline provide the essential raw imagery of Eliade's "axis mundi," lending credibility to the belief that locations such as this are natural centers of sacredness and power characterized by an increased accessibility to the magic and influence of the supernatural world. (Photo by Michael Lemons.)

share of offbeat adventurers hoping to settle into an alternative "good life." Individuals arriving with this mindset find compelling reasons to equate lower Puna's environment and geography with their vision of paradise. As Grove points out, islands in general and tropical islands in particular have, in the Western mind, long been associated with the location of paradisiacal "gardens of Eden"—isolated and peripheral refuges harboring the essence of a lost, innocent, pristine Other that is both dear and fragile while at the same time offering possibilities of environmental abundance and benevolence.[14] Simultaneously, the flowing lavas of the Kilauea and law and infrastructure exemplify an image of nature found in lower Puna that confounds mainstream visions of Hawaiian paradise as it attracts spiritually oriented individuals seeking to experience *communitas* within an apparent *axis mundi*.[15]

The Actors: Vignettes of Nature/Culture Dualism and Nature Spirituality

Scene 1: Willow is mulching a taro patch and rolling her eyes at the voice on her portable radio. It's a public radio announcement reminding everyone to wear sunscreen, since the sun's rays are harmful and can result in skin cancer after prolonged exposure. "Don't they understand that UV and the sun are natural?" she says. "There's nothing wrong with sunshine!" I could take that statement as representative of a basic latent nature/culture dichotomization that permeates every organic farm, health retreat, and permaculture community I've worked at here in Hawaii. Vilifying the beliefs of the modern techno-"culture" while simultaneously identifying and glorifying a realm of "natural" beliefs, behaviors, and objects that is at odds with this "culture" is a consistent ideological pattern that I've seen expressed at each of these various alternative living experiments. It is an inherently countercultural ideology that can easily take on utopian dimensions within circles of the most faithful. Rules of right and wrong, and beliefs about whom to trust as authoritative holders of knowledge can become reversed. Within these circles, discussions of the possibility of breatharianism are likely to go unchallenged, as are diatribes declaring that the need for sunscreen products is a lie promoted by a conspiratorial corporate-government industrial complex. Noni fermentation, composted human

Images from the websites of two communities in lower Puna that use the permaculture trope as part of their self-description. Understanding the post-materialist, egalitarian, and mystic orientations of the typical permaculture participant helps to explain why themes of nature communion, community, spiritual connection, and personal healing become emblematic of the permaculture experience. (http://permaculture-hawaii.com/.)

manure, raw meat—these things are "in," as are many things considered naturally stinky, gross, unacceptable, and/or unthinkable by the majority culture. If there is an indigenous tradition somewhere in the world that can be interpreted to be in alignment with the latest trending belief, then that belief may find itself promoted into the higher echelons of bona fide alternative knowledges, a subversive realm of sorcerers' stones and panacean paths of escape from the closet of mainstream lies, home to lost truths that bring the believer one step closer to the salvation and freedom that exist on the Other Side.

Scene 2: Angel is driving me crazy. He is a twenty-something haole from urban, middle-class New Mexico who arrived in Hawaii with the belief that he wouldn't need much money since it would be easy to grow his own food. Now he's waiting for his parents to finance his plane flight home. Angel started out at the La'akea Permaculture Community but didn't like the fact that he had to pay $360 a month for room and board even though he was working for the community twenty-eight hours each week. His next stint, at Josanna's Organic Garden, had promised him as much free food from the land as he could eat, but he wasn't satisfied with the sparse raw diet that the promise entailed—"a lot of things just aren't in season right now"—nor with the workload that came along with growing food in paradise. At the moment I'm giving him a ride into town so he can spend his recent allotment of food stamps at the local health food store. He is looking out the window of my car at a green expanse of ferns and ohia trees and trying to convince me that a few acres of this land has everything you need to survive, "if only we knew how to tend the jungle like the original Hawaiians did." I try to tell him that no traditional Hawaiian with a choice would ever attempt intensive crop production on the soilless lava rubble from which this ohia-fern forest sprouts, and that as far as anyone knows such areas were typically used mainly for hunting or the gathering of herbs, nonfood ornamentals, and construction materials. But Angel remains certain of his convictions. Plots of land in the ohia-fern forest sell for cheap, and he wishes he could persuade his parents to lend him the money to buy a small piece of land here. Then, he says with confidence, he could start his own permaculture community in which the food grown would be free to anyone willing to work.

Scene 3: Riox has just returned to the garden. She's a thirty-something office professional from Texas who recently left an unfulfilling marriage and "felt a call" to go to Hawaii. Now, absolutely naked, she's pulling weeds next to me at a raw vegan community called Pangaia. After asking me a few questions about my research, she shifts gears and tells me with a wide-eyed smile how incredible it feels to be pulling weeds without any clothes on—"cleansing" and "liberating" are two of the words she uses. Like many newcomers from the mainland who arrive fresh on Puna's permaculture scene, Riox at the moment seems open to just about anything. Personally, I'm feeling very square at the moment, clothed and closed. If I take my clothes off too and go along with Riox's ecstatic flow, will that be part of cultural immersion?[16] I decide to keep my clothes on and work my way over to an isolated corner of the garden, validating my distance on the basis of ensuring sound research that doesn't cross some ethical boundary between researcher and subject. At the same time, I am in full knowledge that such boundaries are exactly the kind that communities like this are trying to break down.

Although Riox is a few years older than the typical permaculturist who arrives in Puna from the mainland for a few weeks, months, or years of back-to-the-land living, Willow, Angel, and Riox all fit a certain profile that tends to prevail among the crowd of permaculture participants in Puna: they are nearly all Caucasian, well educated, intellectually oriented, left of center in political beliefs, and from urban/suburban middle-class backgrounds. That pattern becomes apparent to anyone who spends a few days or even a few hours at any one of Puna's permaculture communities. However, the majority of Americans fitting this profile aren't rushing off to join permaculture communities. In the end, the most important differences that separate Puna's permaculture participants from the majority culture seem to be along the finer lines of a nature-oriented spiritual belief system, coupled with a tendency toward mystical experiences and combined with an open/experimental and altruistic/other-oriented value system most aptly summarized as egalitarian.[17]

These particular differences became apparent as a result of time spent in participant observation at five permaculture communities, and were further supported by the results of in-depth interviews with twenty-five of Puna's permaculture participants spread over six communities. These observations were tested statistically through seven surveys administered for the purpose of comparing Puna's permaculture participants to three control groups used to represent majority culture. While space limitations prevent me from presenting and expounding upon the results of all seven surveys, I provide here a short summary of the results of one of them in order to demonstrate my thesis that the variables measured by these surveys can shed light on the differences in nature spirituality and nature/culture dualism between Puna's permaculture participants and members of the majority culture.[18] My goal is twofold: to show how these variables catalyze, drive, steer, and reinforce dualized nature/culture beliefs, and to highlight how these variables demonstrate and elucidate some common patterns among Puna's permaculture participants that may be at odds with permaculture's basic pragmatic goal of sustainable food production and consumption.

Performance Catalysts: Postmaterialist Ideals Driving a Materialist Pursuit

Not everybody finds inspiration in Puna's permaculture communities. Each semester, I offer my Hawaii Community College anthropology classes a chance to participate in a field trip to the La'akea Permaculture Community in lower Puna. The students are mostly young local adults fulfilling course prerequisites for acceptance to the college nursing program. While many experience La'akea's endeavors as positive, healthy, and honorable, others view the community as negative, dangerous, and even a little bit foolish:

> The community college needs to be very careful because that field trip we went on the homes were not safe to be in, they had no final inspection, and was totally illegal. They could have been a major law suit if someone was injured. . . . The La'akea project was worth 4 total extra credit points, but that place was filthy, and would not recommend this to other students. . . . I think the community college needs to take a second look at this instructor who potentially could have put our lives in danger, especially because we were down in [lower Puna] which is in a lava flow zone. thanks.[19]

To understand the basis for such differences in opinion, one good place to start is by looking at the basic differences between materialist and postmaterialist worldviews. The quote above (taken from an anonymous end-of-semester course evaluation), expressing physical safety concerns posed by a lack of institutionalized standards and the presence of natural hazards, presents a typical if extreme example of a "materialist" worldview as defined by Inglehart.[20] Permaculture participants, on the other hand, are predominantly "postmaterialist" in outlook. In fact, survey results showed the single most significant difference separating Puna's permaculture participants from the majority culture was the degree to which Inglehart's postmaterialism survey demonstrated among Puna's permaculture participants a preoccupation with postmaterialist concerns.

Inglehart has administered his postmaterialism survey to over sixteen thousand people in fifty countries to test his theory that individuals raised in conditions of insecurity and material want spend their adult lives concerned with matters of material security and wealth, while individuals raised in relative safety and comfort tend to have "postmaterial needs" that focus on nonmaterialist forms of morality, pleasure seeking, and aesthetics. Postmaterialists, as a result, give high priority to nonmonetary ideals, the importance of community input, and the need to maintain clean and healthy environments, rather than to values associated with materialist concerns such as economic growth, a strong national defense, and "law and order."[21]

Inglehart's theory of postmaterialism thus sees the novel value changes associated with environmentalism and other substantive concerns as the result of having grown up with a sense of material security.[22] As materialist concerns have been met in earlier years through economic and physical security, attention is increasingly focused on aesthetic, relational, and quality-of-life issues.[23] The postmaterialist leaning of Puna's permaculturists is particularly interesting considering that the ultimate goal of permaculture system design is long-term material security and autonomy—a materialist concern—while the ethical imperatives that drive Puna's permaculturists and serve as the key means to this end are postmaterialist at heart.

The graphs below show the postmaterialism score results for permaculture participants in Puna versus three control groups in Hilo, Hawaii, representing majority culture (Hawaii Walmart shoppers, a University of Hawaii, Hilo, SOC100 class, and a Hawaii Community College SOC100 class). Higher scores reflect a greater degree of postmaterialism.

In essence, a postmaterialist focus means that, for the typical Puna

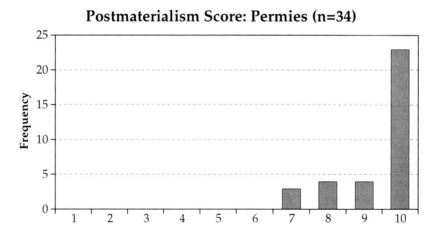

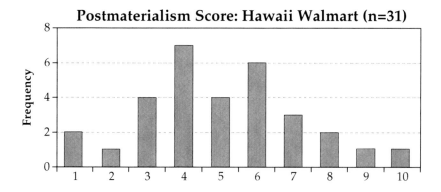

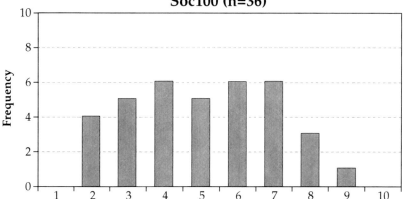

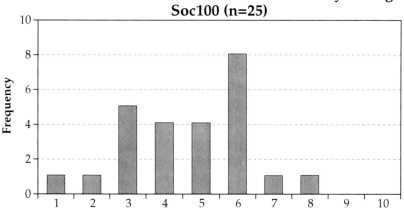

permaculturist, the overall goal of life does not necessarily begin with, and certainly doesn't stop at, a concern for food, shelter, water, and other forms of needed security. Following Inglehart's logic, Puna's permaculturists as a group are among those whose focus on material concerns is substantially less driving than for most people—they are exactly those who are the least likely to be satisfied with the simple, low-impact, pragmatic pursuit of a permanent supply of food, shelter, and water that was originally intended to constitute the prize at the end of the permaculture rainbow.[24] Something else underlies these individuals' drive for permaculture participation, something else that led them to Puna in the first place. Attitudes and actions oriented toward food, sustainability, and the natural environment in these communities likely arise more from social, spiritual, and existential needs than from material pragmatism; likewise, the drive to perform farming activity within Puna's permaculture scene is likely being catalyzed more by a desire to engender a social experience of communitas and a sense of sacred connection with the natural environment than by a primary preoccupation with securing a long-term source of future food and shelter needs.

If the postmaterialist pursuit of material sustainability presents an irony for the typical permaculture project in Puna, it may ultimately pose a problem for the spread of the permaculture paradigm as a solution to worldwide social and environmental problems. On a planet with a burgeoning population, in which most material resources are becoming increasingly overutilized, one can project a future that will witness an increasing number of individuals born into and growing up in circumstances far less materially secure than those that theoretically bred the postmaterialist orientations of Puna's permaculturists. Such circumstances, according to postmaterialism theory, tend to breed a materialist outlook on life—and materialist beliefs and actions are likely to be at odds with permaculture's postmaterialist approach to long-term material sustainability and social/environmental equity.

As the survey results in the postmaterialism scores suggest, the majority of Puna's permaculturists tend to demonstrate postmaterialist leanings of the most extreme nature. Comparably, even the typical resident of Hawaii—already relatively likely to be far more materially secure than the average member of the worldwide population—tends toward the materialist end of Inglehart's material/postmaterial spectrum.[25] Hence, Hawaii residents mirror the typical worldwide results of the postmaterialism survey, in which even nonextreme forms of postmaterialism tend to remain a minority viewpoint, even in the most industrialized countries.[26] Thus, one of permaculture's greatest obstacles may be that its intentional, spiritualized approach to the need for an environmentally sustainable lifestyle, focused ideally on the autonomous and sustainable production and consumption of agricultural products, may continue to be a postmaterialist endeavor of substantial concern only to a select and relatively materially privileged portion of the population.

Study Questions

1. What, according to Lemons, is paradoxical about the relationship between current academic theorizing on nature/culture dichotomization and observations of permaculture participants on the Big Island of Hawaii?
2. In what ways, according to Lemons, does postmaterialism conflict with the basic goal of permaculture?

Notes

1. All names in this chapter have been fictionalized.
2. Many scholars consider the tendency to conceptualize nature and culture as separate categories to be a uniquely Western way of thinking, or at least particularly pronounced in Western culture. For the former view, see Clarence Glacken, *Traces on the Rhodian Shore: Nature and Culture in Western Thought from Ancient Times to the End of the Eighteenth Century* (Berkeley: University of California Press, 1967); Stephen Horigan, *Nature and Culture in Western Discourses* (London: Routledge, 1988); Bruno Latour, *We Have Never Been Modern* (Cambridge, Mass.: Harvard University Press, 1993); Max Oelschlager, *The Idea of Wilderness: From Prehistory to the Age of Ecology* (New Haven, Conn.: Yale University Press, 1993); Vandana Shiva, *Close to Home: Women Reconnect Ecology, Health, and Development* (London: Earthscan, 1994). For the latter position, see Elizabeth Croll and David Parkin, *Bush Base, Forest Farm: Culture, Environment and Development* (London: Routledge, 1992); Philippe Descola and Gisli Palsson, *Nature and Society* (New York: Routledge, 1996); Roy Ellen and Katsuyoshi Fukui, *Redefining Nature: Ecology, Culture, and Domestication* (Washington, D.C.: Berg, 1996); Helaine Selin, *Nature across Cultures* (New York: Springer-Verlag, 2003). Modern critics of nature/culture dualism see its most basic flaw as stemming from the way it historically misrepresents the interdependent and seamless

relationship that truly exists between humans and the ecological environment. See Lynn White, "The Historical Roots of Our Ecologic Crisis," *Science*, March 1967, 1203–7; Bruce Winterhalder, *Historical Ecology* (School of American Research, 1994); William Balee, "The Research Program of Historical Ecology," *Annual Review of Anthropology* 35 (October 2006): 75–98. This resulting misrepresentation causes dualized nature/culture thinking to play a central role in the emergence and persistence of various social and environmental problems, including historic Western patterns of environmental conquest and the use of nature/culture dichotomization to justify uneven power relationships between men (representing refined "culture") and women (representing "nature") as well as between humans (higher "culture") and other animals (lesser "nature"). For environmental conquest, see Richard Grove, *Green Imperialism* (Cambridge: Cambridge University Press, 1995); Roderick Nash, *Wilderness and the American Mind* (New Haven, Conn.: Yale University Press, 1967); Mark Spence, *Dispossessing the Wilderness* (New York: Oxford University Press, 1999). For uneven power relationships, see Sherry Ortner, "Is Female to Male as Nature Is to Culture?" *Feminist Studies* 1, no. 2 (1972): 68–87; Carolyn Merchant, *The Death of Nature: Women, Ecology, and the Scientific Revolution* (San Francisco: Harper and Row, 1980); Carol MacCormack and Marilyn Strathern, *Nature, Culture, and Gender* (Cambridge: Cambridge University Press, 1980); Donna Haraway, *Primate Visions: Gender, Race, and Nature in the World of Modern Science* (New York: Routledge, 1989); Rosemary Ruether, *New Woman, New Earth: Sexist Ideologies and Human Liberation* (Boston: Beacon, 1995); Bruno Latour, *Politics of Nature: How to Bring the Sciences into Democracy* (Cambridge, Mass.: Harvard University Press, 2004). These critiques have led many scholars to begin exploring ways to rethink the nature/culture relationship as a means of overcoming the limitations and problems posed by dualized thinking. See Donna Haraway, *Simians, Cyborgs, and Women: The Reinvention of Nature* (New York: Routledge, 1991); William Cronon, *Uncommon Ground: Toward Reinventing Nature* (New York: Norton, 1995); Arturo Escobar, "After Nature: Steps to an Anti-essentialist Political Ecology," *Current Anthropology* 40, no. 1 (1999): 1–30; Tim Ingold, *The Perception of the Environment* (London: Routledge, 2000); Val Plumwood, *Environmental Culture: The Ecological Crisis of Reason* (New York: Routledge, 2002); Philippe Descola, "Beyond Nature and Culture: Radcliffe-Brown Lecture in Social Anthropology, 2005," in *Proceedings of the British Academy* (Oxford: Oxford University Press, 2006), 139:137–55; James Proctor, "Environment after Nature: Time for a New Vision," in *Envisioning Nature, Science, and Religion*, ed. James Proctor (West Conshohocken, Pa.: Templeton Foundation Press, 2009), 1–14.

3. Roy Rappaport, *Ritual and Religion in the Making of Humanity* (Cambridge: Cambridge University Press, 1999).

4. Dramaturgy is a sociological perspective that looks at daily human interaction as similar to staged theatrical events; two of its main arguments are that (1) people's beliefs, actions, perspectives, and motivations are dependent upon loca-

tion, time period, and the audience; and (2) people's daily "performances" during everyday interactions are meant to create/cause/manage a particular impression or dramatic effect (typically self-serving and/or dependent on acceptance by and amenability with the audience) that requires the suppression of beliefs, actions, perspectives, and motivations that do not contribute to the intended impression or effect. These performances typically follow loose "scripts" that are meant to provide the "actor" with some guidance and direction regarding what is and is not an appropriate belief/behavior for purposes of acceptable performance.

5. Millenarianism typically refers to a belief pattern in which the core belief entails the coming of a major transformative moment or event. This core belief is accompanied by one or more secondary beliefs that may include an apocalyptic moment associated with the transformative moment or event; the arrival of a supernatural or preternatural entity or force in association with this transformative moment or event; belief in a current period of malevolence, corruption, or evil leading up to the transformative moment or event; belief in a time period previous to the current period that was relatively free of corruption and/or evil; and/or a belief that particular actions associated with group membership will allow group members to experience a positive post-transformative or postapocalyptic period that will not be experienced as positive by nonmembers.

6. By referring to permaculture as a trope, I intend here to recognize the fact that the rapid rise and widespread popular use of the term *permaculture* has resulted in various groups, communities, and individuals utilizing the word in ways that vary in respect to meaning, intention, and associated action. What may be understood by one group, community, or individual to be an example of a permaculture-related action, concept, or motivation may not fit into another group's, community's, or individual's understanding of what the word means or encompasses. The result is that the word *permaculture*, as a popular trope within the alternative lifestyle scene, ends up having a much wider range of meanings, associations, and uses than that originally intended by its founders and/or those falling into the category of strict traditional practitioners, for whom the term implies a much more specific and therefore narrow range of agreed-upon meanings. "Permaculture" thus demonstrates a memetic quality that contributes to its increasingly widespread use as a signifier of participation in the alternative agriculture scene and alternative lifestyle pursuits in general. One consequence of this memetic quality is that there is little consensus on whether this use of the trope in association with an increasingly varied range of circumstances is or is not appropriate. For purposes of fairness and objectivity in this chapter, I do not make any claims about the authenticity of any single group's, community's, or individual's definition or understanding of the word. Instead, as mentioned in the text above, I allow my working definition of a Puna "permaculture community" to include (1) any group or community in Puna that utilizes the "permaculture" trope as part of its self-definition, self-advertisement, and/or self-description; and (2) any group or community in Puna that is claimed to be an example of permaculture

by members of other groups or communities in Puna, including those groups or communities that self-define themselves as such.

7. See Michael Van Patrick Lemons, "Lifestyles of the Down and Prosperous: Nature/Culture, Counterculture, and the Culture of Sustainability" (Ph.D. diss., University of Florida, 2012) for an in-depth discussion of these aspects.

8. In other words, they tend to be characterized by a relatively high prevalence, acceptance, and outright expression of spiritualized New Age themes.

9. Patricia Leigh Brown, "Life on a Lava Field," *New York Times*, July 2, 2005.

10. Big Island Visitor's Bureau, "Pull over and Park It in Puna," http://www.bigisland.org/daytrips/228/day-two-pull-over-park-it-in-puna (accessed April 9, 2007).

11. Rod Thompson, "Ice Storm: Epidemic of the Islands," *Honolulu Star Bulletin*, September 12, 2003, http://starbulletin.com/2003/09/12/news/story4.html (accessed April 10, 2007).

12. Ibid.

13. George Hicks, *Experimental Americans: Celo and Utopian Community in the Twentieth Century* (Urbana: University of Illinois Press, 2001); John Edgerton, *Visions of Utopia* (Knoxville: University of Tennessee Press, 1977).

14. Grove, *Green Imperialism*.

15. Mircea Eliade used the term *axis mundi* to describe a physical geographic location that is interpreted to be a communication point and/or point of accessibility between the earth and supernatural realms located above or below the earth. Key to Eliade's use of the term was the concept that the physical and geographic characteristics of these locations lend themselves to such interpretations. Eliade, *Images and Symbols: Studies in Religious Symbolism* (Princeton: Princeton University Press, 1991).

16. When using the "participant-observation" approach to immersion in another culture, the field researcher faces the ethical problem during participation of deciding which activities are and which are not acceptable means of accessing the native worldview. Anthropologists (for example, Bronislaw Malinowski) have been known to marry into the tribes they study and even participate in tribal warfare against neighboring villages. Sociologists studying gang mentality have participated in brutal and illegal initiation ceremonies; others have had themselves arrested in order to study the sexual habits of prison inmates. Efforts within the modern social sciences increasingly attempt to define and standardize the limits of participant-observation as a means of protecting both researcher and subject; nonetheless, the basic act of cultural immersion into the unknown Other will continue to remain an act that requires crossing those very same physical, cultural, and ethical boundaries that seem to distinguish "Us" from "Them" in the first place.

17. Egalitarianism in this case is used in Mary Douglas's meaning, in which a sense of purity, harmony, and intrinsic concept of fairness and righteousness during interactions and exchanges is the idealized experience within the defined body. This body can be the physical body itself, a social body, and/or an envi-

ronmental body that includes interactions between humans and nonhumans. Crucial to this definition is a sense that this egalitarian body is different than some other nonharmonious, nonegalitarian body that becomes contrasted to the perceived ideal body of which one is part, trying to become part of, or trying to create or re-create. Mary Douglas, *Natural Symbols* (New York: Routledge, 1970).

18. See Lemons, "Lifestyles of the Down and Prosperous" for an in-depth look at the results of each of the seven surveys. The seven surveys were: Inglehart's materialism/postmaterialism survey (Ronald Inglehart, *The Silent Revolution: Changing Values and Political Styles among Western Publics* [Princeton, N.J.: Princeton University Press, 1977]); Dake's grid/group survey (Karl Dake, "Orienting Dispositions in the Perception of Risk: An Analysis of Contemporary Worldviews and Cultural Biases," *Journal of Cross-Cultural Psychology* 22, no. 1 [1991]: 61–82); the Schwartz values survey (Shalom Schwartz, Gila Melech, Arielle Lehmann, Steven Burgess, Mari Harris, and Vicki Owens, "Extending the Cross-Cultural Validity of the Theory of Basic Human Values with a Different Method of Measurement," *Journal of Cross-Cultural Psychology* 32, no. 5 [2001]: 519–42); the Hood mysticism survey (Ralph Hood, "The Construction and Preliminary Validation of a Measure of Reported Mystical Experience," *Journal for the Scientific Study of Religion* 14, no. 1 [1975]: 29–41); Jacob's back-to-the-land survey (Jeffrey Jacob, *New Pioneers: The Back-to-the-Land Movement and the Search for a Sustainable Future* [Philadelphia: Pennsylvania State University Press, 1997]); the connectedness-to-nature scale (F. Stephan Mayer and Cynthia McPherson Frantz, "The Connectedness to Nature Scale: A Measure of Individual's Feeling in Community with Nature," *Journal of Environmental Psychology* 24, no. 4 [2004]: 503–15); and the modified new ecological paradigm / dominant social paradigm (NEP/DSP) survey (Helen La Trobe and Tim Acott, "A Modified NEP/DSP Environmental Attitudes Scale," *Journal of Environmental Education* 32, no. 1 [2000]: 12–20).

19. Direct quote (anonymous) from a Hawaii Community College student's ANTH200 course evaluation of a field trip to the La'akea Permaculture Community.

20. Inglehart, *The Silent Revolution*.

21. Ronald Inglehart, Miguel Basanez, Jaime Diez-Dedrano, Loek Halman, and Ruud Luijkx, *Human Beliefs and Values: A Cross-Cultural Sourcebook Based on the 1999–2002 Values Surveys* (Mexico City: Siglo XXI, 2004).

22. Inglehart, *The Silent Revolution*.

23. Ronald Inglehart, "Aggregate Stability and Individual-Level Flux in Mass Belief Systems: The Level of Analysis Paradox," *American Political Science Review* 79, no. 97 (1985): 116.

24. Inglehart, *The Silent Revolution*.

25. See Lemons, "Lifestyles of the Down and Prosperous," appendix A, for an explanation of why Hilo Walmart shoppers were used to represent majority culture in Hawaii.

26. According to Inglehart et al. (*Human Beliefs and Values*), the countries with the highest worldwide percentage of postmaterialists as of the year 2000 were Australia (35 percent), Austria (30 percent), Canada (29 percent), Italy (28 percent), Argentina (25 percent), and the United States (25 percent). From the graphs above, using scores from 7 through 10 as the range indicating a postmaterialist worldview, survey results show that 22 percent of the Hawaii Walmart control group comprised postmaterialists (very close to the US average) while 100 percent of the permaculture test group comprised postmaterialists.

5

Hindu Traditions and Peasant Farming in the Himalayan Foothills of Nepal

Jagannath Adhikari

This chapter examines how Hindu concepts of life and the cosmos shaped human relationships with the environment—particularly the human use of resources for agriculture in a subsistence farming system prevalent in the Himalayan foothills of Nepal. Even though Hindu societies now are no better than other societies following different faiths in conserving the environment or reducing pollution, whether on farms or in cities, certain concepts contained in the Hindu religious belief system based on *dharma* (merit), *paap* (sin), and *karma* (deed) encompass moral principles that may help in the conservation of resources like soil, water, and forest; in the maintenance of biodiversity; and in the preservation of community cohesion. These spiritual concepts can be useful in conserving resources if applied in a way to suit the present context. This is especially so in rural farming communities. After all, the resource management system in Hindu society is a long-established one, clearly indicating that it has some elements that could help stabilize the ecological system while growing food in a sustainable manner. In this chapter, I attempt to identify those elements.

In this essay, I will analyze Nepal's efforts to modernize agriculture, efforts that have led to deteriorating environmental conditions. I bring insights from my own birthplace, Baidam in the Pokhara Valley in midwestern Nepal. Next, I will examine traditional agrarian ways of life based on Hindu religious worldviews to explore the ecologically prudent ways and beliefs inherent in this system. I look at their relevance to today's world in reducing the sustainability crisis of present-day chemical agriculture in a farm-based economy. It is not my aim to romanticize the traditional farming system or the Hindu religion, as the society following these traditions has also undergone tremendous stress in the past fifty years caused by population growth and the new desires of people to participate in the consumerism perpetrated through the globalized economy. There is also controversy as to whether the whole Hindu religion is conducive to nature conservation. This leads to the question of why those ecologically prudent practices were not applied to solve environmental problems. I also attempt to answer these questions.

Situating the Context

Hindu religious texts like the Vedas as well as daily practices contain references to the earth in general along with guidelines regarding how to manage agricultural fields.[1] The general guidance addresses the purpose of farming as well as specific components of the farming system. A handbook on sustainable agriculture written in Sanskrit in verse, *Krisi-Parashar,* dates back to as early as the sixth century CE.[2] Saint Parashara wrote this work to benefit farmers in their farming practices. The book illustrates the basic principles of farming, including how astrology—the positions of the sun, moon, and other stars—guide farming practices. The book gives prominence to how the early Aryans, followers of the Hindu belief system, accorded importance to the gods of natural and atmospheric elements like the rain, the wind, and agriculture in general. This treatise covers all aspects of farming—weather and climate, management of land, management of cattle, agricultural tools and implements, seed collection and preservation, and all agricultural processes from preparation of fields to harvesting of crops and storage of yields.[3]

As these guidelines were developed long ago, questions have often been raised as to their relevance solving present-day ecological problems. In this regard, it should be noted that the ancient system of farming was managed in line with principles to which we do not give much value now. On the other hand, they could still give us useful guidance. The basic purposes of farming or producing food in ancient times (about the first to the tenth century CE) were subsistence and to fulfill religious functions, including sharing food with others. In an encompassing worldview, "others" included guests, pilgrims passing by any village, temple, or sacred place, and more-than-human species. The traditional farming system that still survives in the mountains of Nepal remains subsistence oriented. Such a long-established system has become complex, characterized by high species diversity in terms of the varietal composition of plants and what people grow. The interconnection among the components in this type of farming system, such as recycling, self-pollination for seed production, and selection for seed improvement, was and is emphasized. The culture stresses sharing food and other resources. Moreover, these management principles underscore the interconnection of human and more-than-human natural elements. Ancient Hindu agricultural texts accented the belief that natural agencies like rain, clouds, and sun respond to the behavior of humans, and thus humans should not take these entities for granted but behave responsibly in order to assure their cooperation in efforts to produce food. This ultimately means the recognition that a reciprocity, an exchange between natural agencies and humans is needed for sustaining the environment and for growing food.

In a long-established system like agriculture informed by core Hindu religious values, there are certainly various elements that are well adapted to the environment and are resilient. They are sure to have experienced many perturbations in the system in the course of their long history, but they had the ability to adjust and remain viable. This resilience is what is required in the agricultural system at the present time in the face of environmental uncertainties, including the impact of climate change. As I discuss below, unlike the Hindu-based systems of agriculture and food, the modern agricultural system emphasizes monoculture and the greater use of energy inputs in the form of chemical fertilizers, pesticides, hormones, and other chemicals. Even within a short period of fifty to sixty years, this system, based as it is on external inputs, has caused instability in the ecosystem in the hills and mountains of Nepal. The external chemical-based system has no self-organizing properties like adaptability, biodiversity, stability, nutrient recycling, and conservation. In this context, it is important to learn lessons from the time-honored resilient system which, despite facing various ups and downs and crises, has demonstrated in-built capabilities to regenerate through self-organizing principles. Sadly, these principles have been lost in the modern agricultural system.

As I refer to it here, Hindu religion, or tradition, denotes diverse types of faiths and practices that broadly follow certain beliefs considered to be the core principles of this religion: the principle of reincarnation, the existence of one absolute being of multiple manifestations (supreme being), the law of cause and effect (karma), the need to follow the path of righteousness (dharma), and the desire for liberation from the cycle of birth and death (*moksha*). Different groups within this religious tradition follow different core ideas, and thus it is difficult to define Hinduism as a religion. It is largely considered a way of life or tradition that has to be experienced in order to know it. At the grassroots level of the geographical area that I study in this chapter, there is an intermixing of Hinduism, Buddhism, and shamanistic and animistic faiths. Agricultural practices in village society are informed by a combination of all these faiths and practices.

The Vedic model of agriculture could integrate all the elements of a stable farming system producing quality food for self-sustenance and feeding others who do not farm. I offer a closer look at these dynamics in my own birthplace, Baidam—a community by Phewa Lake within the metropolis of Pokhara in midwestern Nepal.

Nepal's Deteriorating Agriculture: The Case of Pokhara Valley

In the Pokhara Valley of Nepal, I have witnessed drastic changes in the tradition of farming—not only in the values attached to land, fauna, flora, and domestic animals, but also on the traditional concept of human membership in the community of life in the village.[4] Until about thirty to forty years ago, these resources were considered, especially by elderly people, as living creatures capable of cursing humans for misusing them. Farmers were constantly worried about what their resources would do if they were not used properly. While doing farming or any other work related to natural resources, the conceptual trilogy of dharma (merit), paap (sin), and karma (deed)—based not only on the traditional worldview of the Hindu religion but also on local faiths like animism and Buddhism—was invoked. Dharma, paap, and karma also informed the objectives of farming and the way it was to be organized. (I elaborate on the properties of these concepts below.) These beliefs also contained strong elements of social justice—like the importance of distribution and the avoidance of accumulation—and of moral ecology—reciprocal exchange between the human and more-than-human worlds.

Practicing Hindus accepted the notion that there was unity in diverse forms of life and that in all species there is the presence of one supreme power.[5] Such an ontological view guided people to view themselves as merely one part of the ecosystem, equal to other life-forms, thus enhancing the value of diversity and encouraging its preservation. This helped to promote localism; the flora and fauna found in the local environment were linked to human rituals and were also important for achieving a balanced food system as well as maintaining health. It is notable that certain foods were considered medicines, used for the healing of the human body as postulated in Arurvedic tradition.

In recent times, Pokhara has urbanized and modernized with the influx of people from within Nepal and abroad. In what farming has remained in the area, modern chemical inputs (fertilizers, pesticides, and hormones) and the hybrid seeds of only a few crops are used for the production of food. This type of farming has contributed to environmental pollution (for example, pollution of local bodies of water, leading to a rapid decline in dissolved oxygen, the growth of various plants, and the death of aquatic species like fish) as well as the produc-

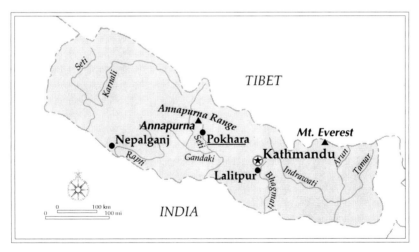

Location of Pokhara in Nepal. (Based on http://www.infoplease.com/atlas/country/nepal.html.)

tion of food unsafe for human consumption, which has caused various health problems.

The pursuit of modernized farming—through government interventions as well as the people's new orientation to adopt modern ideas—seen in Pokhara is common throughout Nepal nowadays except in some pockets where modernization forces have not yet reached. Nepal's agriculture has still remained subsistence oriented, intended primarily to fulfill the requirements of the family, although as of now, about 60 percent of families who rely on agriculture as their main occupation cannot meet their food and nutritional requirements by their farming. Unlike in the Western developed countries, farming in Nepal is done on small landholdings, basically family farms. On average, a farm consists of only about 0.7 hectare; very few farms are larger than 4 hectares.[6]

The traditional farming system in the middle hills region of Nepal, and particularly in Pokhara, was mixed: forest, farm, and animals were integrated for growing a diversity of crops and foods. A typical farm household would produce different types of cereals, pulses (lentils), herbs/medicinal plants, vegetables, and fruits, supplementing the family diet with wild foods collected from the forest. A few domestic animals were kept—cows, buffaloes, chickens, pigs, goats, or sheep—depending upon the specific cultural practice of the people. These used the crop waste and converted it into milk/milk products, eggs, draft (animal) power for plowing and transporting, and manure. Domestic animals also used forest resources like pasture and fodder; these animals, in turn, produced foods for human consumption and other products like manure to increase soil fertility. Forest, pastures, and water sources were managed by the communities and were, and still are, common properties.

A diversity of crops was grown on the farm, and some crops were intermixed. Generally, cereals and lentils were intermixed with crops so that the pulses would capture the nitrogen from the air and enrich the soil, at the same time providing a protein supplement to cereal-based diets like rice, corn, and millet. Vegetables, both seasonal and perennial, were cropped near the house to meet the family's nutritional and ritual requirements. Many other plants were grown and conserved as required for the rituals. All the required inputs—seeds, manure, knowl-

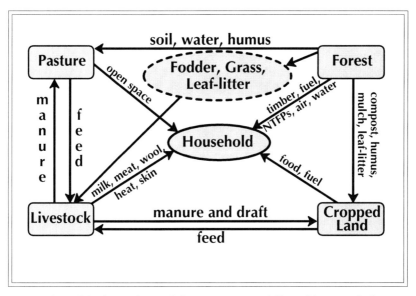

A typical model of a traditional farming system followed by Nepali farmers that integrated different ecological niches like farmland, forest, pasture, and livestock. Mixed farming conserved biodiversity and provided necessities to households including medicinal herbs, fruits, vegetables, and NTFPs (non-timber forest products). (Figure by Jagannath Adhikari.)

edge, and labor—were local. Farmers did not have to purchase seeds or know-how, as both were shared freely. Seeds were open-pollinated and improved in each generation through selection. Farmers used to take special care in the selection and storage of seed. This is evident from a popular saying: "Anikal ma biu joganau hulmul ma jiu joganau" (One has to protect seed during famine and one's life in a mob).

Chemicals like fertilizers, herbicides, hormones, and pesticides were not available in those days. In cases of pest or disease attacks, people depended again on local biological materials, such as herbs. The basic advantage of this integrated, mixed, and diverse farming system, even practiced on the small unit of the family farm, was that there was recycling of resources and thus the sustainability of the system was maintained. Humans were provided with a diversity of foods as required for a balanced diet and a healthy life.

Attempts to modernize agriculture in Nepal started in earnest in the 1950s, when the country was opened to the outside world. The concept of modernization followed the Western urban-industrial model of development, which was imagined and then imposed as a better course to meet the challenges posed by population growth and new expectations of a high standard of material living. Chemical fertilizers and pesticides were promoted; initially they were given to farmers as gifts, but later on prices were charged—and regularly increased. Traditional or indigenous seeds were considered "backward" and unproductive. Even the government agricultural extension programs discouraged farmers from using local seeds, urging them instead to plant either modern seeds that came from abroad or a few improved varieties released from research stations. Since the 1990s, the use of hybrid seeds has been increasing. The soil has become so poor that it needs a strong dose of fertilizer, pesticides, hormones, and other chemicals to sustain the production of crops. Although recently there has been some resistance to these new seeds and the new style of farming based on chemicals, their use has been growing because of a perception that whatever is modern and developed (the Nepalese call that *bikase*) is better. The traditional social and cultural values guiding land use and growing food have declined. Current agrarian values and practices, including monoculture and the excessive use of external inputs, are guided by market demands.

Farming and the Concepts of Dharma, Paap, and Karma

Dharma in the Hindu religion/tradition refers to the duties of people as defined by the society, depending upon the position of the individual. Doing one's dharma (*svadharma*) means not only acting ethically but also assuming the duties that are proper to the class or caste into which one was born (due to one's past karma) and to the stage of life one is presently in. Hindu philosophy defines four stages of life— *brahmacharya* (study leaving the family), *grihastha* (householder), *vanaprastha* (retreat from household chores and living in the forest), and *sannyasa* (renouncing the world and withdrawing from material life), which represent a progression from lower knowledge to higher knowledge and, ultimately, to moksha (salvation). The first three stages are connected to the household and living in this worldly life, whereas renouncers become ascetics by denying the materialistic life. There is also a connection between these two broad categories. The householders are supposed to look after the renouncers, and thus food production is meant to be shared with such people. Food producers therefore have some moral responsibilities and are guided by religious requirements. There are two models in Hindu religion—householders-cum-farmers and ascetics. The former are called to follow the "devotional" model because they have passions and devotions to worldly affairs and to material wealth. But this way of life is not isolated from the ascetic model—those who follow this model reduce the pressure on available resources because they live with only a few basic necessities and they do not have accumulative attitudes.

Farmers are supposed to follow what is called *kisan dharma* (or farmer's dharma), and this dharma means virtue. Virtue was understood as right ecological action and right ritual activities related to farming communities and farming resources. Farming is a secular (*sansaric,* devotional or worldly) world, which is again opposed to the religious world. In religious dharma, there is more emphasis on the transience of life, which calls for withdrawal from the world (asceticism). But in sansaric dharma and kisan dharma, farmers are to work on the land, producing more food as they develop harmony and balance in the ecosystem, considering themselves as a part of all life forms in the farm ecosystem.

Kisan (farmer) is the second stage in the life of people following the farming profession, and to embrace this occupation is to become a householder. In this stage of life, a farmer needs to properly pursue not only *dharma* (religious duties) but also *artha* (material prosperity) and *kama* (passion and reproduction). The farmer/householder has the responsibility not only to sustain his or her household but also to support needy community members, including ascetics, through alms and gifts and organizing rituals. Rituals were aimed at benefiting nature and society as well as oneself. Accordingly, many rituals to be followed were traditionally aligned with agricultural cycles. Remnants of a few of these rituals can still be observed in rural areas. The rituals include worshipping other life forms, which essentially means showing reverence toward them. Among these rituals are the worship of serpents (on a special day the virtues of snakes are extolled), frogs, land, vultures (*garuda*), trees, and crops; the marriages of trees are also celebrated.

These rituals are aimed at creating a balance in the ecosystem and producing more food. They serve the practical purpose of biologically controlling insects and other pests that might cause damage to the crops. For example, rituals related to snakes and frogs would help in maintaining a balance of rats, snakes, frogs, and vultures to avoid an overpopulation of rats.

Like land, water, too, has a special place in Hindu tradition. It is regarded as one of the five primal elements (*pancha-tatva*) of the universe. Water is also considered a source of purification and expiation. The sources of water were sacred and were worshipped with respect. Polluting water sources was considered sin (paap).[7] Accordingly, rivers were/are also sacred. The Himalayan mountains, considered a water tower, are worshipped and respected, believed to connect earth and heaven. Hindu scriptures clearly mention that the sight of the Himalaya in early morning takes away one's sins, and these mountains are to be admired from a distance—not to be conquered by humans. Adherence to this philosophy might have maintained the rivers and other water bodies clean because the Himalaya is the source of water. Aptly, one of the mountains in the Himalayan range just north of Pokhara valley is called Annapurna—the storehouse of grain. The snow collected here was traditionally considered grain because it supplied water to rivers and other water sources. Some rural people, especially the elderly, still regard natural lakes as having sacred waters and feel cleansed if they bathe in these waters. For some time, people had no scientific knowledge of the cause-and-effect relationship of throwing modern inputs such as chemicals, fertilizers, and plastics into the water and ecological pollution.

Many of the resources in the villages were kept communally, considered "common goods." People who expanded, enhanced, and enriched these common goods would have more dharma. There were no patent rights on innovations. A farmer would freely share or exchange good seed developed on his or her land. Giving seeds to places and people struck by famines was a great dharma, an act earning merit, for the farmers. People would help each other mutually by donating free labor. Maintaining a common communitarian life was itself a dharma—dharma as a neighbor or as a fellow villager.

The fear of paap, which is the opposite of dharma, was invoked to restrain people from doing *adharmic* things. There is paap in not following the moral obligation to help or support people in need, the poor, and the destitute. There is paap in harming one's fellow human beings. The concept of paap is equally invoked in doing harm to nature and not following one's dharma to use it properly. The mirror-image concepts of dharma and paap in Hindu religion/tradition were aimed at making people identify with nature to maintain harmony with all of creation.

Karma: Source of a Caring Attitude

Karma is a concept closely related to dharma. Using the idea of karma, peasants used to connect their actions and the consequences thereof not only to the present but to the past of their ancestors and to the future of their descendants. It would also encourage them to do right karma and to accept natural events without complaint. Whatever benefits or harm people reaped from their actions, they would be content, saying it was the *karma ko phal* (the fruit of karma). In fact, according to scriptures, once karma (action) is started, it continues without a break. Even when a person dies, his or her karma survives in the form of memory and carries over into the next life. Sometimes the consequences of karma are indecipherable to individuals—an individual's family members and later descendents may be forced to pay for past crimes or may benefit from past good actions.[8] The concepts of karma and reincarnation could help make people compassionate and caring for others even as they encouraged them to practice self-restraint.

Unity and Diversity in the Farm Ecosystem: Making Connection with the Cosmos

The mixed, diversified, and integrated farming methods that were followed under the subsistence system described above and the rituals contained in this system were aimed at nurturing unity among diverse life forms. This was achieved in a self-contained farming system geared toward producing what was required for the family and community not only in terms of human needs but also to connect human life with the natural world. This entailed a diversified farming system with emphasis on preserving various species of plants and animals and developing interconnections among them for reuse and recycling of resources and avoiding waste. The major emphasis was on maintaining the dynamic balance between the five basic elements of the universe—*prithiwi* (land), *jal* (water), *tej* (radiation, energy, or fire), *bayu* (air), and *aakas* (cosmic space)—as they relate to human activities. That dynamic balance formed the structure and infused the life of all creatures. It was believed that if this balance were not maintained, then both nature and the human body would become sick and dysfunctional.

Farm and forest were viewed as interconnected spaces for the transfer of energy as well as for the shifting of people's residence. According to Hindu philosophy, the farm or agrarian life needs to be connected with the forest life for the regeneration of energy and creativity. Forests were seen as spaces where teaching and thinking could be conducted. The remnants of these sacred forests are still found in many villages in Nepal. Many books were composed when the sages stayed in the forest. Many kings in the past, like King Rama of the epic Ramayana, stayed in the forest, becoming philosopher-kings, in order to understand the complexity and dynamics of natural life, which would make them wise.

Intimately linked to biological diversity are cultural practices and customs, such as myths regarding the virtues of sacred plants and trees, festivals for the germination of seeds, and customary rights setting limits on the use of natural resources. The practice of *navadhanya* (nine seeds to be used for rituals) embodies a link between cosmic influences and life forms on earth since each of the nine seeds is associated with a planet or cosmic force—like chickpea representing Jupiter, sesame Saturn, rice Venus, and barley the sun. In Durga Puja (worship of the goddess Durga), barley, rice, and the leaves of various plants are required. People need to germinate, grow, and maintain these plants. As food rituals (systems) also change according to the seasons in order to cope with the challenges of particular seasonal patterns, diversity in food production is required. Such a repertoire of rich diversity was based on local fauna and flora cultivated in gardens and farms as well as found in the nearby forests. Thus biological and cultural diversities reinforced one another.

For example, seed was considered not only a source of food but also an element of culture and rituals. One such ritual is called *satbeej charne* (hundred-seed dispersing), wherein people have to disperse at least one hundred types of seed in a sacred place, some of them located in high mountain areas. This ritual is to be performed by someone within a year of the death of a family member. In some places the seeds could propagate naturally; in cold areas in the mountains, they could be stored for two or three years, providing a storehouse that could be accessed in times of famine. The practice was also a basis of maintaining biodiversity.

The traditional system of farming was so effective in preserving agrobiodiversity that there were as many as two thousand landraces (varieties) of rice in Nepal alone, growing in areas from 60 meters in altitude to 3,050 meters.[9] However, a study revealed that of the known eighteen hundred varieties of indigenous landraces of rice, including wild species in Nepal, only one variety is popularly cultivated in Tarai, where rice is widely grown.[10] In Pokhara Valley, fifty-three local varieties of rice were recorded as late as 2007 in the small village of Begnas.[11] A few popular varieties of this rice are still grown, but most others are replaced by modern varieties. This evidence suggests that local species and landraces are disappearing. Modification of cropping patterns and the expanded use of hybrid seeds and imported seeds of a few improved varieties are the main causes of this decline in agrobiodiversity.

Localism: Exchange and the Moral Economy

The value of local self-sufficiency in traditional farming historically helped to avoid long-distance trade in food. The local rituals required that crops and animals be produced in the local context. As a result, farming preserved biodiversity. Some exchange was practiced, but the food mileage was very low. In fact, the exchange of produce, seeds, manures, labor, and ideas to preserve the land and produce more food was emphasized. Donations of seeds and food were considered sacred and were done for free, and this act of distribution was required if there was surplus production in a year. The purpose was to earn religious merits (dharma), not money. This ideology contrasts sharply with the present-day practice of selling the seeds and patenting them. Traditional farmers constantly improved their seeds, but they did not receive any patents—it was done for their kisan dharma. Seed exchanges also took place between communities, but this exchange was done not with the idea to compete with others or to destroy another seed. Such practices merely added biodiversity.

As high-yielding varieties and chemical pesticides are widely used now, the practices associated with traditional varieties have fallen into oblivion. Hence, the erosion of crop diversity is accompanied by a loss in cultural diversity. Biodiversity has been reduced because of homogenized growing conditions, such as using the same few high-yielding crops, the same few high-yielding seeds, and the same kind of surface irrigation. As a result, the erosion of genetic diversity has accelerated.

The decline of biodiverse cropping systems I have seen in Nepal is not an isolated phenomenon. Rather, it is linked to broad processes of change shaped by dynamic political and socioeconomic forces—the introduction of the Green Revolution, liberalization of the economy, and other aspects of globalization. The increased emphasis on export-bound crops and the advent of agribusiness justify the expansion of monocultures of corn, cabbage, tomatoes, and other crops. At the same time, the penetration of corporate food industries' highly processed and packaged food (biscuits, canned food, noodles, and other products having a long shelf life) is destroying local food culture, biodiversity, production, and human health.

Reflection on the Future of Agriculture

The traditional farming system, informed by the Hindu ethos and principles, was influenced by a distinct worldview. As I have shown above, within this worldview, nature was valued differently—not merely as an object or a commodity but as an embodiment of spiritual values with a deep connection to divine processes, or the realm of the sacred. In this chapter, I have examined a community in which the Hindu religion is dominant but, as practiced at the grassroots level, is a confluence of Hindu principles, Buddhism, shamanism, and animism. As argued by Nepali anthropologist Pramod Parajuli, the coeditor of this volume, despite their religious affiliations, locally specific, place-based people depending on nature for survival and sustenance could be considered "ecological ethnicities."[12]

Some of the principles of Hinduism have been embedded in rituals that contain wisdom for the conservation of nature. The concepts of dharma, paap, and karma as understood by people socialized in Hindu communities encouraged people to do the right thing for nature, the land, and one another, without regard to the generation of wealth. The right act was to produce food for the sustenance of human beings and the regeneration of nature. The ultimate goal was the well-being of both the community of human beings and of more-than-human species through food consumption and ritual activities. As a result, biodiversity was maintained through rituals, and localism was emphasized in the production and use of the produce. Dharma and rituals also defined what level of consumption is righteous and the criteria for the right accumulation of wealth. The pursuit of dharma and other achievements like artha (wealth) and kama (passion) did not lead to conspicuous consumption. Today's problem in Nepal, and for that matter in all developing countries, is not only decreasing production but also conspicuous consumption and the wrong kind of consumption.

The Green Revolution model of agriculture in conjunction with the urban-industrial model of development as pursued in developed countries emphasized "productivity" at the expense of biodiversity. Biodiversity became a burden and an obstacle rather than an asset. The Green Revolution put emphasis on a few select crops like wheat, corn, and rice. The so-called high-yielding varieties (HYV) of seeds of wheat and rice were brought from developed countries and disseminated in Nepal. Vandana Shiva has argued, "HYV seeds are misnamed because the term implies that the seeds are high yielding in and of themselves. The distinguishing feature of these seeds, however, is that they are highly responsive to certain key inputs such as fertiliser and irrigation. They are actually, *high response varieties.*"[13] The continued use of chemical fertilizers and pesticides has altered the ecological balance in many parts of the world. Despite these interventions, yields are stagnating and the growth rate of production is on the decline. Other side effects include waterlogging, soil deficiency in micronutrients, and an increase in soil salinity.

In the context of today's ecological crises, exacerbated by modern chemical- and fossil fuel–based agriculture, recovering and regenerating some central elements of the traditional Hindu principles of farming could be useful. The concept of two-pronged dharma (one emphasizing the moral economy in the distribution of food and feeding others, and another focusing on giving back to nature to maintain its reproductive capacities) seems very relevant in today's world. Such an approach would reduce social inequality, malnutrition, and food crises, and at the same time maintain the sustainability of the ecosystem. The ideas of food sovereignty and localism are also contained in this philosophy, as explained above. In its simplest form, dharma is a mechanism that creates respect for nature and improves its capacity to regenerate.[14] Such a relationship of respect is imperative if we want to achieve sustainable agriculture.

Reflecting back on the resiliency of what I considered a Vedic system of agriculture, I certainly do not think it advisable to go back to ancient times. However, concepts emphasizing a balanced and harmonious relationship in the ecosystem between the human and more-than-human worlds as practiced in the past may sensitize the consciousness of present-day farmers and policy makers. To this end, I propose that building a locally based, diverse food production system could be a highly effective way forward.

Study Questions

1. Do ancient texts, beliefs, rituals, doctrines, and customs that were generated long before concepts such as evolution and soil science arose have something to contribute to contemporary sustainable agriculture? What elements of rural Hinduism does the author suggest might be relevant, and why?
2. What are examples of management principles that "underscore the interconnection of human and more-than-human natural elements," and how do these differ from industrial agriculture?
3. What was locally exchanged in rural Nepalese farming, and how did this contribute to a sustainable agriculture as well as the creation of place-specific Hindu concepts and rituals?

Notes

1. Suchitra J. Sarma, "The Circle of Existence and Interdependence: Ecopoetry of the Vedic Ksetra or Field," *Interdisciplinary Studies on Literature and Environment* 11, no. 2 (2004): 79–106.

2. *Krishi-Parashara* (*Agriculture by Parashara*), trans. Nalini Sadhale, Asian Agri-History Foundation, bulletin 2 (Secundarabad: Asian Agri-History Foundation, 1999).

3. Manikant Shah and D. P. Agrawal, "*Krishi Parashara:* An Early Sanskrit Text on Agriculture," n.d., www.indianscience.org/essays/24-%20E-F-Krishi%20 Parashar.pdf (accessed August 1, 2010).

4. Pokhara metropolis is now an urban society, but I will discuss farming practices that obtained when it was still a rural society, that is, the situation in the 1960s and 1970s. This place, a popular tourist destination now, has been

modernized and urbanized at a rapid rate. Detailed description about the changing nature of this town can be found in Jagannath Adhikari and David Seddon, *Pokhara: Biography of a Town* (Kathmandu: Mandala Book Point, 2001).

5. Ranchor Prime, *Vedic Ecology: Practical Wisdom for Surviving the 21st Century* (San Rafael, Calif.: Mandala, 2002).

6. Central Bureau of Statistics, *Nepal Living Standard Survey*, vol. 2 (Kathmandu: Central Bureau of Statistics, 2012).

7. Sudhindra Sharma, "Water in Hinduism: Continuities and Disjuncture between Scriptural Canons and Local Traditions in Nepal," *Water Nepal* 9–10, nos. 1–2 (2003): 215–47.

8. O. P. Dwivedi, "Dharmic Ecology," in *Hinduism and Ecology: The Intersection of Earth, Sky and Water*, ed. Christopher Key Chapple and Mary Evelyn Tucker (Cambridge, Mass.: Harvard University Press, 2000), 3–22.

9. Bal K. Joshi, "Rice Gene Pool for Tarai and Inner Tarai Areas of Nepal," *Agricultural Research Journal* 6 (2005): 10–22.

10. G. L. Shrestha and M. P. Upadhyay, "Wild Relatives of Cultivated Rice Crops in Nepal," in *Wild Relatives of Cultivated Plants in Nepal: Proceedings of National Conference*, ed. R. Shrestha and B. Shrestha (Kathmandu: GEM, Nepal, 1999), 72–82.

11. R. B. Rana, C. Garforth, D. Jarvis, and B. Staphit, "Influence of Socioeconomic and Cultural Factors in Rice Variety Diversity Management on Farm in Nepal," *Agriculture and Human Values* 24, no. 4 (2007): 461–72.

12. Pramod Parajuli, "How Can Four Trees Make a Jungle?" *Terra Nova* 3, no. 3 (2004), republished in *Terrain Magazine*, http://www.terrain.org/essays/14/parajuli.htm.

13. Vandana Shiva, Afsar H. Jafri, and Ashok Emani Manish Pande, *Seeds of Suicide: The Ecological and Human Costs of Globalisation of Agriculture* (New Delhi: Research Foundation for Science, Technology and Ecology, 2002), 2.

14. Pankaj Jain, *Dharma and Ecology of Hindu Communities: Sustenance and Sustainability* (Farnham, U.K.: Ashgate, 2011).

6

Dharma for the Earth, Water, and Agriculture

Perspectives from the Swadhyaya

Pankaj Jain

This chapter is about the Swadhyayis, Swadhyaya practitioners, in the Indian states of Gujarat and Maharashtra. The Swadhyaya movement arose in the mid-twentieth century in Gujarat as a new religious movement led by its founder, the late Pandurang Shastri Athavale (1920–2003). In my research, I discovered that there is no category of "environmentalism" in the "way of life" of Swadhyayis living in the rural areas of western India. Following Weightman and Pandey, I argue that the concept of *dharma* can be successfully applied as an overarching term for the sustainability of ecology and agriculture and the religious lives of Swadhyayis.[1] By developing reverential relationships with trees, water, cows, and other ecological resources, Swadhyayis strive to develop their dharmic teachings into practice.

Athavale often cited the definition of dharma from the *Mahābhārata* (12.110.11): that which sustains both the personal order and the cosmic order. Swadhyayis, like many other Hindu communities, use "dharma" to describe their ethos as it relates to their religion and the natural order. For them, the distinction between the religious ethos and the ecological order is negligible, so they express both with the common term *dharma* or *dharam*. Dharma as "virtue ethics" has served as a moral guide for Indians for millennia.[2] In my research with the Swadhyayis, I found that their inspirations were the Hindu epic heroes and their guru, whom they see as role models practicing dharma to attain *moksa*, spiritual liberation. Athavale repeatedly stressed that only actions done with a devotional motive can be considered dharmic actions leading

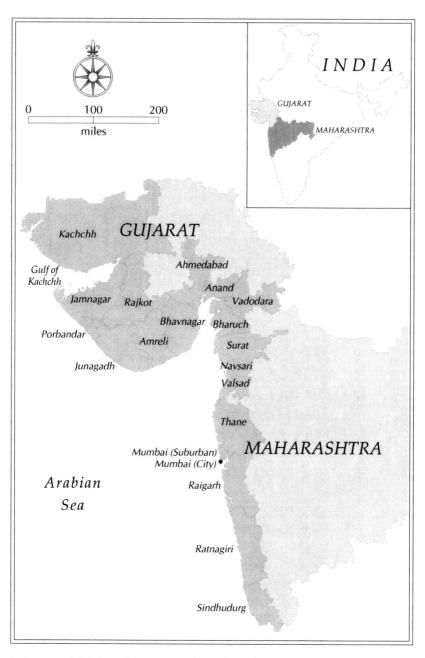

Gujarat and Maharashtra coast. (Based on http://joshuaproject.net/people_groups/16464/IN.)

to moksa. He correlated the motive of the action with the potential for moksa. Based on my case study of Swadhyayis, an ethical framework based on dharma and karma that is also integrated with moksa can serve as an important step in the development of a comprehensive Hindu environmental ethics.

McKim Marriott has suggested that dharma can be an *ethnosociological* category to study and analyze the Indic world that frequently transgresses the world of religion, environmental ethics, and the human social order, as is evident from my case studies of Swadhyayis.[3] Swadhyayis, like other Hindus, use dharma to mean both their religious practices and their social duties. Thus, I suggest that dharma can function as a bridge between the ecological notions and environmental ethics of local Hindu communities and the ecological message related to the planet earth. The word *dharma* can be effectively used to translate the ecological awareness to reach out to local communities of Hindus based on its meanings related to duties, ecological order, sustenance, virtues, righteousness, and religion.

Swadhyaya's Dharmic Ecology

Swadhyaya is one of the least-known new religious movements that arose in the mid-twentieth century in the western states of India. Although this movement now has some presence in several Western countries, including the United States, Canada, and the United Kingdom, it has not received the attention of scholars of Hinduism with the exception of a few introductory essays.[4]

Having heard about the new religious movement Swadhyaya during my trip to India in 2006, I called the organization's office in Mumbai to investigate. Soon I found myself on my way to Valsad in Gujarat, where I arrived at the home of a Swadhyaya volunteer, Maheshbhai, who took me to some of the farms and water-harvesting sites managed by local Swadhyayis.[5] All of them welcomed me warmly and enthusiastically explained their activities and the ideologies of the Swadhyaya movement. As the Swadhyayi farmers told me about the way they perceive nature and the vision of their guru Athavale, I asked questions related to environmentalism and sustainability. What I present below is based on several such interviews with Swadhyaya followers. I have also extracted relevant information from the literature of Swadhyaya that is based on the video-recorded discourses of Athavale.

Swadhyaya (literally, self-study) was inspired by its founder, Pandurang Shastri Athavale, known as Dādājī (meaning "elder brother" in Marathi and "grandfather" in Hindi). He was born in 1920 to a Maharashtrian Brahmin family. His father had founded a traditional school in 1926 called Gītā Pāthshālā (Bhagvad Gita School) to teach the Hindu scriptures in Mumbai. In 1942, when suddenly his father fell ill, young Athavale took over responsibility for the school. Although a Maharashtrian by birth, he developed fluency in Gujarati since most of his listeners were Gujarati. Not much is known about the early years of Athavale's work, which seems to have been limited to preaching based on traditional Indian texts. However, his focus soon changed when he participated in a religion conference in Japan in 1954. This proved to be a turning point in his life. Apparently, he was challenged there to show any village in India where people live based on the Bhagvad Gita's message.[6] After his return from Japan, he decided to develop a role model based on the traditional dharmic teachings. His work first took shape in 1956 when he launched Tattvajñāna Vidyāpīṭha, a residential school with a mission to study the Indic traditions and philosophies. This event can be considered one of the early milestones of his work.

For the first few decades after starting his work, Athavale's focus was to teach and spread the doctrine of the "Indwelling God," the divine presence within humans and other beings. For Athavale, *asmitā* (self-esteem) was an important virtue to be developed based on the awareness of Indwelling God.[7] Although the ancient Vedantic texts identify Ātman within each living entity as identical with Brahman—the all-pervading soul of the universe—such Brahminical concepts have largely been the subject of research and discussion among scholars rather than practiced in the daily lives of ordinary Indians living in villages. I should note here that during this period, Indian villages were largely undeveloped, and the fruits of Indian economic progress had not yet reached these rural areas. Thus, Athavale must have faced an uphill battle to infuse a sense of self-esteem and faith in Indian cultural traditions in villagers struggling to make ends meet. When Western societies were already in their "postmodern" phase of facing the environmental problems that arose due to industrialization, Indian villagers were still in their "premodern" phase of agricultural economy. It is noteworthy that Athavale succeeded in his mission even under these circumstances.

Before I introduce the ecological work of the Swadhyayis, it is important to mention that they deny their environmental significance. In fact, one was taken aback when I told him about my topic of research. In his own words, "You might misrepresent Swadhyaya if you choose to research it from an ecological perspective. Swadhyaya and its activities are only about our devotion to the Almighty; sustainability is not our concern. Environmental problems are due to industrialization and the solution lies beyond Swadhyaya's activities. Swadhyayis are not environmentalists!"

Based on my observations of Swadhyaya's several activities, I tend to agree with him. Athavale repeatedly emphasized that the main goal of Swadhyaya is to transform human society based on the Upanishadic concept of Indwelling God.[8] According to him, since the Almighty resides in everybody, one should develop a sense of *spiritual* self-respect irrespective of *materialistic* prestige or possessions. In addition to expressing one's own dignity, the concept of Indwelling God also helps transcend the divisions of class, caste, and religion, and Athavale exhorted his followers to develop the Swadhyaya community as a global family. The activities of Swadhyaya, woven around this main principle, are aimed at Indian cultural revival.

Although sustainability is neither the means nor the goal of Swadhyaya's activities, natural resources such as the earth, water, trees, and cattle are revered and nurtured by Swadhyayis. I argue that a multivalent term like *dharma* can comprehend and describe this kaleidoscopic phenomenon and the way it relates to ecology. Swadhyaya followers do not regard environmentalism as their main duty, their dharma. Alternatively, from the outside, one can regard their dharma, their cultural practices, as ecologically sustainable, as I show below. I also want to note that my observations are based on their activities in the rural parts of India; the urban and the diasporic Swadhyayis do not have such ecological projects yet.

Swadhyaya's "Earth Dharma"

Athavale's reverential discourses about the earth and water took a constructive shape when his followers launched several projects related to groundwater. In a country like India, where the agriculture in large parts of the country is dependent on rainfall, a lot of work is being done to raise public awareness of rainwater harvesting. Although various Indian organizations have started harvesting the rainwater, Swadhyaya's work in this regard is different because of its underlying inspiration based on devotional reverence for the water and for the earth.[9] According to Athavale, "If you quench the thirst of Mother Earth, she will quench yours." This "earth dharma" became the driving force for Swadhyaya, especially after three successive droughts between 1985 and 1987 in Gujarat. Some Swadhyayi villages started trying out well-recharging experiments. By 1992 the efforts started taking the shape of a movement, and all the nearby farmers began collecting as much rainfall as they could on their fields and in the village, canalizing it to a recharge source. The Swadhyaya "water ethic" is "Rain falling on your roof stays in your house; rain falling in your field stays in your field; rain falling in your village stays in your village." Swadhyaya's work demonstrates a great potential to mobilize the community to undertake ecological restoration at the grassroots level by invoking the cultural and dharmic paradigms. N. R. Sheth describes similar observations:

> Some friends and I observed in June 1994 an outstanding effort in recharging of water resources in a village where a project of constructing a simple check-dam to impound waters flowing in a rivulet was in the final stage of completion. A group of young men constituting the local unit of the Swadhyaya youth centre had conceived and undertaken this project to store the rivulet water in a tank created within it by means of the dam. The stones and soil dug out to make the tank were used to build a 100-metre long dam. The main purpose of this project was to preserve water in the river tank to raise the water table for wells in the neighboring villages. The entire project was executed by an extremely innovative deployment of collective labors. Over a thousand men and women from villages around the dam site had offered their time and energy for the purpose in the spirit of *śramabhakti* (devotional labor). Typically, the males worked on the dam site for about six hours from 7:30 pm in the evening after they had finished work on their own farms. The females worked during the day in neighborhood groups. The people had followed this pattern of work for nine months when we visited the place and they were expecting to finish

the work within a week. Barring small amounts of money spent on explosives to break stone, no cash expenditure was involved in this work. Had the project been executed in a conventional manner, it would have cost around Rs. 400,000. A visitor to a *Nirmal Nīr* work-site cannot fail to be moved by the picnic-like social atmosphere as well as a crusader-like spirit exuded by the volunteers in spite of very hard physical labor they perform at a time when they normally relax or sleep. The social bonding of love and goodwill, which is created among Swadhyayis as well as between them and others is truly remarkable.[10]

One of the most remarkable features of this work is its consistent denial of any outside financial, political, or social help. It has continued to depend solely on the devotional inspiration of its own volunteers. According to Jitendrabhai, a senior Swadhyayi who has supervised several Nirmal Nīr projects in Gujarat in his volunteer work for Swadhyaya for more than two decades, it is important to note the dharmic perspective in *Nirmal Nīr prayog*:[11]

> In the morning, we ask for her [the earth's] forgiveness since we touch her with our feet as we wake up. We bow to her and call her as the wife of Visnu; thus the earth is our divine mother. Just as *Śiva Linga* is worshipped by pouring clean water on it, *Nirmal Nīr* is also our way to worship the earth with clean water. It is an *abhiseka,* ritual of sprinkling, done for Mother Earth. In this prayog, water is cleaned in different stages using sand, brick powder, and coal, and this clean water is stored in wells and ponds. Water enters the wells through the small holes and splinters in their walls. Existing ponds are also deepened in this prayog. People work together to deepen the wells, which also deepens their mutual relationships.

Recalling the visit of water conservationist Rajendra Singh, winner of the Magsaysay Award, to the Nirmal Nīr, Jitendrabhai continued:

> Swadhyayis of one village worked for other villages selflessly with great motivation. Everybody brings his or her own food and shares it with others. This working and sharing with each other builds and develops relationships among all castes and classes of the village. Muslims work

together with Hindus, Harijans work together with Brahmins. Muslims recite their Qur'an verses while Hindus recite Sanskrit verses at the time of regular prayers. After the Gujarat riots, Muslims of Lunāwādā village in Anand District had told the district collector that only Swadhyayis should be trusted to deliver government aid to the Muslim victims of the riots since most other Hindus had turned against Muslims. Earlier farmers used to quarrel if water from neighboring farms would enter their own farms, but after working together in Swadhyaya prayogs they now welcome water drained out from neighboring farms to come to their farms. This reflects increased cooperation and unity among the farmers. In another example, farmers of Maganpura village share each other's losses if anybody's cotton crops fail to yield sufficiently. Another benefit of Swadhyaya prayogs that villagers experienced is that the water tables of all nearby wells increased.

The above ideas expressed by a lay Swadhyayi clearly show the zeal and the spirit of the Swadhyaya work. Although communal tensions have flared up several times in Gujarat, Swadhyayis show a different attitude existing in this state. In addition, most people would be surprised to note the selfless enthusiasm of Swadhyayis. Their motive is not just to do the utilitarian work of the village, or even to protect the ecological resources of the village. Rather, the underlying objective of Swadhyayis is to bring about constructive social manifestation of their spiritual understanding toward the divine that is dwelling in them, their village society, and the village ecology. Indeed, Jitendrabhai succinctly summarized this in his own words:

> In ancient times, Indian sages also were aware of agriculture and ecological resources. For instance, Kaśyapa is referred to as "the father of Indian soil." Similarly, Parāshar is attributed as the writer of *Krsi Pārāshar*, a Sanskrit text for agriculture. Similarly, several community leaders and wealthy people had built water tanks called *Bāvadi* in many villages and towns with Hindu idols near them. However, Swadhyaya prayogs are based on *Karma Yoga*, not *Karma Kānda*. Nirmal Nīr prayogs do not have any ritualistic idol worship. They are prayogs for implementing devotion in one's behavior instead of limiting it to rituals. This openness invites people of all castes, sects, and religions to come and work together.

Overall, Swadhyayis' attitude toward water and the earth shows their long-standing relationships with their ecology, which is revitalized by the thoughts and teachings of Athavale. I now present the agricultural work undertaken by Swadhyayis, a widespread experiment undertaken mostly in Gujarat and Maharashtra because most followers come from farming communities there.

Yogeśvara Krsi: Swadhyaya's Experiments with Agriculture

According to Athavale, agriculture includes four components: trust of neighbor, love for animals, faith in God, and respect for nature.[12] Obviously, the second and fourth components affect ecology directly. In several Hindu festivals, oxen used in Yogeśvara Krsi are bathed and beautifully decorated, and their procession is arranged in the entire village.[13] Ladies smear the foreheads of their husbands and the oxen with the sanctified red color and welcome them. Later, the farms are tilled together, all the oxen participating. Often, the response is so overwhelming that more oxen are brought to a particular piece of land than are required for farming. Thus, this experiment also develops unity, harmony, and goodwill in the entire village.

A related experiment, *Śridarśanam,* involves twenty villages. Twenty people from different villages periodically come together to live on a monthly or yearly basis. One cow also lives with every twenty people to supply their dairy needs. This cow is dearly cared for by. Since the people participating in this devotional experiment treat this cow reverentially, their attitude becomes reverential toward their own personal domestic animals when they return home to perform their routine farming. Thus, in this mechanical age of tractors and other machines, the relationship of farmers with their farm animals is strengthened and revitalized. Yogeśvara Krsi also inspires farmers to develop respect and love for nature. One of the Swadhyayi volunteers, Jitubhai, explained: "One who goes to farms just for work is a farmer, but one who goes to enjoy and respect greenery goes with reverence for and gratitude to God. This reverential perspective inspires to make the entire world green."

Another volunteer explained that they cannot avoid the bare minimum of violence required during the harvest, but they try to avoid harming the nearby plants, trees, and bushes. Vermiculture compost has also been taken up by the villagers on a large scale.[14] The products generated from this devotional farming are treated as the *prasāda* (divine gift) from the Almighty. This helps develop the understanding that nature is not just for anybody's consumption or exploitation but should be revered since it is God's creation. Similarly, the inherent love for animals in farmers is strengthened by such experiments. To recount an example from Yogeśvara Krsi, a Swadhyayi narrated this incident. Swadhyayis go to offer their efficiency at Śridarśanam periodically, as I noted above. When one's turn came, his wife was extremely sick and his crop was ready to be cut. He was in a dilemma whether to go or not. Eventually, the couple decided that the husband should give priority to "God's work," and the wife stayed alone at home. When the husband came back, he found that his farming needs were already taken care of. Later, they discovered that other Swadhyayis had secretly helped the couple by neatly collecting their crops in their warehouse. These secret helpers had criminal records but now, with Swadhyaya experiments, they were so deeply transformed that they were moved by the feeling of divine unity to help others.

It is important to note that the Swadhyaya agricultural experiments are different from the purely "natural farming" pioneered by Masanobu Fukuoka in Japan in the 1940s.[15] Although natural farming is also being promoted by Mohan Shankar Deshpande and other Indians, Swadhyaya's approach is different.[16] In the latter case, there is very little mention of *Kriśipārāśara,* the ancient Sanskrit text about farming. Unlike natural farming, or *Rishi Krishi,* Swadhyayis do not hesitate to till the ground or use natural fertilizers. Rather than rejecting the commonly used tools and methods of agriculture, the Swadhyaya approach is simply to bring about the ethical, spiritual, and social transformation of the farmers (and of the rest of the society). This transformation seeks to develop a familial interrelationship among the farmers of a small village and a reverential perspective toward Mother Earth.

The noted Indian environmentalist the late Anil Agarwal has mentioned that Hindu beliefs, values, and practices, built on a "utilitarian conservationism, rather than "protectionist conservationism," could play an important role in restoring a balance between environmental conservation and economic growth.[17] Swadhyaya experiments do not fall in either category. In fact, when I interviewed some Swadhyayis, they vehemently denied both utilitarian and protectionist motives behind their experiments, underscoring instead the devotional motive.

In line with anthropologist David Haberman's theoretical framework and his observations from fieldwork on the banks of the Yamuna, these Swadhyaya experiments are Indian counterparts of what could be called "environmental activism."[18] Similar to Haberman's examples, Swadhyaya experiments are inspired by Indic traditions. I visited one farm in Gujarat in 2006 where local Swadhyayis had told me that devotional farming has inspired the farmers to grow more trees even on their personal farms. They also started using more organic and traditional fertilizers, such as earthworms. Their perspective toward trees was changed from exploitative to reverential. When I asked them about the practical challenges or difficulties related to Swadhyaya experiments, they noted several. The biggest challenge is to be able to sustain the transformation based on Swadhyaya's teachings. Without the dharmic perspective, the work can become "mechanical." Another challenge is to take these experiments and replicate them at a more far-reaching level. So far, these have remained smaller local initiatives at the district level rather than projects at the regional or national level. They also confessed that the number of volunteers available to work at different farms varies according to the intensity and depth of belief in Swadhyaya's principles in the surrounding villages. Since the spread of Swadhyaya is not uniform across the different villages and towns of Gujarat and Maharashtra, the number of volunteers working at such experiments is also not uniform.

The practitioners of these experiments do not label them as "environmental projects," and yet they have succeeded in sustaining natural resources in many Indian villages. As in other such work, the challenge now is to maintain the projects and to develop new such experiments, especially since Athavale, the motivating force behind the movement, passed away in 2003. To my knowledge, the current leadership has not developed new ecological experiments. However, it seems focused on strengthening existing experiments by inspiring more villagers to join the movement.

Swadhyaya is still a nascent movement with the charisma of its founder yet fresh in the memories of its followers. Will they become more active environmentalists in the new century in the absence of Athavale? Swadhyaya is also emerging as a global movement. When Swadhyayis migrate to different parts of the world, will they mobilize their environmental sensibility to respond to the problems of climate change in their diasporic environments?

Overall, we can conclude that the dharmic ecological work done by Athavale and his followers can be compared with ecological work done by environmental NGOs. However, for the Swadhyayis, their work is simply a reflection of their *krtibhakti,* activity inspired by their devotion to the divinity inherent in themselves and in nature around them. For Athavale's followers, trees and plants merely symbolize the divine force that works to connect human society and nature. As Swadhyayis told me: "*To be is to be related.*" By developing reverential relationships with trees, water, cows, and other ecological resources, Swadhyayis strive to develop their dharmic teachings into practice.

Study Questions

1. How might the practice of sustainable agriculture be impacted if there were no separate category of "environmentalism" within a culture/worldview? How does the concept of environmentalism inform your own understanding about the goals of sustainable agriculture?
2. How does understanding the importance of dharma help us better understand what we would call ecological practices, especially as these occur in Swadhyayi communities?
3. How might basing farming culture on devotion influence farming activities?

Notes

1. S. Weightman and S. M. Pandey, "The Semantic Fields of *Dharma* and *Kartavy* in Modern Hindi," in *The Concept of Duty in South Asia,* ed. Wendy D. O'Flaherty, J. Duncan, and M. Derrett (Columbia, Mo.: South Asia Books for the School of Oriental and Africa Studies, 1978).
2. Bimal Matilal, *Ethics and Epics: Philosophy, Culture, and Religion,* ed. Jonardon Ganeri (Oxford: Oxford University Press, 2002).

3. McKim Marriott, ed., *India through Hindu Categories* (New Delhi: Sage, 1990).

4. Between 1994 and 1996, some observers and scholars visited the Swadhyaya villages. Their observations were compiled in an edited volume by Raj K. Srivastava: *Vital Connections: Self, Society, and God: Perspectives on Swadhyaya* (New York: Weatherhill, 1998). This is a helpful introduction to the movement. In addition, see G. Dharampal-Frick, "Swadhyaya and the 'Stream' of Religious Revitalization," in *Charisma and Canon: Essays on the Religious History of the Indian Subcontinent*, ed. Vasudha Dalmia, Angelika Malinar, and Martin Christof (New Delhi: Oxford University Press, 2001); G. A. James, "Athavale and Swadhyaya," in *The Encyclopedia of Religion and Nature*, ed. Bron Taylor (New York: Thoemmes Continuum, 2005); J. T. Little, "Video Vachana, Swadhyaya and Sacred Tapes," in *Media and the Transformation of Religion in South Asia*, ed. Lawrence A. Babb and Susan S. Wadley (Philadelphia: University of Pennsylvania Press, 1995); M. Paranjape, ed., *Dharma and Development: The Future of Survival* (Delhi: Samvad India Foundation, 2005); Trichur S. Rukmani, *Turmoil, Hope, and the Swadhyaya* (Montreal: CASA Conference, 1999); Devinder Sharma, "Shedding Tears over Failed Watersheds," *Business Line*, June 14, 2000; and Betty M. Unterberger and Rekha R. Sharma, "Shri Pandurang Vaijnath Athavale Shastri and the Swadhyaya Movement in India," *Journal of Third World Studies* 7 (1990): 116–32. Ananta Giri has published a monograph on self-development and social transformation brought about by Swadhyaya: *Self-Development and Social Transformations? The Vision and Practice of the Self-Study Mobilization of Swadhyaya* (Lanham, Md.: Lexington Books, 2009). My own monograph also has a significant chapter on Swadhyaya: Pankaj Jain, *Dharma and Ecology of Hindu Communities: Sustenance and Sustainability* (Farnham, U.K.: Ashgate, 2011), 17–50.

5. In this chapter I have used pseudonyms except for the well-known personalities within the movement.

6. Srivastava, *Vital Connections*.

7. Asmitā is also mentioned in the Yogasūtra (1.17) as a positive attribute to the state of Samādhi and as a negative attribute to be cleansed from the mind (2.3 and 2.6).

8. The Hindu philosophical texts the Upanishads describe "God" as the Absolute or Supreme Soul. The relationship between Supreme Soul and individual soul is likened to the Indwelling God and the soul within one's heart.

9. Sharma, "Shedding Tears over Failed Watersheds."

10. N. R. Sheth, "A Spiritual Approach to Social Transformation," in *The Other Gujarat*, ed. Takashi Shinoda (Mumbai: Popular Prakashan, 2002), 14.

11. In the Indian languages Sanskrit, Gujarati, Marathi, and Hindi, *prayog* means experiment, and the word is used widely in several Swadhyaya activities.

12. Information in this section is based on my interviews with several Swadhyayis in India in 2006.

13. Ann G. Gold, "Story, Ritual, and Environment in Rajasthan," in *Sacred*

Landscapes and Cultural Politics: Planting a Tree, ed. Philip P. Arnold and Ann G. Gold (Aldershot, U.K.: Ashgate, 2001) has observed that such celebration of cattle is linked with the collective management of rainmaking and fertility themes. It also helps develop appropriate harmonies among people and between people and the environment (see especially pp. 128–29).

14. B. K. Sinha, "The Answers Within," *Down to Earth*, May 1998.

15. Masanobu Fukuoka, *The One-Straw Revolution* (Emmaus, Pa.: Rodale, 1978).

16. See http://www.rishi-krishi.com/index.htm (accessed January 20, 2007).

17. Anil Agarwal, "Can Hindu Beliefs and Values Help India Meet Its Ecological Crisis?" in *Hinduism and Ecology: The Intersection of Earth, Sky, and Water*, ed. C. Chapple and M. Evelyn Tucker (Cambridge, Mass.: Harvard University Press, 2000), 165–79.

18. David L. Haberman, *River of Love in an Age of Pollution: The Yamuna River of Northern India* (Berkeley: University of California Press, 2006).

7

Gandhi's Agrarian Legacy
Practicing Food, Justice, and Sustainability in India

A. Whitney Sanford

Almost a century ago, Mohandas K. Gandhi offered a paradigm for food democracy that emphasized sustainability, equity, and social justice regarding natural resources. Since 1959, members of Brahma Vidya Mandir (BVM), an intentional community for women in Paunar, Maharashtra, situated deep in the heart of the Indian subcontinent, have wrestled with the practical implications of translating Gandhian values such as self-sufficiency, nonviolence, and public service into specific practices of food production and consumption.[1] Today, members of BVM and associated farmers illustrate how Gandhi's thought shapes a religious response to the environmental and socioeconomic problems wrought by globalized large-scale agricultural systems.[2] They articulate their responses to failures in contemporary industrial agriculture in religiously inflected language drawn from the *Bhagavad Gita*. They use the *Bhagavad Gita*, a central Hindu text, as a guide to develop agricultural practices that they deem nonviolent. Like Gandhi and BVM founder Vinoba Bhave, the members of the Brahma Vidya Mandir question the dominant narrative of what constitutes the public good and, in terms of food and agriculture, what constitutes good food.

Vinoba Bhave (1895–1982), a contemporary and a follower of Mahatma Gandhi (1869–1948) who is considered Gandhi's spiritual successor, established BVM and five other *ashrams* (spiritual retreats or communities) throughout India to ensure the continuation of Gandhi's lifework on nonviolence and justice for the poor. A Brahman from Maharashtra, Bhave began following Gandhi in 1916 at the age of twenty-one and worked to implement and practice Gandhi's social thought throughout his life. Although Bhave founded six ashrams, he spent much of his life focused on the two in Paunar, Maharashtra: the Paramdham Ashram for men (established 1938), and BVM for women (1959). He established these "experimental laboratories" in remote sites but intended them to be socially relevant, showcasing a nonviolent community that was self-reliant and cooperative. Residents would engage in productive labor, not relying solely on gifts (as is the case with many ashrams), and their existence would demonstrate the possibility of new, more egalitarian social structures. Ashrams, or religiously oriented intentional communities, are a common feature of the Hindu landscape, and these founded by Bhave bear structural similarities with others, including the taking of vows, separation of sexes, and devotional practices. Nonetheless, few ashrams emphasize social change as BVM and Bhave's other ashrams do, making them highly idiosyncratic.

While the women of BVM have removed themselves from mainstream society both geographically and psychologically, they continue Gandhi and Bhave's work both through the examples of their own lives and by educating and training short- and long-term BVM visitors, who are primarily, but not exclusively, women. Nilayam Nivedita and Samvad Farm, two nearby agricultural ashrams, for example, have adopted and enacted practices learned at BVM. The women of BVM and the farmers who work with them enact a counternarrative to the prevalent one that large-scale agriculture is inevitable, necessary, and the sole possibility of feeding the world. They consciously reject a narrative of progress that privileges centralization of knowledge and power and increases reliance on expensive technologies. The narrative of contemporary agriculture, for example, emphasizes productivity, that is, high yields (an arbitrary standard), as the sole measure of value, and progress in the form of improved seeds and inputs such as fertilizers that are priced beyond the reach of most farmers.[3]

In contrast, BVM members grow a diversity of foods and crops that reflect traditional systems of intercropping and small-farm diversity. These women consciously live out ideals of self-rule, nonviolence, and local economies through their methods of farming and food distribution. Their choices of low-input technologies echo Gandhi's call for appropriate technologies. Their lives and language reflect a Gandhian critique of the paradigms of agricultural modernity.

Gandhi's attention to agricultural conditions and distribution of natural resources reflects a form of environmentalism that incorporates equity concerns.[4] Environmental movements in the Global South tend to pair ecological and equity issues, creating a robust environmentalism of the poor, which notes that equity has not always been recognized as an environmental concern due to its emphasis on social concerns. Gandhi's environmental thought addressed agrarian and environmental issues in the context of the social needs of village India.[5] Freedom lay in democratic and broad access to the means of production and survival: for example, community access to forests, water, and healthy soils. His environmental thought offers a paradigm to evaluate the intimate—and frequently overlooked—ties between the environmental degradation and social inequities that inform contemporary concerns about sustainability and the food we eat.

Although communities such as BVM contribute to a growing chorus in India and abroad of those questioning the existing industrialized food system, they by no means constitute a large and powerful movement. Today, India proudly proclaims its sizeable and growing middle class, and while many Indians revere Gandhi as a national hero, they have also, en masse, rejected Gandhian austerity in favor of US-style consumerism. Nonetheless, these communities prompt us to imagine what it means to enact alternative agricultural and social practices, and they help us envision how we might adapt and apply Gandhian ideas in other social and geographic contexts.

The counternarrative BVM's residents practice demonstrates how community members work out the practicalities and trade-offs in their application of self-sufficiency, nonviolence, and radical democracy to their own social and geographic context. Few visitors are likely to join small communities such as BVM, and many features of them, such as consensus-style governance, are not scalable for larger social networks or communities. Even so, small communities such as BVM are necessary for the process of social change—they demonstrate alternate frames of reference. What if values championed by Gandhi, such as nonviolence (*ahimsa*), commitment to the public good (*sarvodaya*), and the Jain concept of nonpossessiveness (*aparigraha),* replaced profit and efficiency as indicators of success in food production?

Gandhi's Thought and the *Bhagavad Gita*

In developing his moral philosophy, Gandhi drew on two traditions: the Hindu, especially the Vaishnava branch, and Jain.[6] Gandhi's home state of Gujarat, while predominantly Hindu, housed significant populations of Jains, from whom he imbibed an ethos that emphasized nonviolence and nonpossessiveness. The *Bhagavad Gita* provides the religious framework for understanding Gandhi's views on sacrifice, duty, and nonviolence. Indeed, the contemporary Gandhi-focused agriculturalists I interviewed stressed the importance of the *Bhagavad Gita* as a moral framework, especially regarding selflessness and reducing ego attachments. The *Bhagavad Gita* (The Song of the Lord), composed between 200 BCE and 200 CE, is included within the *Mahabharata,* one of Hinduism's two great epics.[7] The *Mahabharata* and other epics are especially important to Hindu ethical theory and practice because the situations narrated in the epics present contexts to think through and resolve ethical dilemmas. The *Bhagavad Gita* deeply influenced Gandhi's thought and practices, and he frequently cited the importance of this text and its teachings on *karma yoga,* the path of action to spiritual liberation.

For Gandhi, Bhave, and countless Hindus, Krishna's discourse offers a guide to transform one's actions in the world into devotion to the divine. Bhave also considered the *Bhagavad Gita* a guide in philosophical and practical matters, and he communicated these insights in his *Gita Pravachan,* or *Talks on the Gita.* For the *karma yogi,* one following the path of selfless action, all actions are motivated by love and service first and only then by the practical result; the karma yogi farmer, for example, works to feed society and to establish a basis of love with all beings, not simply for wages. Work and labor become a prayer; and compassion and love render service a joy rather than drudgery.[8] "[The *Gita*] will come to the lowliest of the low, to the poor and the weak and the ignorant, not to keep them in that state, but to grasp them by their hands and lift them up. Its only desire is that man should purify his daily life and reach the ultimate state, the final destination. In fact, this is the very aim and object of the *Gita*."⁹ For both Gandhi and Bhave, the *Bhagavad Gita* provided the frame and authority to restructure society to benefit the poor and oppressed.

Practicing a Gandhian Agriculture

Both Gandhi and Bhave established ashram-agricultural communities, such as Phoenix in South Africa and Sevagram in Wardha, Maharashtra, that elevate "agriculture and artisanry [to a] a spiritual dimension." Ashram residents, for example, gather for religious services either daily or weekly.¹⁰ These utopian farm-ashrams emphasize the dignity of human labor and promote "bread labor," that each person should contribute his or her own labor for goods consumed. Gandhi borrowed the concept of bread labor from Tolstoy, but stated that the third chapter of the *Bhagavad Gita* reflects this principle—that is, food eaten without sacrifice, or bread labor, is stolen.¹¹

Vandana Shiva, founder of the Delhi-based Research Foundation for Science, Technology and Ecology (RFSTE), has become a prominent voice in food sovereignty movements in India and abroad. Her thought, rhetoric, and activism draw significantly on Gandhian ideals and symbolism. The 1998 "Monsanto Quit India" campaign against Monsanto's Terminator technology, for example, echoed Gandhi's "Quit India" rhetoric of independence, and Shiva has framed her continued resistance to GMOs (genetically modified organisms) in India as *seed democracy*, a trope that circulates globally as *food democracy*.¹² Shiva includes seed democracy, food democracy, and water democracy as essential elements of earth democracy.

While Shiva is perhaps the best-known agricultural activist, less familiar farmers and activists also frame their responses in Gandhian terms. Gujarati farmer Bhaskar Save, named "the Gandhi of Organic Farming," has become a spokesman and an early adopter of organic, or natural, farming based on Gandhian principles and has inspired countless others in this endeavor.¹³ "Non-violence, the essential mark of cultural and spiritual evolution," he has contended, "is only possible through natural farming."¹⁴ Kalpavriksha, his fourteen-acre farm, lies just north of the Gujarat-Maharashtra border and yields a variety of fruits, vegetables, and rice, outproducing many chemically based farms.

Save was influenced by the writings of Gandhi and Vinoba Bhave as he developed his agricultural methods, particularly Bhave's work on *Adivasi* (indigenous or tribal) intercropping techniques that rely on symbiotic plant relationships. Save, like other farmers who embrace Gandhi's concept of local economies, has not entered the export market and has chosen to trade or sell his food nearby.[15]

Brahma Vidya Mandir

In the context of this work, in 2008 and 2009, I visited BVM and two agriculturally focused ashrams, Nilayam Nivedita and Samvad Farm, and spoke with Gandhi-inspired farmers who visited BVM. These farmer-activists are located in rural Maharashtra, far west of the urban center of Mumbai, and are proximate to Wardha, where Gandhi established his own ashram, Sevagram. Gandhi chose Wardha because it represented, geographically and symbolically, the heart of India. Today, this agricultural region hosts a growing community of farmers who have established farms based on Gandhian social thought and Hindu practice. BVM and Nilayam Nivedita are in Paunar, ten kilometers from Wardha, and Samvad Farm is approximately two hours by bus north, creating a triangle with Nagpur, the largest urban center in eastern Maharashtra.

These farmers and activists articulated their concerns regarding industrial agriculture, the global financial system, and the overwhelming power of multinational corporations, consistently framing their own actions in terms of engaged critiques of big agriculture, big business, and greed. The women in the ashram are well educated and informed, and have dedicated their lives to service. They represent a small demographic of India's population—a subset living out Gandhian values. A parallel in the United States might be the small population of college-educated, middle-class Americans who have chosen to populate agrarian-focused intentional communities. In short, I interacted with a set of farmers and activists who have the power, education, and means to focus their energies on their social and environmental concerns, so when I use the word *farmer*, it refers to these farmer-activists, not to the social groups, often peasants, to whom the term is usually applied.

Vinoba Bhave established BVM for women to achieve spiritual

liberation and to practice ideals of self-sufficiency, nonviolence, and self-discipline in a community setting. He thought that India needed a class of what he called "social workers" whose work and lives would demonstrate by example how to build a new form of society, in part by living together without regard to caste, language, religion, or nationality.[16] Bhave realized that India had a long tradition of Brahma Vidya, the search for knowledge of the divine, but that few of India's philosophers had reflected on community life. So BVM would maintain India's long philosophical tradition but would do so communally, a collective spiritual liberation—in the hands of women.[17] Approximately twenty-five women resided there, and most have taken vows of celibacy (*brahmacarya*) and thus have the status of religious sisters.[18]

The community emphasizes performing one's duty, practicing nonviolence, and reducing ego to decrease greed and attachment to consumer goods. They are primarily self-reliant in terms of food and water; they spin daily, although not enough to make their own clothes. When I visited, I participated in daily activities, including gardening and food preparation, and observed outreach activities in the form of a conference of Gandhian activists. They have no designated leader, and all decisions are made by consensus. This cohesive, bounded community, with its emphasis on participatory decision making, exemplifies to me the critical role of small grassroots organizations for social change.[19] This ashram has become a hub of agricultural resistance to large-scale industrial agriculture, and the women have trained a number of farmer-activists, men and women, who themselves train others.

BVM sits along the Dham River, which narrows to a trickle in the hot months before the summer monsoon. Upon arriving at the ashram from the main road from Nagpur, one must cross the bridge over the river, then double back onto a smaller access road and cross the river again. Beyond the entrance gates (which are closed for several hours in the afternoon), a short stretch of road, approximately one hundred yards, continues up a small hill to the ashram's buildings. Visitors first see a fenced in garden on the right side of the road, then a set of one-story buildings, including rooms that house male visitors and guests during ashram events and a kitchen that can produce food for large numbers of people.

Most of the ashram's residents and female guests live in the build-

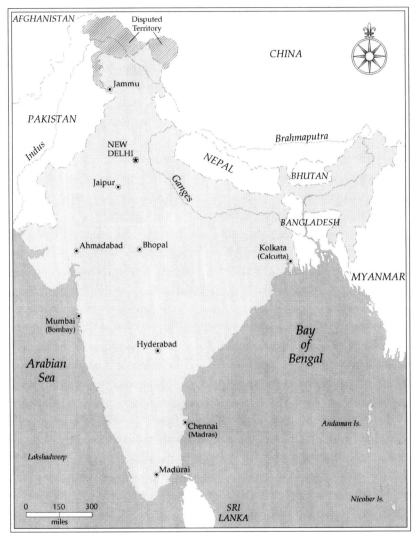

Map of India. (Based on http://www.geography-site.co.uk/pages/countries/atlas/india.html.)

ings that form a quadrangle around the central courtyard. The garden, worked by members and guests, covers approximately half the courtyard. The other half holds long-buried statues of deities discovered by Bhave when he established the ashram as well as several large sundials that Bhave liked. These buildings also house the ashram bookstore, featuring publications by Gandhi, Bhave, and others; the temple; the kitchen and dining area; and a conference room. Many of the buildings feature covered walkways, sitting areas, and a platform for prayers and spinning.

The residents of BVM perform communal prayers, reciting from the *Ishavasya Upanishad* at dawn, the *Vishnu-Sahasranama* (One Thousand Names of Vishnu) at midmorning, and the *Bhagavad Gita* in the evening. During midmorning prayers, they make homespun cloth. For Gandhi, spinning was a dense and critical symbol that encapsulated his views on the multiple facets of independence, self-sufficiency, and inner control. Gandhi believed home-based work such as spinning offered women unprecedented autonomy over their economic lives. As a political symbol, the spinning wheel, or *charkha*, represented freedom from Great Britain's cotton industry that had blocked India's attempts to develop its own manufacturing base. Even today, virtually every village in India holds a Khadi Emporium that sells cloths and other village-produced items. While many Indians and Europeans saw technology, urbanization, and industrialization as forms of progress, Gandhi critiqued these trends because he feared that the need for goods and the accompanying technologies would enslave people.[20]

Gandhi considered personal transformation necessary for meaningful social change, and BVM community members stress mindfulness and a disciplined self as critical for the practice of nonviolence. This disciplined self is able to control desires that lead to greed and violence. As people adopted the discipline and practice of spinning, reciprocally, their identities changed, and this practice changed their ideas of what constituted appropriate clothing—from British to local homespun. Similarly, food, like clothing, is central to our identities, individual and social, and our intimacy with food and clothing offers possibilities to enact choices that are socially and environmentally sustainable at the personal, community, and political levels.

While the women do not make their own clothes, they wear *khadi*, either saris or *salwar-khameez* (a long tunic with baggy, pajama-like pants). Wearing these clothes maintains and enlivens the symbolic nexus popularized by Gandhi, but it also integrates contemporary controversies surrounding cotton production. Today, the seeds and necessary technological packets, such as herbicides and pesticides, incur both massive debt and high ecological costs, and many cite the nutritional deficits that resulted when monoculture cotton crops replaced traditional systems of intercropping food and cotton. The journalist Aparna Pallava, for example, has described efforts in rural Maharashtra to reintroduce traditional (and nearly extinct) intercropping methods in which women plant vegetables and pulses amid rows of cotton.[21] BVM is able to source a small percentage of its cotton locally, from an ashram between Wardha and Paunar, thus supporting regional self-sufficiency and local economies, and this cotton is organic. Most of BVM's cotton is local but not organic, so the members, perhaps not consciously, have had to choose between the violence of a globalized distribution system to obtain organic cotton and the violence of conventional agriculture.

In the morning after prayers, residents and visitors work, performing such tasks as sweeping, food preparation, and gardening. Visitors generally do approximately thirty minutes of sweeping and thirty minutes of work in the garden in addition to food preparation. Equitable distribution of agricultural and other tasks mean that those who might romanticize this labor actually perform it. Bhave believed that three to three and a half hours of productive and well-planned labor, which is also considered a form of worship, would be sufficient to support the ashram and that all labor should be considered equal.[22] Several sisters emphasized to me that this practice reflected Bhave's philosophy of voluntary labor—work, especially agricultural work, is better than meditation for spiritual development because it helps to open up the mind.

Bhave considered physical labor, including cleaning, a form of "faith in action," and bodily labor must be voluntarily undertaken with love.[23] Recounting his first meeting with Gandhi, Bhave noted that his "initiation" into service and ashram life was peeling vegetables, a task that he had never before performed. Seeing Gandhi himself, a national leader, preparing food taught him a lesson about bodily labor and illustrated the meaning of karma yoga, which, in Bhave's eyes, was embodied by Gandhi.[24] BVM's food practices, consensus-style decision making, and shared labor practices help the sisters reduce ego attachments and personal inclinations that would otherwise disrupt the community.

The *Bhagavad Gita*'s discussion of sacrificing the fruits of one's labor to the divine provides philosophical and practical resources in the difficult task of subverting one's own ego and needs to that of the larger group, never an easy process. Decisions about what vegetables should be grown and under what conditions are made by consensus so that all residents have a voice in the process; however, the trade-off is that this process diminishes the influence of those who might have the most agricultural experience. So, in one case, one sister with strong gardening skills wanted to mulch, but others objected, concerned that mulching would lead to increased mosquito populations, a significant concern in a region where malaria is prevalent. This instance of competing values, consensus versus self-sufficiency through agriculture skill, demonstrated the difficulties of actually living a set of values.

The sisters eat a vegetarian diet and generally avoid addictive substances, including tea and coffee, although they serve tea during conferences. The food was pure and *sattvik,* a category describing light foods that do not arouse the passions. Their food reflects a traditional simple Indian meal, based on rice and dal, vegetables, chapattis (unleavened wheat flat bread), yogurt, and milk. When I visited in December 2009, we prepared and ate seasonal vegetables such as radish, eggplant, and *lauki,* a mild gourd. When cutting these vegetables, I was admonished to do so with *ahimsa,* that is, to avoid harming the worms that inevitably appear in organic produce. Their food was locally sourced: the milk, ghee (clarified butter), and yogurt came from their own cows, and the remaining milk was made into *pedas,* a popular dessert. The cow manure was used for fertilizer, and the ashram has a biogas digester for the cow manure that provides some of their energy needs.

The biogas digester illustrates what we now think of as an "appropriate technology" (a term popularized by E. F. Schumacher in *Small Is Beautiful*).[25] Gandhi argued for human-scale, village-based technologies that enhance agricultural productivity and return the benefits to village populations.[26] To evaluate appropriate technologies he simply asked, "Who benefits?" This question was rooted in neither an antiscience nor an antitechnology view; its point was that appropriate technologies diffuse knowledge, fit local conditions, and benefit local economies and so represent alternative and more equitable paradigms for development.

The sisters grow their own vegetables without inputs such as herbicides and pesticides. Water from the kitchen's gray-water system provides some fertilizer; the sisters use ash to first wash the pots and dishes, and the ash water then fertilizes the garden. The sisters use only hand tools in the garden, a practice established by Bhave. Wheat and pulses grow in the fields surrounding the ashram, and villagers provide the labor in these fields (along with the labor for the biogas), using teams of oxen. As Bhave found in his own experiments with food, like giving up milk, the sisters have to balance competing principles when considering food: spiritual health; bodily health; *swadeshi,* or local; and cost.[27] Reflecting Gandhi's emphasis on experimentation, their grappling with translating broad principles such as nonviolence into the realities of everyday life demonstrates that their chosen practices are not inevitable or singular solutions but the result of consciously evaluating the trade-offs and benefits within their specific context. Their chosen practices illustrate one particular instantiation of Gandhian values— these practices are not necessarily generalizable or scalable, but they encourage visitors to consider how these values might be translated to their own circumstances.

Nilayam Nivedita and Samvad Farm

The farmers of BVM, along with those of Samvad Farm and Nilayam Nivedita, draw significantly on a set of interrelated Gandhian ideals, especially regional self-sufficiency, nonviolence, and appropriate technologies, to rethink food and food production in the context of contemporary agrarian challenges. Residents at these agricultural ashrams who have learned the philosophical approaches in tandem with agricultural techniques have established their own farm-ashrams and train students there. They embody Gandhi's emphasis on praxis over theory. Nonviolence, for example, is a call to action, not simply a lack of action or withdrawal, and these farms are actively rethinking and consciously approaching all aspects of food production, including distribution, using the rubric of nonviolence.

A BVM student founded Nilayam Nivedita twenty-five years ago. The farm relies on animal traction and grows bananas, pulses, barley, and wheat. Only four acres are suitable for farming, and this soil has been heavily amended. The local soil, heavy with red clay, is poor, and the amended soil in the four acres under production is visibly distinct.

This farm-ashram primarily trains students who leave to start their own farms after approximately six years.

In May 2008, after a two-hour bus ride, I arrived at Samvad Farm, a ten-acre farm with six acres of orchard, in Amarvati, Maharashtra. Samvad Farm's bioregion is wetter than that of BVM and Nilayam Nivedita, and during the bus ride, I was struck by the difference in fertility and vegetation. Karuna and Vasant Funtane founded the farm in approximately 1984, and now run it with their two adult sons. Karuna Futane lived at BVM from the age of four and has maintained close ties with the organization. Samvad Farm also has an active educational and outreach component, hosting local schoolchildren, foreign students, and others who wish to learn about its farming, water, and building practices. As stated in its brochure, Samvad Farm is a "group of volunteers working in the Gandhian way for *sarvodaya*, the 'upliftment of all'" and is open to visitors and volunteers.

Samvad Farm practices what is described as "need-based" approach, cultivating plants and trees to fill its members' food and shelter needs. When Vasant showed me around the farm, he pointed out how the bulk of their needs, including medicinal, are fulfilled by materials from the farm. In addition to crops and orchards, the two-acre "food forest" provides plants for food, medicinal needs, and building. The buildings, including the house, a cow stall, and a hostel for visitors, were all built from local mud and materials. The term *need-based* reflects Gandhi's statement that "the world has enough for everyone's need, but not everyone's greed" and his idea of "trusteeship," that it is one's duty and service to nurture the land. Today, these farmers and others have argued that agriculture and the contemporary consumer-driven society have moved from "need-based to greed-based" lifeways, confirming Gandhi's fears.[28]

Using only what they need and being self-sufficient offers the members of Samvad Farm a freedom not enjoyed by farmers trapped by debt. Instead, Bhave argued that restoring the land and soil fertility should be considered a ritual sacrifice (*yajna*); all production, including that of food and clothing, should function as a ritual sacrifice to replenish what has been taken from nature.[29] These farmers draw upon the *Bhagavad Gita's* emphasis on performing one's duty and reducing ego to reduce greed and attachment to consumer goods.

Both Samvad Farm and Nilayam Nivedita have developed and reintroduced local varieties of crops such as mangos and sorghum (*jowar*) to ensure the continuation of local knowledge and control of agricultural techniques. Those at Samvad Farm, for example, have developed new varieties of mangos appropriate for the particular soil and weather conditions and maintain varieties in danger of being lost. In line with Gandhian ideals, however, they sell and trade the produce only locally even though they have the opportunity to export their mangos to urban markets. In choosing to serve their immediate vicinities, these farmers consciously reject the neoliberal model proposed (and sometimes mandated) by international trade organizations that sees export-driven trade as the best means to global food security.

These sites of agricultural activism maintain nonviolent farming practices, broadly extending the reach of what might be defined as nonviolent. It is not a stretch to consider toxic pesticides and herbicides as violence, given their harm to human and nonhuman communities. Yet the violence of institutionalized inequity, unjust economic relations, and food systems designed around the needs of the wealthy are less obvious. Cultivating inner nonviolence through attention to the *Bhagavad Gita* and using a consensus-based decision-making process to establish nonviolence is an interior practice related deeply to personal transformation.

These farmers' practices reflect Gandhi's broad understanding of violence and nonviolence and the principle that self-reliance and self-rule are intimately linked with—and rely upon—concepts of nonviolence and decentralization. Centralization, and the resulting need to enforce laws and regulations, was to Gandhi a form of violence.[30] Moreover, what many deem "progress" might inflict violence on marginalized populations. Gandhi implicated the railway system, which has often been lauded as Britain's great contribution to India, in famines because trains transported grains from rural areas to cities and markets. Producing food locally, he believed, reduces food waste and enhances diversified agricultural knowledge of local seeds, landraces, and soils.[31]

In this vein, BVM, Samvad Farm, and Nilayam Nivedita maintain seed banks of local varieties that protect indigenous agrarian knowledge and are adapted to local drought conditions. Echoing activist Vandana Shiva's condemnation of the violence of the Green Revolution and the promised "gene" revolution, these farmers characterize GMOs as violence toward nature and human communities. Although none of these farms are of a size or scale to use these seeds and their associated technologies, they have witnessed the social and economic consequences to farmers in Maharashtra and nearby Andhra Pradesh.

As a result, Samvad Farm and Nilayam Nivedita have crafted their market practices in accord with Gandhi's call for swadeshi—localized economics. According to the Gandhian Satish Kumar, the village-based approach is inherently environmentally sustainable and community enhancing, in part, because it encourages villagers to "take care of themselves, their families, their neighbours, their animals, lands, forestry, and all the natural resources for the benefit of present and future generations."[32] Swadeshi, Gandhi argued, cannot be legislated or enforced with violence. Instead, the transition to a local economy lies in a personal cultivation of nonviolence and sarvodaya (benefits for the many); moreover, persuasion and education encourage peaceful transition.[33] Unlike his Marxist contemporaries, Gandhi did not advocate class conflict or violence in land reform. Instead, he called for large landowners, or zamindars, to become "model landowners" and use property for the good of the peasants.[34]

Lessons from BVM, Samvad Farm, and Nilayam Nivedita

These farm-ashrams apply a Gandhian framework—not a dogma—to fashion equitable alternatives to existing narratives about food, agriculture, and society. Gandhi referred to his own work as a series of experiments and rejected the term *Gandhism,* which would imply a particular ideology, although many have continued to use the term.[35] Gandhi's social thought offers a flexible framework that can be adapted to a variety of conditions and even used in ways that Gandhi himself might not have wanted. What, then, does it mean to follow Gandhi?

BVM, Samvad Farm, and Nilayam Nivedita enact their values in food production and consumption and demonstrate a balance of reflection and practice. Their focus on practice as well as consensus decision making places these communities in a context of process and experimentation, rather than adherence to fixed dogma. The reflexivity of engagement and assessment ensures that this process is not a simple application of a Gandhian platform or an ideological absorption in which theory is divorced from practice, a persistent problem for intentional communities seeking social change.

Like Gandhi, these communities are not nostalgic for a premodern utopia, nor do they seek to reclaim a romanticized past, concerns raised by Nanda and Mawdsley, among others.[36] While Gandhi's *Hind Swaraj* lauded India's traditional values and villages, he did not advocate a nostalgic return to feudal or premodern styles of living in which decision making was limited to very few people. Instead, he proposed novel forms of village development and participatory democracy based on his concept of *swaraj*, which offered unprecedented levels of autonomy to populations such as women and the poor.[37] Similarly, the members of BVM have sought new ways to promote democracy and to avoid replicating existing repressive hierarchies in their own community. In experimenting with innovative and integrated forms of governance, sustainable agriculture, and religious practice, members reenvision concepts of and relations between self, nature, and community. In these ways they demonstrate how relations within small communities can contribute to broader and positive social and environmental change.

These practical and religious responses are situated in a holistic paradigm that integrates material, economic, social, and religious realms. These holistic worldviews reflect a new agrarianism that rejects reductionist, scientific agriculture with its modern separation of religious, economic, and scientific realms.[38] Instead, this new agrarianism privileges a holistic understanding that integrates sustainable agriculture and religion and firmly acknowledges the material dimensions of agriculture and social equity, including health and nutritional benefits. Similarly, BVM, Samvad Farm, and Nilayam Nivedita are not antiscience, but they explicitly consider the financial and social dimensions of agriculture.

Since independence, India has steadily retreated from a rural, agrarian-based society and moved toward an urban-focused, consumer-based society. To some, a Gandhian platform emphasizing agrarian values, nonviolence, and regional self-sufficiency appears quaint and a step in the wrong direction. One might reasonably wonder what small, intentional communities such as BVM can contribute to an increasingly corporate, violent, and unsustainable world: their agricultural production will not feed a hungry world, nor can more than 7 billion of us engage in consensus-style governance. Vinoba Bhave predicted that the world would face two competing narratives: commitment to the public good, or what Gandhi warned of—corporate tyranny and the seductive lure of consumer goods.[39] The intentional communities I have analyzed in this chapter provide counternarratives and lived counterexamples of radical democracy to the tyranny Gandhi warned of and resisted.

BVM presents Gandhi as a frame within which to redirect our thoughts about food and community and, in focusing visitors' attention on the source of food and its production, offers a new awareness for many who have not questioned these issues. Scholarly studies and pop culture phenomena both suggest that in many ways, concerns about food sources and safety are growing in both developed and developing countries.

Our food choices would look significantly different if we grounded our decisions in a Gandhian framework. BVM offers that framework and demonstrates one possible application of Gandhian values in the context of a Hindu-oriented intentional community. Bhave himself used the example of yogurt, stating that the ashram's work, like yogurt, could be mixed with milk to make more yogurt that could be spread to other villages.[40] Visitors, then, can make changes appropriate to their own situations, as did the founders of Nilayam Nivedita and Samvad Farm, and then demonstrate specific applications of Gandhian values to ever-widening circles.

Study Questions

1. How would our food choices look different if they were grounded in a Gandhian framework?
2. What is the role of narratives in shaping food choices?
3. How might living in an intentional community aid one's ability to make more sustainable food choices?

Notes

This research was originally published in *Journal of the International Society for the Study of Religion, Nature, and Culture* 7, no. 1 (2013): 65–87. I wish to thank the editors of the journal.

1. Brahma Vidya Mandir can be translated as the temple (*mandir*) to seek the knowledge and wisdom (*vidya*) of the absolute (*brahma*).
2. Shahid Amin, "Gandhi as Mahatma: Gorakhpur District, Eastern UP, 1921-2," in *Subaltern Studies: Writings on South Asian History and Society*, ed. Ranajit Guha (New Delhi: Oxford University Press India, 1984), 25; Pramod

Parajuli, "Revisiting Gandhi and Zapata: Motion of Global Capital, Geographies of Difference and the Formation of Ecological Ethnicities," in *In the Way of Development: Indigenous Peoples, Life Projects and Globalization*, ed. Mario Blaser and Harvey Feit (London: Zed Books, 2004), 2–3. The gender implications of the organization, although fascinating and critical, are the subject of a separate work.

3. Whitney Sanford, *Growing Stories from India: Religion and the Fate of Agriculture* (Lexington: University Press of Kentucky, 2011).

4. Readers interested in a more comprehensive discussion of Gandhi's environmental and agricultural thought are directed to M. K. Gandhi, *An Autobiography; or, My Experiments with Truth*, trans. Mahadev Desai (Ahmedabad: Navajivan, 1927); M. K. Gandhi, *Village Swaraj*, ed. H. M. Vyas (Ahmedabad: Navajivan, 1962); Ramachandra Guha and Juan Martinez-Alier, *Varieties of Environmentalism: Essays North and South* (London: Earthscan, 1997); Ramachandra Guha, *How Much Should a Person Consume? Environmentalism in India and the United States* (Berkeley: University of California Press, 2006); David Haberman, *River of Love in an Age of Pollution: The Yamuna River of Northern India* (Berkeley: University of California Press, 2006); David Hardiman, *Gandhi in His Time and Ours* (Delhi: Permanent Black, 2003); Vinay Lal, "Too Deep for Deep Ecology: Gandhi and the Ecological Vision of Life," in *Hinduism and Ecology*, ed. Christopher Key Chapple and Mary Evelyn Tucker (Cambridge, Mass.: Harvard University Press, 2000); Rudrangshu Mukherjee, ed., *The Penguin Gandhi Reader* (Delhi: Penguin Books, 1993); and Larry Shinn, "The Inner Logic of Gandhian Ecology," in Chapple and Tucker, *Hinduism and Ecology*.

5. Hardiman, *Gandhi in His Time and Ours*, 75–76; Guha and Martinez-Alier, *Varieties of Environmentalism*.

6. Vaishnavas are those Hindus who worship the deity Vishnu or one of his earthly descents such as Rama or Krishna.

7. The *Bhagavad Gita* recounts the dialogue between the warrior Arjuna and his charioteer, the deity Krishna, on the eve of the great battle between cousins, the Pandavas and the Kauravas. Arjuna asks how it could possibly be right to fight his family members on the battlefield, and Krishna offers a discourse about duty, selfless action, and devotion to God. This massive war ushered the earth into the *Kaliyug*, the era of immorality and decay, and, illustrated in the Hindu epic *Mahabharata*, the bloodied fields of Kurukshetra lay as a testament to its devastation. Gandhi interpreted the *Bhagavad Gita* as a paean to nonviolence, an idiosyncratic reading at odds with historical Hindu understandings of the text. Bradley Clough, "Gandhi, Non-violence and the Bhagavad-Gita," in *Holy War Violence and the Bhagavad-Gita*, ed. Steven J. Rosen (Hampton: Deepak Heritage Books, 2002), 61.

8. Vinoba Bhave, *Talks on the Gita*, trans. Parag Cholkar (Wardha: Paramdham Prakashan, 2007 (1940), 48–50, 58.

9. Ibid., 82.

10. Gandhi, *Village Swaraj*, 94–96; Hardiman, *Gandhi in His Time and Ours*, 76.

11. Gandhi, *Village Swaraj*, 43.
12. Ian Scoones, *Science, Agriculture, and the Politics of Policy: The Case of Biotechnology in India* (Hyderabad: Orient Longsman, 2005), 305–6.
13. This information is compiled from Martin Khor and Lim Li Lin, "Water-Efficient Trench Irrigation for Horticulture," in *Good Practices & Innovative Experiences in the South*, vol. 2, *Social Policies, Indigenous Knowledge and Appropriate Technology*, ed. Martin Khor and Lim Li Lin (London: Zed Books, 2001), 178–95; Bharat Mansata, *The Vision of Natural Farming* (Kolkata: Earthcare Books, 2010); and Anahita Mukherji, "Gandhi of Organic Farming Honored," *Times of India*, September 4, 2006.
14. Mansata, *The Vision of Natural Farming*, 235.
15. Khor and Lin, "Water-Efficient Trench Irrigation for Horticulture," 190.
16. Brahma Vidya Mandir pamphlet; Marjorie Sykes, *Moved by Love: The Memoirs of Vinoba Bhave* (Wardha: Gram-Seva Mandal, 2006), 119.
17. Sykes, *Moved by Love*, 222–24.
18. Men participated in programs and conferences organized by BVM, and several of them resided as brothers at BVM on a permanent basis.
19. Gustavo Esteva and Madhu Prakas, *Grassroots Postmodernism: Remaking the Soil of Cultures* (London: Zed Books, 1998); Anna Peterson, *Seeds of the Kingdom: Utopian Communities in the Americas* (New York: Oxford University Press, 2005).
20. Mukherjee, *The Penguin Gandhi Reader*, 16–18.
21. Aparna Pallava, "What's for Lunch, Mother?" *Hitavada*, June 8, 2008.
22. Sykes, *Moved by Love*, 225.
23. Ibid., 218–21, 246–47.
24. Ibid., 71, 73.
25. E. F. Schumacher, *Small Is Beautiful: Economics as if People Mattered* (London: Blond and Briggs, 1973).
26. Gandhi, *Village Swaraj*, 26–27.
27. Sykes, *Moved by Love*, 244.
28. Priti Agrawal, "Shaswat Yogic Farming," *Times of India*, July 19, 2011.
29. Bhave, *Talks on the Gita*, 245–46.
30. Gandhi, *Village Swaraj*, 34, 39.
31. Ibid., 125; Mukherjee, *The Penguin Gandhi Reader*, 23.
32. Satish Kumar, "Gandhi's Swadeshi—The Economics of Permanence," in *The Case against the Global Economy and for a Turn toward the Local*, ed. Jerry Mander and Edward Goldsmith (San Francisco: Sierra Club Books, 1997), 420.
33. Gandhi, *Village Swaraj*, 35, 58; Mukherjee, *The Penguin Gandhi Reader*, 41–46.
34. Gandhi, *Village Swaraj*, 47, 53, 98–99; Hardiman, *Gandhi in His Time and Ours*, 83–84; Mukherjee, *The Penguin Gandhi Reader*, 34–38.
35. D. G. Tendulkar, *Mahatma* (Ahmedabad: Navajivan, 1960), 4:4.
36. Meera Nanda, *Prophets Facing Backwards: Postmodern Critiques of Sci-*

ence and Hindu Nationalism in India (New Brunswick, N.J.: Rutgers University Press, 2003); Emma Mawdsley, "Hindu Nationalism, Neo-traditionalism and Environmental Discourses in India," *Geoforum* 37, no. 3 (2006).

37. M. K. Gandhi, *Hind Swaraj and Other Writings*, ed. Anthony J. Parel (New Delhi: Cambridge University Press, 1997).

38. Frederick Kirschenmann, *Cultivating an Ecological Conscience: Essays from a Farmer Philosopher* (Lexington: University Press of Kentucky, 2010); Norman Wirzba, *The Essential Agrarian Reader: The Future of Culture, Community, and the Land* (Lexington: University Press of Kentucky, 2004).

39. Sykes, *Moved by Love*, 123–24.

40. Ibid., 125.

8

Thailand's Moral Rice Revolution
Cultivating a Collective Ecological Consciousness

Alexander Harrow Kaufman

Thailand possesses a rich history embedded in the cultivation of rice varieties adapted to diverse biogeographical zones. Rice is more than the staple food of the Thai people; it is the cornerstone of the Thai culinary arts. Rice is a life-giving force that plays a crucial role in social systems and religious practices. For thousands of years, varieties of *Oryza officinalis* (wild rice) were cultivated at subsistence levels to feed farmers' families.[1] Farmers relied upon "embodied knowledge" that was passed on from generation to generation to crossbreed the most resilient and tasty strains of rice.[2] As farmers were dependent on natural weather patterns, the introduction of formal irrigation systems (700 CE) was a defining moment in Thailand's economic history. Improved water access contributed to greater food security and the advancement of Thai civilization.[3] As such, rice cultivation in Thailand is laden with political, economic, and sociocultural values.

The religious practices of Thailand's first human inhabitants (predominantly Lao, Mon, Khmer, Shan, and Tai) governed the process of rice cultivation. Although Theravada Buddhism was installed as the official religion in Thailand, changes in religious practices were slow to take hold.[4] Animist and Brahmanic rituals were entrenched in the local culture and closely tied to rural community life.[5] Farmer households paid reverence to Khwan Khao (Rice Soul), Mae Phosop (Rice Mother), Mae Thoranee (Earth Mother), and Mae Khongka (River Mother). Rituals were performed on behalf of specific deities at each stage of the cultivation process: plowing, sowing, transplanting, harvesting, and threshing.[6] For example, an auspicious day was chosen for plowing,

Organic rice paddies in northeastern Thailand. (Photo by Alexander Harrow Kaufman.)

and the fields were approached "from the South in the belief that [this] would burst the stomach of the Naga (serpent deity) to spread its excrement and to fertilize the soil."[7] Sowing was the next significant activity in the rice farmers' schedule, and in central Thailand a ceremony was performed to "invite the Rice Mother to come out of the granary for the rice to be sown in the field."[8] In Issan (the northeast region of Thailand), where rice was transplanted rather than sown, a ritual was conducted to venerate the reincarnation of the Rice Mother by transplanting seven stalks of rice (inhabited by the Rice Soul) to the paddy fields. After the grains were formed, it was believed that the rice seedlings had been impregnated by the Rice Mother. Offerings were given to relieve the pains of pregnancy, and the farmer recited kind words "to soothe and flatter" the Rice Mother. Before the seasonal harvest, the Issan people reaped seven stalks of rice. Next, the rice grains were released from the seven stalks in what was called *pong khao,* which was meant "to invite the Rice Goddess (or Mother) down to the floor for threshing."[9] This ceremony aimed to appease the Rice Mother and to avoid frightening the Rice Soul. Lastly, farmers took offerings of boiled eggs, rice cakes, liquor, and bananas along with the seven rice plants as a representation of the Rice Soul. After these numerous acts of reverence were completed, the rice was ready for human consumption.

Rituals also were conducted on behalf of domesticated farm animals, in particular the Asiatic water buffalo. Thai farmers relied upon buffalo to produce natural fertilizer and pull plows. They served as a source of meat and nutrition as well. Thai buffalo were venerated for their contributions to rice farming through two traditional rites. *Su Khwan Kwai,* a ritual whereby a Buddhist monk performed a blessing on behalf of the buffalo, was one. The second rite, not identified by name, was conducted by the farmer to honor the buffalo for their role in the cultivation of rice. Despite the arduous tasks performed by the buffalo, they were afforded adequate rest and considered trusted members of Thai agrarian society.[10]

Although the aforementioned ritual practices appear to signify a deep respect for Mother Nature, some researchers claim these practices were nothing more than a reaffirmation of the economic benefits of animal husbandry. Along these lines, some experts argue that the rituals found in many early agrarian societies were merely attempts to control the unpredictability of the natural environment.[11] Nonetheless, ethnographers show that up until the mid-twentieth century, Thai rice farmers practiced traditional forms of agriculture and took active steps to conserve natural resources in their respective communities.[12]

Yet in spite of the value of this so-called local knowledge, most farmers discarded traditional methods in favor of the productivity gains offered by modern agricultural technology.[13] Although some farmers benefited from increased productivity, the use of synthetic fertilizers and pesticides contributed to high rates of soil erosion and a reduction in soil quality.[14] Not only has the misuse of pesticides played a part in the degradation of natural resources, but farmers have also suffered from associated health problems; research shows that in some cases, Thai farmers have died from pesticide exposure.[15] Moreover, the capital demands of these innovations in agriculture have contributed to the exodus of rural folk to the cities in search of paid employment.[16] This dynamic mirrors the migration of Mayan farmers from their ancestral lands discussed in chapter 1 of this volume.

Buddhist Environmental Values

Responding to the failures of the Green Revolution, scholars have sought to restore the dignity of Thai farmers and transform rural society through a traditional way of life based on Buddhist principles.[17] Some scholars have called upon the Buddhist tradition to argue that the Thai people, and in particular rural dwellers, have a predisposition to pro-environmental values.[18] Researchers cite the example of sacred trees ordained with a saffron cloth wrapped around them as an indicator of environmental stewardship. Stories of powerful Buddhist monks acting as custodians of community forests have become part of contemporary Thai folklore.[19] Along these lines, rural dwellers are said to draw upon a pro-environmental Buddhist vernacular in their daily life. For example, Thai farmers commonly use the word *dhammachart* (nature), which has its roots in the Pali language and the Pali Canon (Buddhist scriptures), while Bangkok residents tend to employ the more modern term *singwaedlom* (environment), which was introduced by central Thai authorities.[20]

In the last few decades, the concept of "Buddhist agriculture" has been advanced by Thai environmental activists, farmer leaders, and socially engaged Buddhist monks.[21] While Prawes Wasi, a social activist, popularized the term, the idea of Buddhist agriculture is said to have originated from the teachings of Buddhadasa (former abbot of the Suanmokkh Temple in southern Thailand). Buddhadasa sought to bring back dignity to the profession of farming: "The Dharma in rice farming, with its dutiful ploughing in the hot sun behind a buffalo, is enjoyable and conducted with a felt smile. The early rice farmers knew such satisfaction because they felt their duty as their most important moral responsibility and action. The most important thing we can do is our duty. We call it 'farming' using the ordinary materials—natural resources—to produce the harvest, which when we commit our minds and spirits, is nirvana."[22] Buddhadasa's words imply that through "moral" work, the farmer has the potential to achieve nirvana, or "a state of absolute calm or enlightenment."[23] His sermon glorified the union of farmer, animal, and the natural environment in the process of making rice.

Building on Buddhadasa's teachings, alternative agriculturalists and scholars have borrowed the first of the Five Precepts to contend that conventional farming methods (the use of synthetic agrochemicals) contravene the teachings of the Pali Canon: "The concept of morality manifests itself in the five basic precepts underlying the rules for monastic life and for laypeople's conduct respectively: 1) not to kill any living being (often interpreted as 'not to harm'); 2) not to take what is not freely given by the owner (stealing); 3) not to indulge in sexual misconduct; 4) not to lie; and 5) not to consume intoxicants that lead to carelessness."[24]

Despite the "merits" of a shift to organic farming methods, some scholars have found fault with what they argue are "Western" interpretations of the Buddhist scriptures.[25] While scholars labor over the alleged misuse of the First Precept, the majority of Thai Buddhist farmers have embraced conventional, chemical-based agriculture methods as a panacea for their economic woes. Conversely, scholars have found that conventional agriculture methods have contributed to farmer debt and eroded community-based social safety nets.[26] Although experts are critical of the social and environmental costs of unabated synthetic agrochemical use, governmental programs and multinational corporations continue to support productivity gains through conventional agriculture methods.

The Thai Alternative Agriculture Network

In the 1970s, Asian and Western alternative agriculture groups extended their reach into Thailand with assistance programs aimed at reducing agricultural debts and rejuvenating soil productivity. Foreign experts collaborated with their local counterparts in the delivery of organic extension programs in the north and northeast regions, where poverty levels were reported to be the highest in Thailand.[27] Civil society organizations (CSOs) reintroduced traditional/organic farming methods to empower rural farmers. Academic institutions, CSOs, and select government agencies lobbied for policies directed at "smallholder agricultural development."[28] In 1997, His Majesty the King of Thailand (HMK) called upon Thai farmers to embrace the "Sufficiency Economy" philosophy and adopt New Theory Agriculture (NTA), a flexible set of guidelines based on the diversification of landholdings through low-input, sustainable agriculture methods.[29] The Agri-nature Foundation developed a national-level program to teach Thai farmers ways to apply NTA to diverse agroecosystems. The Green Net Cooperative / Earth Net Foundation launched assistance programs to support farmers with organic certification requirements and to provide access to niche markets.[30] The controversial Santi Asoke Group established agricultural communes that linked Buddhism and alternative agriculture through a formal doctrine of practice.[31] As a way to lobby for changes in government policies, CSOs and farmer groups came together to form the Thai Alternative Agriculture Network (AAN).

Despite the efforts of sustainable agriculture proponents and consumer demands for safe food, less than 1 percent of Thai farmers have shifted to organic farming methods.[32] While researchers have shown some of the financial advantages to making this shift, fewer studies have examined what prevents Thai farmers from adopting organic agriculture methods.[33] In this chapter I explore the development of a Buddhist-inspired farmers' group in northeast Thailand and exhibit the socioecological benefits of the farmers' collective work.

The Dharma Garden Community and Temple

In the early 1970s, two schoolteachers from Patiew Village in Yasothon Province donated an old jute plantation to Monk Khammak as a place for Buddhist practice and meditation. Guided by the Five Precepts, Monk Khammak and his followers chose to become vegetarians and designated this land as a conservation area. Monk Khammak decided to name this community the Dharma Garden. At this time, he took the name Luang Poh Thammachart (Nature Abbot). To support his vision, followers grew rice, fruit, trees, and vegetables using only natural agriculture methods. Gradually, the Dharma Garden Community and Temple became a center for farmers seeking to learn natural agriculture methods. In 1987, the Dharma Garden Foundation was registered as an official institution under the direction of Luang Poh Thammachart and local farmers. Rice cultivated under the foundation's auspices gained International Federation of Organic Agriculture (IFOAM) certification in 2004.[34]

Currently, the Dharma Garden Community covers 150 *rai* (1 rai = .16 hectares), one-third of which is designated for religious activities, and two-thirds for agriculture and training purposes. There are fifteen regular resident volunteers, eleven monks, and thirty laypersons who reside at the temple for different periods of time. There is a cen-

Map of Thailand with detail of Yasothon Province. (Based on *National Statistics Office Statistical Yearbook,* 2010, http://web.nso.go.th/eng/link/solink.htm; United Nations Cartographic Section, Thailand map 3853, revision 2, July 2009, http://www.un.org/Depts/Cartographic/english/htmain.htm.)

tral temple, huts (for monks and nuns), a residence hall, rice paddies, vegetable gardens, a mill, a fertilizer factory, a holistic health center, learning centers, cooperative stores, a radio station, and a vegetarian restaurant on the property.

Temple activities are managed by a committee of monks and elders with help from resident and nonresident volunteers. Daily activities are overseen by a layperson, Nikom Phetpha. Previously, Poh Nikom was a soldier in the Thai army and held a senior position in the Bank of Agriculture and Agricultural Cooperatives (BAAC).[35] Poh Nikom's prior experience helped to secure BAAC funding for a program to assist farmers with debt rehabilitation through organic agriculture. Poh Nikom also acts as a facilitator, a spokesperson, and an extension specialist. Once I found Poh Nikom thirty meters up a tree removing decaying branches to avert damage to temple property. As temple volunteers do not receive salaries, Poh Nikom also maintains an organic rice farm to help support his wife and two children.

A joint interview with Poh Songkran, a part-time volunteer (and working member of the fertilizer cooperative), and Khun Grasehboon, a full-time volunteer, provided further details about temple activities.[36] Poh Songkran described himself as a former Bangkok taxi driver who was once a drug abuser and gang member. He claimed to have made a full recovery through his association with the temple and work as an organic farmer. Khun Grasehboon was formerly employed with other nonprofit organizations that provided social services. Both volunteers also ran their own organic farms when not occupied with temple activities.

Through the temple's community radio station, 91.5 MHz FM, monks and followers broadcast news, music, training programs, and Buddhist teachings. Poh Songkran and Khun Grasehboon serve multiple roles at the temple, including as radio disc jockeys. They take great pride in their broadcasting work:

> The radio station was started in 2002 by SIF (Social Investment Fund), and it serves as a center to connect Dharma Garden Temple projects.
> The station now reaches five other provinces: Ubon Ratchathani, Sisaket, Amnat Charoen, Roi Et, and Mukdahan. I believe that there are at least twenty-one thousand people listening to our radio station, and 80 percent are farmers. The station has volunteers who assist with work all year

Mother Earth ritual. (Photo by Alexander Harrow Kaufman.)

round. Our objective is to deliver radio programs that cover Buddhism, art, culture, way of life, and self-reliant agriculture ventures. The radio station talks more about organic farming on weekends, and we often invite local wise men to talk about or discuss their experiences.

Poh Nikom, who also serves as the station director, provided further information about the radio programs: "It is estimated that [our] member-listeners produce three hundred thousand kilograms of rice per year. There is no advertising on the radio station, and efforts are funded mainly through donations. Various community residents and the temple's abbot serve as the announcers. Many members reported that they were attracted to the temple through the radio station. As such, the radio station has become an important tool in the Dharma Garden Foundation's organic extension and spiritual development outreach programs."

The Moral Rice Network

In 2005, the newly appointed abbot Luang Poh Supatto and lay members initiated a specialized program to support their mission, calling it Kow Khunatham (Moral Rice). The Moral Rice mission is to impart Buddhist teachings, expand organic agriculture, reduce farmer debts, and encourage the consumption of healthy and vegetarian food. The Dharma Garden Temple invites farmers from around the nation to participate in organic agriculture support programs and sell rice through the temple's specialized marketing channels. The organization's training programs include topics such as leadership skills, team building, detoxification, soap making, Effective Microorganisms (liquid fertilizer), and dry organic fertilizer.[37]

Farmers who join Moral Rice are encouraged to work in closely knit groups to produce natural fertilizers and other inputs used to enhance rice quality and output. To be accepted into the program, farmers are asked to take seven vows:

1. Adhere to the Five Precepts and quit all types of vices (for example, drinking, gambling, and smoking).
2. Commit to self-reliance. For example, members must pay for their own organic agriculture certification at a cost of 360 baht.
3. Commit to exchanging knowledge at least once a month and participating in quarterly meetings each year at the temple.
4. Gain knowledge of the management process after harvest to maintain the quality and standard of husked rice of at least 38 percent (per gram), a moisture content of not more than 15 percent, and contamination of not more than 2 percent.
5. Strive to build the Moral Rice brand as a way to help society with good and low-cost rice.
6. Manage the Moral Rice market in two directions: help Moral Rice and its satellite centers process and sell milled rice to the marketplace, and gather together with all members to increase negotiating power in the sale of husked rice.
7. Strive to build food security for themselves, their families, and society.

The "accreditation" process was carried out through a "circle of trust" whereby members monitor each other's progress against the Five Precepts. Members also are encouraged to conform to IFOAM standards. Through their network, members sell a diversity of organic rice products

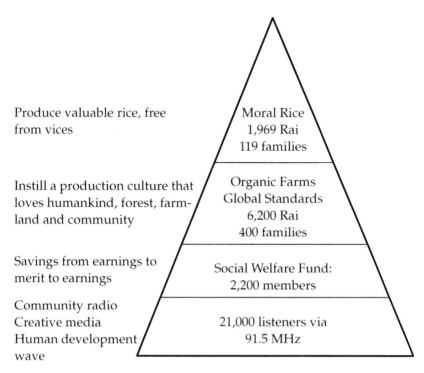

Moral Rice: a pyramid of relationships. (Moral Rice Network, 2012, http://www.moralrice.net.)

to both domestic retail shops and international exporters. Registries in early 2015 showed a total of 119 Moral Rice Network members and 400 certified organic members. Despite a shortage of Moral Rice farmers, over two thousand people subscribed to the temple's social welfare fund.

Wijit Boonserng, Moral Rice chairperson and farmer, explained the benefits of what members call the "One Day, One Baht" fund to farmers and their families: "The way to start becoming a member of Moral Rice is to have them join the Dharma Garden Temple, learn to sacrifice themselves, and give to the others by becoming a member of the savings cooperative. They give one baht a day [1 baht = 1 US$] to support other members when they are in trouble. The fund is separated into two parts: one is for members' welfare—helping those in the hospital—another is to support Moral Rice activities like buying rice from members, buying machines and tools for the mill, and making fertilizer."

Poh Wijit is a key decision maker in temple affairs and was a founding member of the organic training center. I asked Poh Wijit to talk about the origins of Moral Rice:

> Organic farming was started at the Dharma Garden Temple in 1972 by Monk Khammak, and he expanded this knowledge to the villagers—for these actions he was named Luang Poh Thammachart. Luang Poh is the role model for organic farming, as he practiced and did it himself. He allowed homeless people to live on temple property under the condition of helping with farming, and then they got some land to farm for themselves. Luang Poh made the example of organic farming for people to have good, safe food aligned with nature while producing excess food for sale. They had the IFOAM certificate when they opened the temple, but the objective was to move away from the idea of only thinking about selling. You have to change yourself first. Organics is part of good Buddhist practice, which is training by combining Buddhism and organics together, thus leading to Moral Rice. To make rice you must have morality, and farmers must be moral, make sacrifices, and not only talk about it.

According to Poh Wijit, Moral Rice farmers not only acquired new agricultural methods, but they also underwent a spiritual shift through their involvement with the temple: "The Four Types of Lotus is about changing people's ways.[38] If the new member can accept this point, that means they truly want to change their way of thinking. If they think this change is good, then they can decide to become a Moral Rice member later on. However, to be a Moral Rice member one needs to strictly practice the Five Precepts."

Poh Wijit described some of the social and ecological benefits of Moral Rice to farmers, families, and consumers:

> Moral Rice happened because I gave them a way to solve the problem and a protection method from the economy, to solve the farmers' flaws and that of their families, by using dharma as the basis, the Five Precepts to support their mind. Farmers should have safe food to consume—this is the most important point about joining Moral Rice. Next the farmer should live in a proper environment to keep the environment safe. Keep the consumers safe. Manage marketing as a means for leftover production

rather than a need to sell to the market, which is why they need IFOAM as a guarantee for the customers worldwide.

Although organic-certified rice offered growers a higher market price, Poh Wijit conveyed the value of Moral Rice in terms of spiritual and physical health: "The benefit of Moral Rice is that it can raise the farmer's level of spirituality. Which brings good spirits to the rice, and it does not produce toxins. The producer does 'good' by making pure food for himself, his family, and for consumers. The aim is for consumers to have good health. To make the rice pure, the farmer's spirit must be pure first and then good things will come back to him." While to some extent Moral Rice teachings emulate the rhetoric of Buddhist agriculture, members also engage in traditional rites to pay reverence to spirits (see below).

Poh Suvit, a former chairperson of Dharma Garden Temple and president of the Nong Yoh Rice Mill in Kudchum District, explained the ethical basis of Moral Rice: "Moral Rice is a second-stage objective of Dharma Garden Temple created through the Buddhist religion, such as the Four Noble Truths and some other teachings that equate with Moral Rice.[39] Most people do not look at morality as having an application to farm work, because they just want to make more money, with a focus on high production only. But for me, morality is the thing that one can implement in their daily life and it does not have a negative effect on others in the community, the environment, and finally visible and invisible matter."

The Moral Rice program set itself apart from other alternative agriculture initiatives through an emphasis on ecospiritual benefits, as Poh Suvit made clear: "Before I joined Moral Rice I lived in the village, but now I have moved to live in my rice field. At the beginning, we felt alone, then we realized that we were living in symbiosis with the environment, as we grow plants and trees for air and also animals—we raise the cow, the cow eats our plants, and the cow manure is good for the plants and the soil. When the soil is healthy, it helps to manufacture my products. You know the value of food and you know how all living things can live together." As Poh Suvit explained, the methods he used to boost soil fertility led to a series of broader physical changes in the ecology on his farm. As a result of both physical and spiritual work, he developed a deep connection with the natural environment on his family's farm. This new perception of the world carried with it a distinctive set of ecospiritual values.

Buddhist Environmental Values Revisited

To examine the environmental values of Moral Rice farmers, I conducted structured and unstructured interviews with temple leaders and thirty-six members of the Dharma Garden Temple from 2007 to 2009. My data analysis showed that Moral Rice farmers' worldviews are significantly correlated with their religious beliefs, values, and collective practices.[40] These findings were similar to the research of Narong and Nuntiya Hutanawat, which revealed that the "adoption of a new vision" is a critical factor in Thai farmers' ability to sustain organic agriculture.[41] While my findings also highlighted the importance of technical assistance in staying the course of organic agriculture, farmers' decisions to adopt organic methods are associated with Buddhist environmental values.

To explore the concept of Buddhist environmental values, I asked participants to define "nature" in their own words. Many mentioned the Five Precepts in their responses:

> I believe organic farmers must believe in themselves to produce organic food in order to protect nature; the Five Precepts help farmers to be honest (Poh Peng).

> I believe the Five Precepts are a good path for organic farmers, because they do not kill other organisms (Mae Pikul).[42]

> Without the Five Precepts in a farmer's heart, they can't be a true organic farmer, because they might easily break the rules (Mae Boontham).

While Moral Rice members linked organic agriculture to concepts of environmental stewardship, many showed a more spiritually based connection to nature. Poh Nikom explained some of the ways that traditional rituals support organic farming: "Before the rice is sown the family performs a ritual. First we say a prayer, *satoo duu* [a Pali-Sanskrit prayer]: we want to grow rice and fruit in the soil, we wish that the Earth Mother will take care of the soil and help the plants to provide a bounty for eating, giving, and sharing with others." Reverence to the Earth Mother resembles the belief in Mama Allpa of the Kichwa-Lamista farmers based in the high Amazon region of Peru (see chapter 2 of this volume). Poh Nikom further

Organic farmer providing offerings to the Earth Mother. (Photo by Alexander Harrow Kaufman.)

describes the role of the Earth Mother, the Rice Mother, and the River Mother in the Thai conceptualization of nature (dhammachart): "No matter where you fall, or where you are, you fall into the soil. Rice represents all kinds of food that come from the earth. And you need water. Every single drop of water comes from the Earth Mother."

Despite responses that showed Moral Rice farmers espoused a Buddhist ecological worldview, many participants spoke of enhancing nature as a duty to society. Moral Rice farmers also linked the integrity of the natural environment to better health and food security. According to Mae Kluay: "Better soil: soil is much healthier. Better ecosystem: more natural food such as fish, crab, snail, and shrimp. More natural vegetables occur. Less shopping because you can grow anything such as vegetables and fruit in healthy soil."

In several queries, I prompted members to discuss the differences between themselves and conventional farmers. Participants revealed that for the most part their farming practices were guided by Buddhist teachings:

> To be a devout Buddhist, Buddhists must live in harmony with other lives, protecting the environment and microorganisms to keep the environment in

balance; so farmers must keep the rice field in balance (Mae Kampoon).

Organic farmers are more devout Buddhists because the organic farmer has more morals in order not to kill lives and is more honest (Mae Jansri).

Organic farmers are more devoted to Buddhism because to be a true farmer one must study and practice Buddhism in order to be able to talk about organic farming with others (Poh Si).

Organic farmers are more devout Buddhists because we are not greedy (Mae Wanee).

However, some participants felt uncomfortable making comparisons between themselves and conventional farmers:

Organic farmers are not more devout than chemical farmers—it all depends on each person, but it doesn't have anything to do with farming (Poh Ratri).

I disagree that organic farmers are more devout Buddhists because any Buddhist can be devout (Mae Ngern).

Buddhism teaches people to love nature, but that does not mean that chemical farmers are less devout (Mae Malaa).

While Moral Rice farmers expressed an acute understanding of the ecological benefits of organic agriculture, their perspectives of nature were also influenced by family members, colleagues, and spiritual leadership. Members nurtured soil ecology as set out in IFOAM guidelines, and in the process were rewarded for their actions through naturally occurring foods in their rice paddies. Ways of practice were influenced by both knowledge acquisition and availability of the factors of production. Knowledge was acquired and put into practice through formal training courses, sharing experiences, and a process of trial and error on the farm. Significantly, the production of organic fertilizer depended on both technological innovations and the natural resources accessible through their local collectives. However, collectives were also highly valued for their social capital, rather than merely as an access point to the factors of production.[43] Reliable marketing channels were also an important means of sustaining Moral Rice membership.

Moral Rice as a Business Development Strategy

In 2009, the Dharma Garden Temple initiated some changes in its marketing strategy through an alliance with TV Burabha (a Thai television production company), which purchases almost 70 percent of the temple's rice products. TV Burabha helps to market Moral Rice through four key channels: (1) the Symbiosis program, (2) special events, (3) green shops, and (4) large retail outlets. Under the Symbiosis program, products are delivered to consumers in what resembles a "farm to table" scheme.[44] In addition to the benefits of consuming Moral Rice, urban members also have the opportunity to participate in organic farmer home-stay programs. The temple also established a relationship with the Kasetsart University Farmer Shop, where it sells 20 percent of its rice production. The final 10 percent of products are sold to local residents through the Dharma Garden Store. Hence, Moral Rice has become more than a brand name; it is a producer/consumer community composed of monks, farmers, and both rural and urban dwellers.

For Moral Rice farmers, the adoption of organic agriculture is more than an ethical decision; it signifies the acceptance of a new way of life. Members build diverse food systems that afford them additional sources of income and greater food security. The decisions they make regarding the use of natural resources are influenced by Buddhist teachings, a scarcity of natural resources, and the necessity to feed their families. Members substantiate organic agriculture practices as part of being "good" Buddhists. Through the act of nurturing the soil, members come to see themselves as interconnected to an extended community of life. Moral Rice farmers work in close-knit groups to monitor each other's progress in relation to the Five Precepts. By working in groups, members are able to access key natural resources and the factors of production. The Buddhist scriptures guide members' actions vis-à-vis each other and the natural environment.

Based on my research, staying the course of Moral Rice is influenced by a multifaceted set of variables. Moral Rice farmers' worldviews, knowledge systems, and practices affect their relationship with the natural environment and bear upon their quality of life. Members report improvements in their well-being through greater self-reliance, mental and physical health, and the development of social capital through farmer collectives. Similar to the findings of Opart Panya and Solot Sirisai, I found that the religious beliefs of some Thai rural dwellers contribute to ecocentric values.[45] As Dharma Garden Temple members realize the extrinsic and intrinsic benefits of organic agriculture, ecocentric values are reinforced. Nevertheless, only a small percentage of temple members have qualified under the strict requirements of Moral Rice. While the Moral Rice Network struggles to build a future for organic agriculture in Thailand, external socioeconomic and environmental factors continue to bear upon the decision making of members and their families.

Study Questions

1. What influence does religion have upon the decision making of farmers in your country?
2. What strategies could the Dharma Garden Temple employ to attract more Thai farmers and consumers to the Moral Rice Network?
3. How could the Moral Rice program be adapted for use in Western countries?

Notes

1. Paul White, "Rice: The Essential Harvest," *National Geographic*, May 1994, 48.

2. Shigeharu Tanabe, *Ecology and Practical Technology: Peasant Farming Systems in Thailand* (Bangkok: White Lotus, 1994).

3. Lindsey Falvey, *Thai Agriculture: Golden Cradle of Millennia* (Bangkok: Kasetsart, 2000).

4. Prah Payutto, *Thai Buddhism in the Buddhist World* (Bangkok: Buddhadhamma Foundation, 2001). Theravada Buddhism, or the "Teaching of the Elders," was said to have arisen in Ceylon (Sri Lanka), taking hold on the Siam (Thai) Peninsula between the sixth and ninth centuries. The Noble Truths, the Eightfold Path, the Dependent Origination, and the law of Karma are the key principles of Theravada Buddhism.

5. Chatumarn Kabilsingh, *Buddhism and Conservation* (Bangkok: Thai Tibet Center, 2010). *Thewadas* (angels) and *pii* (ghosts) were said to inhabit certain forests, mountains, and bodies of water.

6. Falvey, *Thai Agriculture.*
7. Solot Sirisai, "Rice Goddess and Rice Farming in Thailand," in *Development, Modernization and Tradition in South East Asia,* ed. U Kyaw Than and Pinit Ratanakul (Salaya: Mahidol University, 1990), 168.
8. Ibid., 169.
9. Ibid., 171.
10. Iam Thongdee, "Thai Farmers and Buffaloes: A Cultural Interpretation," in Than and Ratanakul, *Development, Modernization and Tradition in South East Asia,* 178.
11. Terry Rambo, "Conceptual Approaches to Human Ecology" (Research Report 14, East-West Environmental and Policy Institute, Honolulu, 1983).
12. Tanabe, *Ecology and Practical Technology.*
13. Falvey, *Thai Agriculture.*
14. United Nations Development Programme, *Sustainable Human Development and Agriculture* (New York: United Nations Development Program, 1994).
15. Data from 606 farmers in Thailand according to IPM DANIDA pesticide health surveys. See report 62 by the IPM DANIDA Project: "Strengthening Farmers' IPM in Pesticide-Intensive Areas," http://thailand.ipm-info.org/download_documents.htm (accessed October 1, 2010).
16. Opart Panya, "'Community-First' Agriculture: A Search for Thailand's Post-crisis Sustainable Transformation," in *Strengthening Community Competence for Social Development,* ed. Robert Doyle (Phitsanulok, Thailand: Asia-Pacific Perspectives, Naresuan University, 2003), 219.
17. Praves Wasi, "Buddhist Agriculture and the Tranquility of Thai Society," in *Turning Point of Thai Farmers,* ed. Seri Phongphit and Robert Bennoun (Bangkok: Thai Institute for Rural Development, 1988), 1.
18. Falvey, *Thai Agriculture.*
19. Jonathan Rigg, *Southeast Asia: The Human Landscape of Modernization and Development* (London: Routledge, 1997).
20. Opart Panya and Solot Sirisai, "Environmental Consciousness in Thailand: Contesting Maps of Eco-Conscious Minds," *Southeast Asian Studies* 41 (June 2003): 59-75.
21. Apichai Puntasen, "Buddhist Economics: Evolution, Theories and Its Application to Various Economic Subjects," special issue, *Chulalongkorn Journal of Buddhist Studies,* November 2008.
22. Lindsey Falvey, *AgriDhamma: The Duty of Professional Agriculturalists; A Lecture by Buddhahassa Bhikku to Agricultural Teachers and Officials on 25 March 1991 at Suan Mokkhapharam, Chaiya, Surat Thani Province, Thailand* (Adelaide, South Australia: Institute for International Development, 2002), 11.
23. Ibid.
24. Daniel Henning, *A Manual for Buddhism and Deep Ecology* (Bangkok: World Buddhist University, 2002), 37; Chatumarn Kabilsingh, "Early Buddhist Views on Nature," in *Dharma Gaia: A Harvest of Essays in Buddhism and Ecol-*

ogy, ed. Allan Hunt Badiner (Berkeley: Parallax, 1990), 8. Kabilsingh states that the First Precept teaches not only that practitioners should refrain from killing but that they should radiate *metta,* or loving-kindness, to all living creatures.

25. Rigg, *Southeast Asia.*

26. Wasi, "Buddhist Agriculture and the Tranquility of Thai Society."

27. Walaiporn Od-ompanich, Areerat Kittsiri, and Thongchai Manitchara, *Organic Rice Marketing, Organic and Inorganic Rice Production,* 2007, http://www.wiso.boku.ac.at/fileadmin/_/ . . . /2007_Lassmann_DA.pdf (accessed January 20, 2009); *National Statistics Office Statistical Yearbook,* 2010, http://web.nso.go.th/eng/link/solink.htm (accessed October 1, 2011).

28. C. Thongtawee, "Attractors in the Paradigm Shift Process towards Sustainable Agriculture of Farmers" (Ph.D. diss., Mahidol University, 2006).

29. Charan Chantalakhana and Lindsey Falvey, *Sufficiency Economy: An Approach for Smallholder Agricultural Development to Enhance Peace and Stability* (Bangkok: Sermmit, 2008). The Sufficiency Economy Philosophy was introduced by the king of Thailand, King Bhumibol Adulyadej, with the aim to promote moderation, self-sufficiency, and reasonable consumption patterns among Thai farmers and the general population.

30. Panee Samerpak, "A Strategy for Sustainable Agriculture System by Rak Thammachat Club in Kudchum District, Yasothon Province, Thailand" (M.A. thesis, Asian Institute of Management, Manila, 2006). See Rigg, *Southeast Asia.*

31. Juliana Essen, *"Right Development": The Santi Asoke Buddhist Reform Movement of Thailand* (Lanham, Md.: Lexington Books, 2005).

32. FiBL and IFOAM, *The World of Organic Agriculture Statistics,* 2013, Frick and Bonn, http://www.organic-world.net/2413.html?&L=0 (accessed September 13, 2013).

33. Narong Hutanawat and Nuntiya Hutanawat, *Sustainable Agriculture: Vision Process and Indicators* [in Thai] (Nonthaburi: Kledthai, 2006).

34. International Federation of Organic Agricultural Movements, http://www.ifoam.org/growing_organic/definitions/doa/index.html (accessed January 2, 2010). This work employs the term *organic* as defined by IFOAM: "Organic agriculture is a production system that sustains the health of soils, ecosystems and people. It relies on ecological processes, biodiversity and cycles adapted to local conditions, rather than the use of inputs with adverse effects."

35. "Poh" is a title equivalent to "Father" or "Elder" in the northeastern Thai dialect.

36. "Khun" is a title equivalent to "Mr." or "Mrs."

37. Sununtar Setboonsarng and Jonathan Gilman, *Alternative Agriculture in Thailand and Japan,* Asian Institute of Technology, 1999, http:www.solutionssite.org/cat11_s0185.htm (accessed December 11, 2007). Effective Microorganisms is a highly concentrated liquid fertilizer and the trademark of the Kyusei Foundation in Japan.

38. Kabilsingh, *Buddhism and Conservation.* The Four Types of Lotus refer

to different stages of human awareness from a Buddhist perspective. Some lotuses are submerged and thus eaten by animals in the pond, while others rise well above the surface. The height of the lotus corresponds with developed faculties, such as mindfulness, meditation, and wisdom.

39. Payutto, *Thai Buddhism in the Buddhist World*, 7-8. The Four Noble Truths are as follows: (1) the Noble Truth of Suffering, (2) the Noble Truth of the Origin of Suffering, (3) the Noble Truth of the Extinction of Suffering, and (4) the Noble Truth of the Path Leading to the Extinction of Suffering.

40. Alexander Kaufman and Jeremiah Mock, "Cultivating Greater Wellbeing: The Benefits Thai Organic Farmers Experience from Adopting Buddhist Eco-spirituality," *Journal of Agricultural and Environmental Ethics* 27, no. 6 (2014): 871-93.

41. Essen, *"Right Development."*

42. "Mae" is a title equivalent to "Mother" or "Elder" in the Thai northeastern dialect.

43. Alexander Kaufman, "Organic Farmers' Connectedness with Nature: Exploring Thailand's Alternative Agriculture Network," *Worldviews, Global Religions, Culture and Ecology* 16 (January 2012): 154-78.

44. This marketing alliance is no longer as discussed in the text due to recent changes.

45. Panya and Sirisai, "Environmental Consciousness in Thailand."

9

The Seven Species and Their Relevance to Sustainable Agriculture in Israel Today

Elaine Solowey

Sacred to Judaism and at the center of modern Jewish and Israeli holiday traditions is the plant set known as the Seven Species. These plants were also crucial to the agricultural systems of biblical times, when they supplied more than 90 percent of the locally produced food. To this day, the Seven Species have not been surpassed as sustainable and ecologically sound crops for the arid areas of the Middle East. They are an interesting mix of species: two grains, wheat and barley; two deciduous fruit trees, the pomegranate and the fig; one palm, the hardy date; the vine; and a Mediterranean evergreen tree, the olive. From these plants, according to the 104th Psalm, came bread, fruit, oil "to make the face shine," and wine "to gladden the heart"—in essence almost all the food and drink that was produced in the area.

A few words on what sustainability meant in biblical times in an arid area are first needed, as the situation of the people living then is difficult for modern people who gain their calories from Green Revolution technologies to understand. Trade was limited; only luxury goods with a high margin of potential profit justified the risk and labor of importation. Thus, while there was a trade in spices, salt, dyes, cloth, and medicinal and sacred herbs, very little food was imported. Dates, wine, and olive oil were sometimes exported, and the Judean date was famous for its size and flavor until the end of Roman times. But a local failure of a staple crop usually meant that the lack could not easily be made up. Indeed, one of the most memorable stories in the Bible deals with the famine that engulfed the entire Near East, sparing only Egypt. In order for his people to survive, the Patriarch Jacob was obliged to send his sons to Egypt with what silver and gold his tribe possessed to buy grain, which they carried back to Canaan on the backs of donkeys and camels.

Because there was little chance of procuring food from outside, sustainable agriculture in the biblical era was extremely conservative, concentrating on reliable crops that were able to produce with sparse and sometimes irregular supplies of water and other inputs. These crops were also protective and generative of the soil, producing many useful yields to feed both humans and animals.

Following the core belief expressed in the book of Genesis that humans were meant to be the stewards and guardians of lesser creatures, relatively complex laws protecting animals, plants, and the use of water and land were strictly observed. Hence, it was forbidden to muzzle the ox used to thresh the grain, to consolidate farms and drive neighbors away, and to deplete or foul running water. Every seventh year was a sabbatical year in which land lay fallow and debts were canceled; after seven sabbatical cycles, during the fiftieth jubilee year, indentured servants were freed. These rules and customs were so important that their execution was overseen by religious authorities. Disputed matters could even be referred to the king if agreement could not be reached at a local level.

Economically and socially, the Seven Species matched the spirit of the times. It was assumed that people would earn their bread by the sweat of their brow and that food should be stored up for lean years. It was also assumed that the community would care for the widow, the orphan, and the disabled from whatever local surplus it was possible to generate. The Seven Species plants also had a vital and sacred dimension. The Hebrew calendar is to this day lunar (with a correctional leap month that aligns it with the solar year) revolving around festivals of a distinctly agricultural nature. Examining the Seven Species one by one reveals their great value and demonstrates the difficulty of separating the element of "culture" from that of "agriculture."

The wheat of both ancient and modern Israel is *emmer*, the tetraploid "mother of grains." Emmer wheat (*Triticum diccocum*) is versatile in the field as it grows well in both sandy and clay soils, is tolerant of salinity, and has a short, heavy stem that prevents it from lodging in the fierce south winds of the region. Emmer is also the raw material of flour for bread baking, edible oil, and fermented grain products; and unlike many wheat varieties, its straw is suitable for both feed and building material. Wheat bread from tetraploid emmer is an excellent food. Producing a harder and denser loaf than bread made from the hexaploid bread wheat that would appear later, it was also higher in protein. This is the hearty bread referred to in the Bible as the "staff of life." In fact, the modern Sabbath loaf has the shape of a three-strand braid, shaped to resemble an ear of emmer wheat.

Barley, one of the hardiest of grains, is water thrifty and very salt tolerant. Barley was grown in areas that were supplied with water by seasonal flooding, including terraced canyons and salt marshes. It is also resistant to heat and cold, and barley straw is particularly valuable for animal feed and thatch. Barley bread was the mainstay of the poor and was prepared unleavened, baked on a heated stone, and is very much like a chapati or a whole grain matzah. The barley harvest was cause for celebration as barley was hardier than wheat and sometimes did well when the wheat crop was sparse.

Both grains are symbolic of human fertility and courage. The "precious seed" referred to in Psalm 126—"Those who go forth weeping, bearing precious seed, will come again rejoicing, bringing their sheaves with them"—is a direct reference to banished and imprisoned Jews of various captivities, coming out of exile into their own lands again and bringing with them their children who had been born in exile.

The grain harvest was celebrated in two holidays, Shavouth, the festival of the first harvest for the emmer types grown over the winter, and Succoth, or Tabernacles, for the grains grown over the summer. Sheaves of grain are the most common decoration on the Shavuoth holiday, and the book of Ruth is read publicly in the synagogue and at holiday gatherings. Ruth, who was by birth a Moabite, left her land and her people to live with her mother-in-law, Naomi, who lost her husband and her two sons in Moav. She declares to Naomi in what is perhaps the most famous declaration of love and loyalty in the Western tradition: "Wherever you go, I will go, your people will be my people and your God my God and where you die and are buried there I shall be buried as well" (Ruth 1:16). Ruth, childless and widowed in her teens but destined to be the great-grandmother of King David, so becomes the most celebrated of converts, meeting her second husband, Boaz, in a field where she is claiming the right of the poor by picking up the scattered ears of barley left by the reapers. By the act of gleaning she is both proclaiming her and Naomi's poverty and assuming responsibility for supporting her mother-in-law. At the height of the barley harvest, Ruth approaches Boaz and asks that he be the kinsman who "redeems" her from her childless state.

David and Coleen Montgomery, who live near Kiryat Malachi in modern Israel, are cultivating two-rowed biblical barley. They claim, "As a fodder crop it is without compare. It grows well with little water on poor soils and takes easily to organic cultivation methods. The animals love the straw. The grain itself is so fat and beautiful it seems to symbolize abundance itself. The ground grain makes a wonderful bread by itself or mixed with wheat flour. Two-rowed barley malts well for beer and ale, too. We've raised different grain crops, but biblical barley is a special crop."[1] David and Coleen are modern Israeli farmers who are utilizing ancient seed and farming wisdom as they help to create a regionally sustainable food supply. Their farming of barley also suggests that the book of Ruth demonstrates the intertwined connections between grain and fertility, grain and social responsibility, and grain and the notion of real wealth.

To this day, formal and holiday meals in Jewish households begin with the blessing of the bread. In this way the food is blessed and God is thanked at the same time, elevating the meal from a mere stop for bodily refueling into a celebration. The act of breaking bread together became an act of friendship and unity, creating a sacrament, and guests, especially strangers, were traditionally welcomed with food. There is no doubt that bread and wine were considered irreplaceable on the table and in the pantry in ancient times. What is less well known is the importance of bread and wine in worship, for the sharing of these foods was the original "communion," a primal sacred act.

Another contemporary Israeli farmer, Elisheva Rugosa of Ariel, has spent two decades promoting the cultivation and use of emmer wheat both in Israel and the United States. "It is a healthier grain," she said in a 2002 interview. "The hexaploid bread wheats are unbalanced nutritionally. They raise the blood sugar too steeply and sharpen hunger rather than satisfying it. Emmer wheat is more balanced and should be used to make our daily bread." Here is another contemporary Israeli farmer benefiting from the place-specific Seven Species that were the backbone of ancient Israeli farming. For modern-day farming in Israel to move away from Green Revolution technologies and in order for a sustainable farming culture to flourish, it appears that farm management decisions based on the Seven Species can provide food that is healthier for the soil and for our bodies.

The olive tree also had a unique and sacred role. Pure olive oil was the anointing oil of the kings of Israel, a symbol of God's blessing. In the 23rd Psalm it is written, "He prepares a table for me in the presence of my enemies, He anoints my head with oil." King David, the author of this psalm, was chosen by a prophet who ignored his older and stronger brothers. Saul, who ruled before him, pursued him with insane jealousy. King David knew from firsthand experience what it was like to have a table prepared for him in the midst of his enemies and what it was like to feel the sacred anointing oil on his ruddy head.

Olive oil lamps lit the Tabernacle and the Temple in addition to providing illumination for ordinary homes. Chanukah, the Jewish festival of lights, celebrates a particular miracle: after the Temple was retaken from the pagans during the Maccabbean war, only one container of pure olive oil remained. To make more oil for the Temple menorah would take several days. The miracle of Chanukah was that the one cruse of oil lasted eight days, more than enough time for more oil to be made. Given the variety of ways in which the products of the olive tree were used, often in very sacred settings, the olive tree became a symbol of purity.

Good olive oil was considered as valuable as a good reputation. The olive may have been the source of the anointing oil for kings, but olive oil had a much more prosaic purpose, supplying much of the caloric intake in a biblical diet that had few sources of fat or oil. The olive was so vital for health that in the fable of Yotam (Jotham), an ancient tale in the book of Judges (9:8–11), the olive tree refuses the honor of becoming the king of trees. "What would the people do without my fruit and my oil?" the tree in the story asks. "Without me they will suffer and be hungry." Other fruit trees refuse the honor as well. In the end, in a bit of ancient irony, the thorn tree accepts the honor, as this tree has nothing better to contribute than its fierce appearance. The olive branch also symbolizes peace and reconciliation. It was the first plant to emerge from the biblical flood, brought back in the beak of one of the birds Noah released.

So the olive is a tree that makes kings, brings peace, feeds the hungry, and fills both humble and exalted dwellings with light. If Yotam's fable were to be rewritten from the point of view of sustainability, the olive tree would be crowned despite its protests.

Adam of Noam Village is a modern organic grower of olive trees who produces oil using ancient techniques. He shares, "To get the mild virgin oil you can dip your bread in, you have to pick the olives when they are perfect, not let them sit, grind the fruit on stone and press the pulp, unheated, not contaminated by solvents or any modern tricks. Then you have wonderful oil all year that never gets acid, never goes rancid." He continues, "Olive wood from the pruning is great too. I can make handles for my tools and little carvings. The olive is the king of Mediterranean trees—it is so useful and it grows on really rough and dry lands with just the rain that falls." Once again we see that the place-specific domesticated Seven Species of the ancient Israelites are able to form the backbone of a regionally specific sustainable agriculture, one that merges plants and religious heritage.

Another extremely hardy tree woven deeply into the fabric of life and tradition is the date palm. References to the date palm, one of the only sources of sugar in biblical Israel, abound in the Bible. A righteous man in Psalm 92, verse 12 is said to flourish like a date palm. Tamar, or "date tree," is a common woman's name. The male counterpart of this name is Tomer. Another popular man's name, Tamir, means literally as impressive as a palm tree. Many coins from both Israel and Judea carry the symbol of fruit-laden date trees.

From the date tree comes date fruit; date sugar (from collected sap); fiber for rope, mats, and twine; leaves and fruit stalks for basket making; and fronds for thatching. Judean dates were procured by Roman emperors for their table. The Romans, who had nothing else good to say about the inhabitants of ancient Israel, very much appreciated the quality of their dates. The date tree, which grows with irregular and saline water in some of the most barren areas in the Mideast, is also one of the four species of plants celebrated during the Feast of Tabernacles (the original fall feast of thanksgiving) as symbolic of the people of Israel. Its presence is seen on the holiday table in the form of an unopened date frond, or *lulav*.

To Lisa Solomon, an expert on date cultivation who lives in Israel's

The Seven Species and Their Relevance to Sustainable Agriculture   201

Kibbutz date orchard in Israel. (Photo by Elaine Solowey.)

arid and saline Arava Valley, the date is the most majestic of trees. "Our dates give an average of 180 to 200 kilos per tree per season. This yield from one of the hottest and most barren places in the world. We have to take good care of them, of course, to make sure they are irrigated, pollinated on time; we support all the heavy bunches by tying them to the fronds in the crown. We thin our dates and protect them with screen bags. But the result is fine fruit, some of the best in the world, that gets packed locally, then exported all over." Lisa uses modern farming practices that do not depend on chemical inputs in order to produce a key food staple of the region. The resiliency of the traditional date will be vital as Israel becomes even more arid with the onset of climate change, helping form the basis of a sustainable agriculture.

The fig and the vine are two plants that are often mentioned together. A fig tree was considered a necessity both on a farm and in a city garden. In a just and peaceful world, according to the Bible, every man would sit under his own fig and his own vine. The vine, grown over a trellis or arbor, would make a shaded place to rest during the long hot summer days. The grapes, fresh or made into wine, quench thirst and "gladden the heart." The fig, which produces delicious fresh fruit that can also be dried and preserved, provides a source of both beauty and nourishment. Together, the fig and vine came to symbolize the abundance of the natural world as well as peace and prosperity.

The vine by itself has come to symbolize the love and grace of God. The biblical Joseph who saved Egypt and his people from famine is called "a vine taken out of Egypt and planted by a well" in Psalm 80. Worn down by slavery and famine, the children of Israel were told that they would bud like vines in the springtime when they came into the Promised Land.

Chaim Oren from the Jordan Valley works for the Ministry of Agriculture as a consultant for the fig crops in Israel. Chaim explains, "It is a prefect tree, a perfect food. Good for digestion and full of moisture. The fig is wonderful dried and makes a lovely wine when fermented. The sap of the fig is healing and has anticancer properties, did you know that? That is why everyone should have his own fig and his own vine like in the Bible. The fig for health and the vine for happiness." Chaim continues, "The vine is important too. It roots you to the place, to the earth, shades your arbor, and there is nothing better than ripe grapes right off the vine in the summer, like a wonderful drink in little bottles that you can eat. It makes living in a desert almost nice when you have figs there and vines." Figs can be grown with minimal chemical inputs, and there is a growing tendency on various kibbutzim and other farms to grow figs with sustainable practices.

The last of the Seven Species, the pomegranate, is a crop of some importance in modern Israel raised for juice, wine, and materia medica: the same reasons it was cultivated in biblical times. This small, graceful tree grows well on marginal lands, while its wild relatives adorn many rocky hillsides. The distinctive scarlet blossom and the beautiful fruit came to symbolize beauty in both the male and female form in biblical times. The pomegranate is considered the most beautiful of the trees of the field.

Semidomesticated and diverse in height, habit, and fruit quality, the pomegranate probably originated in what is now Iran. It had spread all over the Middle East by biblical times and was a valued source of fruit, wine and other beverages, dye, fine timber, and medicine. Images

Pomegranate trees in Israel. (Photo by Elaine Solowey.)

of pomegranates made of precious metals even decorate the posts of the Torah scroll (the scroll itself is called the Tree of Life).

In the Song of Songs the beloved's lips are like the pomegranate. Other sacred verses find the colors of the pomegranate fruit in rosy cheeks and the tree's graceful shape in the human form. The pomegranate is reputed to have 613 seeds inside it, the exact number of Judaism's major and minor commandments. The edge of the high priest's garment was decorated with a row of bells and pomegranates that rang as he walked. Indeed, one of the few documented and dated First Temple artifacts is a tiny pomegranate carved from ivory that may have graced a priestly garment.

Dan Rimon of the Pomegranate Growers Association in Israel is full of praise for the pomegranate. "It's a beautiful tree that needs little water. It adapts well to most climates. The fruit, when cultivated properly, is big, juicy, and beautiful. A pomegranate tree in flower or fruit is a sight to make you catch your breath. Wine and liqueur from pomegranates is tasty and healthy, too. The fresh juice is considered a drink

Pomegranate fruit in Israel. (Photo by Elaine Solowey.)

good for the heart. Oil from pomegranate seeds is used to control high blood pressure. They knew what they were talking about in the old days when they put pomegranates as ornaments on the Tree of Life."

In addition to food for the people of ancient Israel, these Seven Species contributed much else to the agricultural economy of the day: animal feed, mulch, thatch for the roof, timber, wood for plows and tools, materia medica, dyestuffs, sugar, and coarse fibers. Traditional Jewish agriculture survived the Seleucid conquest and endured into the Roman occupation, adding only a few new crops, such as carobs and saffron. The agricultural traditions continued after the failed Bar Kochba revolt in 70 CE in places like Yavne, Safed, and Ein Gedi, but were unable to survive the Muslim invasion in the eighth century when many remaining Jewish farmers fled into the Diaspora.

Farmers and agriculture suffered greatly in the next centuries as the area was invaded by competing strains of Muslims, Mongols, and then European crusaders, each group looting and laying waste to the orchards, water systems, cisterns, and aqueducts until very little remained of what was once a fertile and productive agricultural area. When the Ottoman Empire ruled the area (from 1299 until World War I), it was so agriculturally depressed and devastated that new crops were brought in from other parts of the empire, including new strains of dates, beans, and grains.

The sixteenth century saw some revival of traditional Jewish crops and settlement outside of the narrow walls of surviving cities, a pattern that continued for the next two centuries. Still, agriculture remained stuck somewhere between the biblical past and the Ottoman system of domination and taxation that reduced Arab peasants, Greek and Armenian merchants, and Jewish craftspeople alike to extreme poverty.

Upon seeing the plight of their coreligionists in the 1800s, wealthy Jews from Europe invested large sums of money to bring new crops, such as citrus, and to revive older crops, such as wheat and vines. Among them were the Baron de Rothschild, still known in modern Israel as "the Generous One," and Moses Montifiore, who brought updated agricultural equipment to the Middle East and built windmills in Jerusalem. With the collapse of the Ottoman Empire, the British became responsible for some areas of the Middle East, and the French for others. The political storms that periodically swept over the region greatly reduced agricultural production until most of the Middle East was an international charity case. Indeed, after Israel declared independence in 1948, absorbing Jewish refugees from Arab countries and displaced persons from World War II, it was necessary to strictly ration food. This was because not enough of any variety of crop, old or new, was being grown to feed the population, and the new country, boycotted by many nations, struggled to survive. This period was called the *Tsena*, or "austerity period," and left such a vivid impression on the soul of the new nation that agricultural research became a priority, leading to an explosion of innovation in irrigation, plant breeding, water desalination, and agroarcheological research.

New and traditional varieties of orchard trees were developed by agricultural pioneers. Modern drip irrigation was invented, and Israel became a world leader in irrigation technology. The ancient Israelite water systems of the Negev were mapped and explored, and Michael Even Ari restored and activated the rain-fed Nabaetan system of water collection by the ruins of the ancient caravan city of Avdat, growing grains and trees in an area with fewer than eighty millimeters of annual rainfall.

Although much of modern Israel's agriculture is geared toward the export market, there have always been farmers who grow traditional species. Dates are the most important crop in the desolate Arava Valley south of the Dead Sea and the arid zones of the Jordan Valley. They are packaged and shipped from these barren valleys all over the world and are one of the more profitable crops for modern farmers as well as one of the few halophytes in the modern crop library. Olives are a widely planted crop in the Galilee and western Negev, supplying most domestic olive oil as well as kosher olive oil for Jewish communities abroad. Grapes for the table and for wine are cultivated in many hill areas from the Lebanese border to the western Negev, with the southernmost vineyard at Naot Smadar in the mountains just north of the Red Sea. In Israel today, figs and vines are the most popular plants in the "house orchard" and the walled garden.

Israel actually has a tiny "wheat belt" between Kiryat Malachi and Beersheva where Jewish and Bedouin farmers sow wheat each year on rain-fed fields. The wheat from this area is particularly well suited for making pasta. Barley is used as a bioremediation plant and for reclamation of saline soils slightly to the south of this region in areas with less annual rainfall. Barley has gone out of favor as a bread grain in the Middle East and is mostly used for beer making, as a whole grain in soup and crackers, and as an animal feed.

A larger theme emerges regarding the Seven Species and contemporary farming in Israel: sustainable agriculture is about not only farming practices but also cultural knowledge. If there is a sustainable culture, as in the case of Israelite identity with local landscapes that stretch back thousands of years, then there is the chance for a sustainable agriculture. Cultural values can play a role in inspiring good, sustainable farming practices.

My Own Experience

I have found a good deal of inspiration and common sense in the agricultural customs described in the Bible and the Talmudic tractate called "Seeds." I will start with the "commonsense" factor. Our modern system of agriculture is based on the continuous abundance of water, fuel, and fertilizer. The biblical systems are based on thrift, scarcity, and an irregular climate. It is an agriculture designed for hard times. However, hard times do not excuse the mistreatment of neighbors.

Water by law must be fairly portioned between upstream and downstream users and cannot be fouled by those upstream. Farms should not be bought up until one owner owns all the visible land around his or her house. Neighbors stopping in an orchard may eat, drink, and refresh themselves with fruit and water but may not carry part of the crop away without permission. The reaper may not turn back for missed stalks or reap around the edges of the field, as these remnants are left for the poor. The first fruits of the trees and vines are also left for the poor. And even in times of war fruit trees must not be destroyed in the territory of the enemy.

The sabbatical years and the jubilee year are fallow years for intensive agriculture. Farmers may harvest from their trees and vines, but they were to let the earth rest and the soil structure re-form. This was especially important in areas with poor soils and limited rainfall.

Scarcity and the need for thrift do not excuse cruelty to animals. Animals must be milked, fed, and put out to graze or brought into the fold regardless of the circumstances of their human owners. Human owners are responsible for damage caused by their animals, not the animals themselves. The ox that treads the grain must not be muzzled, and animals of unequal strength cannot be yoked together. Moreover, in general hunting was discouraged as it was assumed that the wild animals belong to God. These ideas are sensible in an area where everything, even the rain, is scarce and precious. In other words, the Seven Species and the system of farming described in the Hebrew Bible represent a form of traditional ecological knowledge that is both practical and viable for creating a resilient, sustainable agriculture in contemporary Israel.

For example, some of our modern dilemmas can be addressed by the rule of *Kil-i-yim,* or the unwise union or hybrid. It is forbidden to breed a dog with a wolf as the result would be a wolfish dog, a danger to flocks and children, or a doggish wolf, unable to hold a place in the pack. In this age of chimeras, genetically engineered plants, and organisms modified with human DNA, it may be necessary to think about the relevance of this ancient law. A wolfish dog is not what it seems—it is not truly domesticated and is more of a danger than an ordinary dog. In the same manner, a plant that looks like a normal fruit or vegetable but contains the genes of inedible plants or secretes its own pesticides is more of a danger than an ordinary food plant. A doggish wolf has been handicapped in its ability to survive. In the same manner, a plant modified to do something it has not evolved to do or to have an abnormally high harvest index may be hampered in its growth and viability. This is a clear message to those who breed (and modify) both plants and animals: it is unwise and against biblical law to change living things with no thought for the organisms' own needs. I believe these are commonsense rules that could be applied to modern practices, some of which are very damaging to the countryside.

I have been inspired and intrigued by the descriptions of plants in the Bible and the explanation of their uses, from the food of the poor (barley) to the frankincense, myrrh, and basalms necessary to worship in the Temple. The incense trees were as important to the religious and spiritual life of that day as the Seven Species were to the economic life; indeed, it is hard to determine where elements of religion ended and economic elements began. Since many food plants were also used medicinally, one begins to sense a rich and integrated tapestry of law, custom, use, and belief that we moderns are only beginning to understand.

Much of my research now is based on admitting what I do not know and defining what I would like to find out using biblical and Talmudic laws and practices as clues. The fact that five of the Seven Species are so important to today's economy in Israel is a testament to the hardiness and adaptability of these plants. Hardiness, however, is only one reason that the Seven Species have endured. The plants have many qualities that made a sustainable and prosperous life possible in an area with an unstable arid to semiarid climate.

Indeed, the First Jewish Commonwealth, based on the cultivation of these species, endured for eighteen hundred years. These species were all indigenous or had become naturalized early in biblical Israel's history. They required little water or were able to survive on an irregular water supply. They produced multiple products that were easily stored and integrated into everyday life. These species were tolerant of a wide variety of soils and were both heat and cold tolerant in the climate of ancient Israel, which varied from searing heat to freezing winter winds and occasional snowstorms. These plants accompanied the Jewish people on their long historical journeys: they were the vegetation of the spiritual landscape, the species planted in the heart.

Religion, agriculture, and tradition are so intertwined in the cultivation of these plants that it may well be asked if they were important because of their sacred dimension—or if they were considered sacred because of their extreme importance to the very survival of the people who grew them. Other crops were cultivated in biblical times, but none were considered such vital elements in the spiritual and agricultural landscape as the Seven Species. And none of the new crops introduced in the last three hundred years to the Turkish province of Palestine (part of which would become a British mandate and then modern Israel) have attained the status of the ancient crops.

Modern Israel is a small country with limited arable land and little water, but it is home to a vigorous agricultural community that has experimented in growing everything from cotton to bananas. While some crops have flourished for a time, their cultivation is often faddish, and new crops usually vanish in times of recession or drought. That is when the influences of the unstable climate and the arid landscape return to shape the crop library, eliminating all but the most beloved, useful, and hardy plants. In recent decades there has been a renewed interest in biblical crops as sustainable alternatives. Among both new and more veteran agricultural endeavors, the Seven Species reappear in their sturdy glory as the once and future sustainable crops of the land of Israel.

Study Questions

1. Why is it important to understand the "culture" of agriculture? And how was ancient Israeli farming a result of both geography and culture?
2. How might the Seven Species be important to the region in the future, given climate scenarios that suggest the region will be more dry and crowded in the years to come?
3. What might our farming system look like if we observed sabbatical years, and let people glean leftovers from fields? Are there any fields near your house? How might our farming system differ if farmers were forgiven debts so they could invest in their farming practices? Is this fair to nonfarmers? Would it help struggling family farmers?

4. What commonsense rules may help generate sustainable farming in your bioregion? Are any local farmers following such rules?
5. What edible plant/animal species are (or were) indigenous to your area, and is anyone growing these now? How might the commodification of genetic knowledge influence this?

Note

1. This and all subsequent data from Israeli and regional farmers come from interviews conducted by the author.

Tending the Garden of Eden
Sacred Jewish Agricultural Traditions

Yigal Deutscher

As farmers, our invitation is to remember we are tending a story, one so much bigger than our own individual selves, so much larger than our own personal fields. We are never *just* cultivating a field. The "field" we are tending is the story we live in. The landscape is the storybook; the plants are its words. This is a living story, dynamic and growing, changing with the seasons and the years. To dig deep into the soil is to reach into the depth of memory, to remember what we may have forgotten, and to uncover layers of the story we may never even have known. To seed into this soil is to add to it, to enrich it with new memory, new experience. In this way, the soil can dream and the story can evolve.

This canvas of soil, this story, is the foundation of culture. The word *culture* comes from the Latin *cultura*, which actually means cultivation. Culture, like our soils, is something that is tended and nourished. It is the story of society, ancient and new, all together. It is within the fertility of our fields that this story grows. It is within the flavors of our harvests that this story is savored. Similarly, when we erode the soil and disrespect the landscape, we lose our cultural story. These are reflections of one another, and the vibrancy of the two are directly linked.

With our seeds, our pitchforks, our harvests, we might choose to approach farming as businesspeople, educators, laborers, artists, researchers, homesteaders. But we are always storytellers, carrying a story that began long before us. How, as farmers, can we cultivate and tend the majestic vitality of this living story so its seeds can sprout, grow beautifully, flower, and form fruit?

The beauty and mystery of this story is that it has many beginnings.

Each of us can find the indigenous memory of this story, its unique origin, depending on where our farm is planted, who our ancestors are, and in which community we live. The answers can be found in the soils. The land opens our senses, our intuitive knowing, and our remembering. The land cultivates us, and tends to our own growth.

My personal story and ancestry has Jewish roots, so I follow these unique branches and pathways for the answers to my questions. After my introduction to farming, it was the land and soils of Israel that whispered to me, courted me. And I chose this encounter.

Much of my learning happened in the days and nights in this land, in her gentle valleys nestled between rolling hills: a rocky landscape with a sparse Mediterranean shrub forest, deeply green with lush grasses and colorful with wildflowers in the wet winter months, and unforgivingly dry in the summer months. This landscape held a rich and diverse plant community. Sage, marjoram, hyssop, mustard, fennel, and a mixture of wild grains grew below the trees and surrounded the protruding boulders. There were figs, olives, almonds, and carobs growing wildly, as well—remnants of earlier plantings from farmers long before me. This land had been lived in for thousands of years, by many peoples, and during my wanders through the hills I would find the remains of old stone houses, terraces, and carved cisterns, some of which still held the winter rainfalls within. The skies were filled with migratory birds in the spring and fall, making their way over the land corridor linking Africa, Europe, and Asia. In the early dawn and last light of day, you could spot gazelle and fox in the hills. With the coming of evening, you heard the howling of the jackals, in unison, welcoming the approaching darkness. And hours later, with the coming sunrise, the stillness and quiet of the night would be broken by a chorus of harmonic birdsong, the fluttering of wings.

Within this setting, we planted biodiverse gardens of herbs, vegetables, flowers, and fruit trees. We grew for an on-site market and for the members of our CSA (community-supported agriculture program). We hosted an educational farming apprenticeship for students from around the world, similarly called by their own curiosity toward a connection with this ancient, magical, and wild storied landscape. The farm itself was a canvas for learning, an invitation for deep connection, more so than a business for production. We were growing crops, and we were growing our own selves. It was in this land, as I learned from the plants and seasons; as I walked through the fields and wild hills at dawn and dusk; as I watched the migratory birds passing above; as I shared alongside elders and apprentices, asking, wondering, and exploring; as I sat on the Sabbath day of rest in the gentle shade under a tree, with an open Bible on my lap, the ink of the pages mixing with the soils of the field—it was here that I was initiated into the story, and it was here that I truly became a farmer.

Soil and Service

In Israel, the rainy season begins in the fall, after months of dry, barren skies, and this is when the first seeds are planted in the soils. It is during this time that the Torah is opened from its beginning and read once again, each Sabbath a new section.[1] And it is during this sacred reading, during seeding time and calling forth the rains, that we are first introduced to humans and farming.

Early in the biblical story, humans are introduced as *adam*, birthed from *adamah*, the soils of the earth (Genesis 2:7). Human body and earth body are intimately connected, reflections of one another. But the desire, the arousal for our birth came from a lack: "The shrubs of the field were not yet on the earth, and the grasses of the field had yet to sprout, for God had not yet sent rain upon the earth, and there was no human *la'avod* the soil" (Genesis 2:5).[2]

It is the lack of our unique offering that initiates the necessity for our participation in the story of creation. And our unique offering has to do with a particular relationship with the soil. The word *la'avod* is usually translated "to work" and, in this context, has popularly been understood as "to till." In this agrarian reading, there was no human to work the soil, so there was no plant yet growing. Everything at this stage was in a dormant state of potentiality and possibility. All the elements of creation were there, but the synthesis and synergy of dynamic growth had yet to be activated. Why? Because our efforts—our unique relationship with soil—had yet to enter the fabric of the story.

The noun form of *la'avod* is *avodah*. This word appears throughout the Torah, and its most common context does not actually involve agricultural tillage. Rather, the most literal translation of la'avod is "to serve," referencing the priestly service, the sacred work that first began in the Mishkan, the desert tabernacle, and continued with the Beit Hamikdash, the Holy Temple in the land of Israel. This was an intricate and detailed relational courting of Spirit, which involved daily immersions in purifying waters, tending the sacred eternal fire, lighting the incense smoke, and offering the animal, grain, and oil sacrifices on the altar.[3] These offerings are called *korbanot,* a word that comes from the verb *me'karev,* "to draw close, to draw near, to come together." The holistic service of the priests was a sacred invocation and unification, a feeding, revealing, strengthening, and honoring of the interconnectedness and interdependence of all beings. This is the foundational intention of avodah.

After the temple and altar were destroyed, nearly two thousand years ago, this service transformed and evolved into new realms of internal and personal dimensions for each individual, and the rabbis began to refer to this evolution of service as prayer, *avodat ha'lev,* the service of the heart.[4] Our prayers of gratitude, honoring, and reverence are meant to recognize and restore the threads that weave together. Prayer is a sacred conversing, a participation in the living conversation, the mutual story, shared among all creation.

This is the paradise of the Garden of Eden, in which we were placed *la'avdah* (Genesis 2:15). Eden, as told in the narrative of scriptural mythology, was a sacred space, a wild space, a place of abundance, peace, and nourishment. It was a green landscape, filled with the diversity of plant life. The word used for plant in this narrative is *siach,* sharing the same root letters as the word *sicha,* "conversation." This is the conversation of the elements: the roots conversing with the soils, the sun conversing with the leaves, the birds conversing with the flowers, the seeds conversing with the winds and animals.

The soil was the canvas, the host, for this expansive, unfolding conversation of creation. And the soil was the mythic, primal umbilicus of our own physical nature. This was our first relationship to soil. We were entered into the midst of these conversations as pollinators, to nourish and tend this intimate dialogue of life with our own humble gratitude, to honor and recognize these conversations through the beautiful offerings of our voice, the sacred service of our hands and heart: the original essence of avodah.

And this was all before the birth of agriculture. In fact, the first we hear of agriculture is actually a curse. A common reading of the Eden story is that the expulsion from this paradise was humanity's transition to farming and the embrace of the agrarian lifestyle, that Adam and Eve's harvesting and eating from the forbidden "Tree of Knowledge of Good and Evil" was our symbolic remembering of the first cultivation of grains, one of the earliest domesticated crops of agriculture.

In the curse of exile from Eden, the word *avodah* appears once again. But this time, it is far different from its original intention. "The Lord God banished them from the Garden of Eden, *la'avod* the soil from which they were taken" (Genesis 3:23). Now avodah is used in a form meaning burden, toil. What was before an act of ease and simplicity, a natural flow, has become removed and distant. "Cursed be the ground because of you; by toil shall you eat of it all the days of your life. . . . By the sweat of your brow shall you eat bread until you return to the ground" (Genesis 3:17–19). This is tillage: cultivation as control and domination, cultivation as manipulation rather than interconnection. This is a new form of avodah, a shadow and falling from the original sacred service of Eden. This new service is linked with the word *eved*, sharing the same root letters as *avodah*. Eved means a slave, and thus this form of service is imposed, forced labor.

So according to Jewish tradition, is farming a curse? It would appear so. And as a farmer, what is the story I am cultivating? Am I perpetuating this curse each time I till, each time I seed? Luckily, no story is static. And we are crafting the story as we live it. Through the mythic remembering of our past, we can dive into the layers of this story. We can trace its own evolution, its own seeds of transformation.

The Agrarian Change

If we were to place the mythic Garden of Eden on a map, a common landing place would be somewhere in the Fertile Crescent. From the narrative of the Torah (Genesis 2:10), we know Eden was a lush landscape of abundance surrounded by four rivers that flowed perennially, nourishing the land. Similarly, the Fertile Crescent is an ancient area of lush river valleys and rich soils stretching in a crescent shape from the Nile to the Tigris and Euphrates. It covers modern-day Israel, Palestine,

Lebanon, Jordan, Syria, Iraq, and Kuwait as well as parts of Egypt, Turkey, and Iran.

This land is known as the cradle of civilization, as it is the birthplace of agriculture as well as of the ripples of human necessity and desire that emerged from agrarian living: settled tribes, growing populations, surplus and storage of harvests, wealth and economic class distinctions, specializations in the workplace, hierarchical governance, organized religion, writing, money, cities, weapons, war, and empires. Taken together, these filled the canvas of civilization as we know it, with its diverse blessings and curses.

The Hebrew calendar dates the Garden of Eden story to about 5,750 years ago, close enough to this cultural transition that the agrarian civilization would have still been in its early stages of development. The Eden narrative offers a snapshot of a much longer evolutionary period. Historians date the transition to agriculture as beginning around 9000 BCE, fully established in parts of the Fertile Crescent by 6000 BCE. This landscape was a unique and fitting laboratory for the evolution of plant domestication and cultivation. The altitudes of the Fertile Crescent range from the lowest place on earth, the Dead Sea, to the alpine peaks of the Zagros mountain range in modern-day Iraq and Iran. This wide range of ecosystems accommodated a high diversity of plant communities, many of which are the wild ancestors of the domesticated crops we eat today, and also provided a wide range of movement for the nomadic peoples of the area, who seasonally migrated to higher altitudes during the summer months and lower into the valleys during the winter months in search of food and game.

From this landscape, two archetypes emerged: the farmer and the shepherd. The farmer claimed the possibilities of the post-Edenic life, while the shepherd remained more aligned with the earlier, preagrarian hunter/gatherer nomadic lifestyle. In our biblical narrative, the farmer is named Cain, the shepherd Abel. Their names hold their essence, and their characters are embodiments of the energies of their names.

The farmer's name, Cain, comes from the Hebrew *kaniti*; Eve proclaims: "I have *kaniti* a male with God" (Genesis 4:1). Kaniti means "acquire," and all forms of this word have to do with acquisition, ownership, and control. Similarly, sharing the same root letters, a *kinyan* is an object of acquisition, a commodity, something that appears in contractual language.

The shepherd, on the other hand, has the Hebrew name Hevel, which is the word for the out-breath, the exhalation. While many commentators also translate this word to mean "futility," it is only a futility when compared with the dominance of the farmer. Rather, this energy is a powerful form of surrender, of release, of flow. The out-breath is a simple allowance, a giving over in faith and trust to what is beyond our control. And this is the shepherd in his or her role: following the wild path of the animals. Yes, a form of domestication, but much less controllable than the crops we plant.

These two archetypes exist together in tension. In the biblical narrative, Cain and Abel both gather their harvests and make an offering to Spirit. God accepts the offering of the shepherd but does not accept that of the farmer. Cain is enraged with jealousy, and God speaks to him directly. We can understand these words as spoken to every farmer to this day: "Why are you angry, and why is your face fallen? Surely, if you do right, there is uplift. But if you do not do right, sin crouches at the door; Its lust is toward you. Yet you can be its Master" (Genesis 4:6–7).

This is the burden of the farmer. We have chosen a path that is rooted in exile from Eden, work that is full of temptation and amnesia. "Sin crouches at the door." This is the desire of control, of arrogance, of claiming the power as our own, of seeing the crops as commodities, the service as work, and the land as our own property. The act of farming can become a fierce hunger, a need to receive, to take, to claim, until there is nothing left. The amnesia within this hunger is a forgetting of the story of Eden, of connection, of avodah as feeding the interconnected relationships within and between all life.

However, the words of God to the farmer are quite different than the curse given to Adam and Eve. "Surely, if you do right, there is uplift." Ultimately, Cain ignores these guiding words and kills his brother, a murder of archetypal proportions. So today, in a world of agricultural dominance, we are still left with the question: What is the "right" avodah of farming? And how can we use farming to master the desires of control it creates within us?

The Third Path

Throughout the Bible, the Hebrew people chose the path of the shepherd, not the farmer. The families of Abraham, Isaac, and Jacob were all nomadic shepherds traveling the land of Canaan. Arriving in the Nile Valley in the land of Egypt during a severe drought sweeping the western lands of the Fertile Crescent, the Hebrew tribes remained as shepherds, even to their own demise, eventually becoming enslaved to the agricultural empire of Egypt.

The story of their exodus is a story of liberation from slavery and oppression. But it is also a story of unlearning. The Hebrew tribes spent almost four hundred years in Egypt, experiencing the abuses and inequalities an agricultural civilization can create. Their passing through the parted sea was their passage through a birth canal, a cleansing and a rebirth. The period of forty years of travel through the desert was an initiation and a collective spiritual awakening.

The landscape of their travels, the wilderness, is called the Midbar, which means "the Speaking Place." Here they would awaken from their amnesia, remember avodah in the Edenic sense, remember the archetype of hevel, of surrender. In this Speaking Place they would relearn the sacred conversations, the sacred connections between all life, and their service of celebrating this web. This is the place of truth, of clarity, of vision, where layers of separation are peeled back. The Midbar cannot be owned or purchased; it is a place beyond control, beyond the realm of agriculture. It is the place of revelation, the home of the natural rhythm of death and birth, where life decomposes and sprouts forth in a seamless cycle, beyond our powers. It is the home of alchemy and transformation, where the elements merge together in continual co-creation and communication. This place is our teacher.

Here, in this untamed landscape of mountain, soil, and rock, the Hebrew tribes would struggle, sometimes bitterly, confused and lost, longing for the slavery they had become accustomed to, begging for the cultivated vegetables they had known growing on the banks of the Nile in the land of Egypt. And here, passing through this struggle, they would relearn avodah. They would activate the service of the priesthood, offering daily sacrifices, daily courting and connection. Here they would gather the *manna*, wild nourishment of the desert, a food beyond cultivation, collected in equal portions for all peoples—a food that would be gathered daily, without the ability to be hoarded or accumulated in storage (Exodus 16:13–23). Here they would learn the practice of *shabbat,* of rest, of weekly surrender and exhalation. And here they would stand before the mountain of Sinai and receive the Torah, their guidance. And this guidance is one that crafts a path for them, for home. They would journey for forty years to this place, and once they arrived, for the first time in their collective history, they would settle and become farmers; they would begin forming an agrarian society. They would enter into the lineage of Cain, but not before recognizing the shadows of this path, not before relearning the arts and skills and service of Abel. And in this way, they would be initiated into a third path, a merging and blending of farmer and gatherer, farmer and shepherd.

Foundations of Jewish Agriculture

The Torah the Hebrew people received in the Midbar was a blueprint for this third path, offering the foundations for a sacred agriculture, a farming in covenant with the wild. These agreements tell a new story of farming, not as a curse of exile or expression of empire, but as a pathway toward redemption, toward home. Within this blueprint, farming becomes a form of avodah, sacred service, aligned with the service of the Garden of Eden, the feeding and nourishing of the relations between all life. This becomes the new story for the new land, the story we have been invited to tend in our fields, in ourselves, and in our communities. What follows are the foundations for this covenant, the intentions for a sacred agriculture and sacred culture

Rest and Release

The Hebrew story of creation unfolds within a cycle of seven. "Six days you shall work, but on the seventh day you shall rest; you shall cease from plowing and harvesting" (Exodus 34:21). Six days of sacred action followed by a sacred shabbat, a temple in time. The seventh day is a period of stillness, completion, and wholeness. Rather than the end of a linear momentum in time, the seventh is the center, the heart of creation. Welcoming the seventh day during shabbat prayers, within the poetic liturgy of *Lecha Dodi,* we sing: "She is the source of blessing for all creation . . . last in deed but first in thought." This seventh day offers us a gift, a sacred period of rest, of renewal, and of reflection. A time to cease from agriculture, to remember the bigger story we are a part of, in service of. This day is a guide, informing the activity and intention of the six days that follow. The period of nondoing is the teacher for the doing. How we sit in rest, in stillness, will teach us how to engage in action, in creating.

In this relational dance between rest and action, between the days of creation and shabbat, the elements are in eternal, dynamic balance, each feeding the other. The fabric of the entire Jewish cosmology, culture, and story is interwoven with the patterns of this cycle. Just as there is a celebration of shabbat, of rest, every seventh day, there is a collective celebration of release every seventh year, a time known as the *shmita.*[5] "When you enter the land that I am giving you, the land shall observe a Sabbath of the Lord. Six years you shall sow your field . . . but in the Seventh Year the land shall have a complete rest" (Leviticus 25:2–3). For an entire year, farmers and farmland release from the rhythms and practice of agriculture. Our muscles and bodies release our grip on our tools, release our grasp on the land. The land itself expresses herself in her wildness, beyond domestication, beyond cultivation. Her soils are rejuvenated and her nutrients restored. Just as the land re-wilds, so too, we re-wild. We settle into observation, watching the needs of the land, the interactions of land, animal, and plant with sun, water, and wind, without human intervention. We are guided to release and surrender in faith, to learn from this period, to humbly gather the wisdom of knowing the land in this way.

Shmita is the law of the wilderness, the law of the Midbar, the Speaking Place beyond human control. If we are to develop agriculture, we must do so with the seed of the wild planted in our hearts, planted in our tools, in our fields. Shmita, like the weekly shabbat, offers us a delicate edge where our two identities, our civilized and wild reflections, can meet and merge together in sacred balance. When every seventh year passes, and we do return to the tools of tilling and cultivating, the hope is to do so as a sacred act, feeding the wild within us and around us, nourishing that which continues to feed us. The cycles of shabbat and shmita guide us with this question, this riddle, and challenge: What would an agricultural society look like if it ceased from agriculture every seventh day, every seventh year? How can this system be designed for shared nourishment and ecological abundance?

Sacred Land

After forty years of wandering through the Midbar, the Hebrew tribes finally entered the land of Israel, crossing the Jordan River. On its banks they camped and made offerings of gratitude, celebrating once again their exodus from Egypt, their release from slavery. Here, on this soil, they would begin their own story as farmers. During a solo excursion deeper into the land, Joshua, the leader of the Hebrew tribes, received a message from an angel in the guise of a warrior. Joshua fell on his knees, bent low to the earth, and humbly asked: "What does my lord say to his servant?" And this angelic being responded to Joshua, "Take your shoe from off your foot; for the place on which you stand is holy" (Joshua 5:14–15). This story reflects back the same message Moses received at the place of the burning bush in the wilderness. The land is holy; the land is sacred. If you are to begin farming, you must know this in every fiber of your being. Every tool you use, every seed you plant, every harvest you gather: you are interacting with the sacred. Through direct relationship with the soil you will know the divine.

When the people settled in Israel, the land was portioned out to each tribe, the size of each parcel according to the tribal population. Each land had its own soils, its own resources, its own unique crops. And as the people of each tribe began to cultivate their soils, they would develop an intimate relationship with place, extending from generation to generation. The stories of their families would merge with the story of the landscape. Each would be woven in with the other. If land were to be sold, it would be only for a short period. After seven shmita cycles, the fiftieth year would mark the jubilee, the *yovel* (Leviticus 25:8). In this year, all tribal peoples would be invited to return once again to their ancestral lands. If they had lost their family land through debt or had chosen to sell it, they would still have equal right to return. Part of the sanctity of the land was the sacred bond with the particular peoples of the land. For the story of the land to be vibrant, it required the partnership and participation of these peoples. In this way, land never became an object, a resource, or a commodity. Even if it did enter the market, the land itself would have no price. The only "sale value" was on the estimated amount of harvest that would be gathered in the years leading up to the next jubilee.

To further support the wild sanctity of land, in each tribal holding, cities would be surrounded by a *migrash,* an open space (Numbers 35:2). This open space was dedicated land for animals, for wilderness, for shepherding. In our planning of cities and agricultural landscapes, there would always remain the connection to the wild, the uncultivated.

Supporting this multigenerational connection to place and the wild came with the recognition that humans are humble guests, graciously receiving a divine gift. The land ethic of Israel was based on this understanding, as given by God: "The land is Mine, and you are temporary settlers with Me" (Leviticus 25:23).

Perennial and Wild Allies

The celebrated native and heritage plant species of the land of Israel are known as the Seven Species, the *Shivat Minim.* Through their bodies and fruits, these plants tell the story of the land. "For the Lord your God is bringing you into a good land . . . a land of wheat and barley, of grape vines, figs, and pomegranates, a land of olive trees and honey [date palms]" (Deuteronomy 8:7–8; also see chapter 9 by Elaine Solowey in this volume). Besides these crops, the land was rich in a variety of other fruit trees and wild edibles, such as the fruit trees carob, pistachio, almond, and mulberry; the herbs sage, hyssop, marjoram, and mallow; onion and asparagus; the seed crops of oats, lentils, and chickpeas. These plants became the foundation for the earliest forms of agriculture in the land. And the tending of these particular crops created an open portal to a more regenerative, wild, and diverse food ecology, in unison with the culture of the land.

Every seventh year, all soil cultivation—tillage and seeding—ceased. Yet this year was meant to be one of abundance, of food to be eaten and shared with all: "Whatever the land yields during its Sabbath will be food for you—for yourself, your male and female servants, and the hired worker and temporary resident who live among you" (Leviticus 25:6). Without agricultural production of grain, legume, or vegetable, the primary harvests of this year were the wild edibles and perennial plants. The need for widespread availability of these crops during the seventh year required a deep respect and awareness, a stewarding and honoring of these plants, during all years of the shmita cycle.

With this rhythm—six years of farming and one year of complete fallow—the natural food ecology that developed integrated perennials as a keystone species in plant production. And this supported a healthier agricultural system. Perennial crops retain their predomesticated vigor and are less dependent on the agricultural inputs of soil cultivation, irrigation, and fertilization, efforts that often come with costs such as soil pollution, erosion, and salinization. Perennials invest more into their own plant body than annuals do, giving them resilience for survival. Their longer and stronger roots make them more drought tolerant and more able to access nutrients in the soil. Their stronger plant bodies allow for stronger resistance to disease and insects. Furthermore, over seasons, decaying organic matter from dead leaves and roots leads to building organic matter within their own soils.

Beyond the tending of perennial food crops, the shmita tradition also inspired an open relationship with the wild lands beyond the borders of agricultural fields, abundant in foods growing without a dependency on human efforts. And the diet was enriched by this. Wild plants tend to have much higher levels of nutrients than our domesticated crops, as their roots dig deeper for their own survival through soils that have not been cultivated. Also, with no irrigation provided, their bodies and fruits may be smaller, but their flavors are much more concentrated.

An agriculture allied with a diverse community of wild and perennial plants is an agriculture of permanence, in partnership and relationship with the natural ecology. This is an agriculture reflecting the Garden of Eden. It is an agriculture that requires patience and deep observation skills; it requires long-term visioning of land protection and fruit-tree tending; it requires education to know how to identify, properly gather, and prepare wild edibles. It is an agricultural planting and tending not simply for this season's harvest but for the generations ahead, harvests our children will be able to enjoy and be nourished by.

Offerings of Gratitude

The earth provides, and with our inputs of seed, water, and fertility, our agricultural landscapes are one of abundance. Yet, for each taking, there must be a return, a giving back, an offering of gifts. Jewish agriculture is embedded in a relationship of reciprocity and mutuality, with specific instructions for agricultural offerings. These offerings were a way to celebrate all the ways we were being fed and to tender something in return, an alchemy born from shared effort, of the primal elements of sun, water, air, and earth, the life force of plant and animal, the farmer's hands, sweat, and tools. These offerings mentored us as well, cultivating our own sense of self-discipline, generosity, and sacred humility.

With every harvest, the first crop to ripen was not for our own consumption. Rather, this was designated as sacred, as *bikkurim*, first fruits to be dedicated to the Temple in Jerusalem. From the spring harvests of grain to the late-summer harvests of fruit crops, the farmers of Israel would make the journey to the Temple with fresh or dried harvests, depending on their distance of travel. And they would arrive in celebration, in a pilgrimage and parade of thanksgiving and gratitude, with decorated wagons pulled by decorated oxen, loaded with ornamentally displayed harvest baskets, with musicians leading the way, guiding them onward to the Temple (Mishnah Bikkurim 3).[6]

Beyond the seasonal offerings of the bikkurim, there were sacred offerings from each harvest, dedicated to the Kohanim and Levites. Unlike the other tribes of Israel, these peoples did not receive a land-based heritage; instead, their avodah was strictly through Temple service. The offerings of the farmers would link these priests to the land while simultaneously linking the farmers to the Temple. Of every crop that entered the home, 10 percent was separated as a *terumah,* a gift, to these priests.

Once the Temple was destroyed, the rabbis of the Talmud taught that the new altar for service was the table within our home. The table itself became the altar; the food we place on it became the offerings; and the prayers we offer became the avodah. The rabbis understood that to eat a food before offering a blessing would be the equivalent of theft. Regardless of the work we as farmers put into the growth and harvest and preparation of this food, it is not ours at all. It belongs to Spirit, to creation, which we serve through our farming. It is the offering of the blessing that allows us the spiritual capacity to consume these foods and be nourished by them. Each blessing is worded in the present tense: "Blessed are you Creator who is creating." We, as farmers, participated in a part of the process, but the existence of the food itself is beyond our own efforts. The first blessings, before consumption, relate to recognition and awareness of this understanding, an honoring of the life we have tended and taken. The second blessing, after consumption, is of gratitude and thanksgiving. These prayers are the final act of cultivation, which began in seed form, continued through tillage, through weeding, through irrigation, beyond harvest, and finally is completed with the cultivation of gratitude and honoring.

Courting the Rains

Unlike the Nile of Egypt and the lush river valleys of the Fertile Crescent, which allowed for human-controlled and human-directed canal irrigation systems, the land of Israel was an arid landscape, "a land of hills and valleys that drinks its water from the rains of the heaven" (Deuteronomy 11:11). The falling of the rain during the short winter months was the life force for this land and its people. It completely shaped their culture and spiritual worldview. "If, then, you observe the commandments that I place upon you this day, to love the Lord your God and serve him with all your heart and soul, I will send the rain for your land in season, the early rain and the late rain. . . . Watch yourselves that your heart is not seduced, and turns aside to serve other gods and bow to them. For the Lord's anger will flare up against you, and he will shut up the skies so that there will be no rain and the ground will not yield its produce" (Deuteronomy 11:13–17).

Our vulnerable dependence on rain, our total lack of control in its falling, was a critical aspect in cultivating the farmer's faith and humility. Farmer became rainmaker, or rain steward, in this way, through relationships of interdependence rather than through human force or labor. If we love God—if we love the wild ecology, plant life, the connections and conversations between all beings; if we honor and steward them—these ecological systems will remain intact and rains will fall. If, however, we disturb these relationships—these connections among plant, earth, animal, and water—the systems will be disrupted, the rain patterns and rhythms will be broken. There will be destructive floods. There will be drought. In this way, there was deep reciprocity and reflection between the human and the wild.

The cycle of the Jewish year followed the rainfall. Each fall season, the new year would be marked by weeks of purification, of repentance, of blowing a *shofar*, ram's horn, to awaken self, the skies, and the earth for the change of season. This period of prayer would create the bridge, the passing into the new year, and a portal into Sukkot, the fall harvest festival. During these seven days, as the last of the tree crops would be gathered in from the agricultural fields, those who had made pilgrimage to Jerusalem would participate in a festive rain ceremony called Nisuch Ha'Mayim, the Pouring of the Waters, which involved ritualistic collecting of groundwater from a flowing mountain spring, parading it to the Temple, and pouring it onto the altar as a libation and invocation. The morning ritual was followed by a night of festive dancing, singing, and music called the Simchat Beit Ha'sho'eva, the Rejoicing of the House of the Water Drawing, held in the Temple courtyard (Mishnah Sukkah 4–5). The purification process of the new year rituals was to serve as a guide toward this culminating holiday and the moment of becoming a rainmaker for the season ahead. This is a worldview in which humans and rain cycles are directly linked, reflections of one another. As humans return to the source of all life in sacred relationship, so, too, the rainfall returns to the earth in its proper time, for blessing.

From this day onward, until spring and the Passover holiday, courting the precious rains would be a part of the farmer's daily avodah through offerings of prayer, and throughout the year, farmers would perform physical tasks that received and welcomed the rains, such as crafting terraces in their hillsides to allow the earth to drink deeply, digging cisterns to hold the rains beyond the wet months, and planting arid-land fruit trees, shrubs, and herbs. These physical prayers mark the landscape of Israel to this day.

Tending Community

The memory of ancestral slavery in a land of agricultural power and wealth was deeply embedded in the tribes of Israel. While they traveled in the desert in preparation to farm the land they would settle in, the laws they received came with strict guidance for sharing the foods they grew with those who were hungry or without land, as well as for offering tithes to local community members in poverty. The responsibility of the farmer is to cultivate relationships, not just food. And these relationships extend beyond plant and soil. These relationships extend into family, into community. Food is the carrier of nourishment, of story, of sustenance, and it is the nexus of community. In this way, the tending of relationships in the fields was a microcosm of the tending of vulnerable relationships within society.

With each cultivation, a corner of the field would be sown and tended but not harvested by the farmer. This section, *peah,* would be dedicated to community members who were without land of their own (Leviticus 19:9-10). The peah would be open for them to come and freely harvest as needed. Similarly, during the harvest season, gleaners would be welcomed to enter fields and collect from remaining crops, a tradition known as *leket.* This is a remembering of food as nourishment, not as commodity. This is a remembering of land as commons, not as private property.

The peak of this remembering came on the shmita year. The cultural shift of the agricultural fallow recalled the gatherer society, the harvests of Eden, the collection of manna in the desert, equal portions for all people. In this year, everyone—landowners, field workers, and landless peasants alike—would be able to enjoy the harvests of the field as equals. The foods that did grow on their own during this time were considered ownerless, *hefker.* In this year, all fields would be left open for harvest, with fences unlocked or taken down. The crops that were collected could not be accumulated or stored for private wealth. Food was completely separated from commerce and the marketplace. As public commons, the harvests were to be shared and fairly distributed through the village network; collective sustenance depended on families tending to one another's mutual and interdependent needs.

In addition, debts would be forgiven during this time, which had a direct impact on farmers (Deuteronomy 15:2). In any year, a poor harvest would result in the need for a loan, in the form of seeds, animals, or money, for the next growing season. To alleviate the financial risk of taking a loan, no interest was to be charged (Leviticus 25:36). However, if a second season of poor harvest followed, it would create a downward spiral for affected farmers until they would lose their land or become indentured servants to pay off their debt. In the seventh year, at the release of debts, these farmers would have a chance to start over without burden.

The shmita year offered a profound and challenging reminder that we, as farmers, are not owners. And this message became a guiding value for all years. A healthy soil or plant ecology must coexist with a healthy social ecology. The farmer is tending to both of these visions. Beyond any economic value placed on the foods we grow, foods are never just a resource or a commodity. Food is nourishment; it is life force, the foundation for community stability, resiliency, and growth. In avodah, in sacred farming, profit is never the goal. Rather it is tending and stewarding relationships, on the field and beyond.

And it was this tending that was at the heart of the sacred service that Spirit desired. Even if the Temple offerings were being practiced, they were meaningless without the conduct of sacred service outside the Temple walls. "The multitude of your [Temple] sacrifices—what are they to me?" says the Lord. "I have more than enough of burnt offerings. . . . Wash yourselves; make yourselves clean. Remove your evil deeds from my sight; stop doing wrong. Learn to do right; seek justice. Aid the oppressed. Uphold the rights of the orphan; defend the widow" (Isaiah 1:11, 16–17). This message expanded, transcended, the Temple borders and brought the call for sacred service directly to the fields, the home, and the community.

Living the Story of Sacred Service

Each year, in the early summer season, after the last rains have fallen and the fields have turned brown, the land of Israel observes Shavuot, a holiday that celebrates simultaneously the anniversary of receiving the Torah and the start of the annual wheat harvest, a joint remembering of the wilderness of the Midbar and the domestication of grains. During the times of the Temple, two loaves of bread, freshly baked with flour from the new wheat crop, were offered on the altar, embodied representations of the alchemy of agriculture. This is a harvest holiday, symbolically merging the harvest of plant and spirit, honoring the relationship between the seed and the sacred. Each of these two layers, of agriculture and Torah, is woven into the other; they are reflections of one another. They are bound together by the agreements for a sacred agriculture and a sacred culture. Each year, Shavuot offers us a reminder of this covenant.[7]

And this covenant is not easy or simple. These agreements continue to be wrestled with by Jewish people in Israel and beyond, and by all peoples, with their own unique sacred pathways and relationships among plant, people, and place. The vision for sacred agriculture is timeless. It is applicable today, with our advanced farming technology and our global agricultural marketplace. It is needed more now than ever. The challenge of farming, its possibility of being a curse for the land and for humanity, is just as real today, if we neglect such relationships.

This is a story of farmer as priest, in sacred avodah, tending the relations between all lives, in service of the conversation among all beings. I share this story as a reminder. Today, in the modern agricultural landscape, you may not see these practices at first glance. Modern farming in Israel, as in many lands, is guided more by the priorities of efficiency, productivity, profit, and technological innovation than by a desire to facilitate deep connection with land and community. The covenant of sacred agriculture is easily forgotten. The Temple in Jerusalem no longer stands, and the agricultural rituals associated with it are no longer active. The tribal identities and land associations have been lost. The jubilee counting has ceased; shmita is not widely observed. The visions of land and food justice have been overlooked.

While the rituals and service of today may not be the same as in ancient Israel, and while the cultural landscape has changed over the many passing years, the values and intentions of these agreements are as clear as ever. They are very much alive, within us and within our soils, as creative seeds of vision, prayer, and practice. This is the beating heart of the story, and peeling back the superficial layers of farm and marketplace will reveal these threads of indigenous knowing and prayerful intention. Through our knowing them and remembering them, they can be woven back together into a beautiful fabric of culture and agriculture. These principles have been the guiding factors in my own agriculture, in my own relationships with plant, place, and Spirit, and for this I am grateful. And these principles are at the heart of the resurgence and reawakening of Jewish farming practices, both in Israel and around the world. We are breathing this old and new story back to life, slowly, delicately, vulnerably, and beautifully. The Temple is our home and our fields. The altar is our table. And we are the priests.

Today, it is a choice to tell this story, to believe in it, to remember it, as we work the field, cultivate the crops, share the harvests. This is the story of the soil, the story of our own body, told through the generations, remembering all the way back to the Garden of Eden. And this particular story is a piece of our larger human collective story, of our sacred origins, and of our sacred relationship to all life. This is our story to remember, our story to tell, our story to live.

Study Questions

1. What is the indigenous agricultural story of the land you cultivate? How are you feeding and being fed by this story?
2. What are the foundational values at the heart of this story, guiding your farming in a sacred manner?
3. If you no longer remember this story, how can you form a new one with the place you call home, with the soils you cultivate, with the community of friends and neighbors nourished by the sustenance of your land?

Notes

1. The Torah (literally, the Teaching or the Instruction) refers to the five books of Moses, from Genesis to Deuteronomy (in Hebrew known as Chumash; also known as the Greek Pentateuch). The Torah comprises the first books of the Jewish Bible (in Hebrew also known as the Tanakh; in Christian tradition also known as the Old Testament). Each week on shabbat, a portion, or *parshah,* of the Torah is read from the Sefer Torah, literally, the "Scroll of the Torah," the five books of Moses handwritten in Hebrew by a scribe, with quill and ink, on pieces of parchment sewn together into one long scroll rolled around carved wooden handles. This ritual reading cycle takes one full year.

2. All references in this chapter are structured by book, chapter, and verse. Translations are by the author.

3. The Hebrew faith believes in God as one universal, cosmic force within and beyond creation. Spirit is called by many names, including HaShem, which means "the Name," as well as YHVH and Elohim (representing the masculine aspects of Spirit), and Shechina (representing the feminine aspects of Spirit).

4. The First Temple (also known as Solomon's Temple) was destroyed during the Babylonian conquest in 586 BCE. The Second Temple was rebuilt in 515 BCE and stood until 70 CE, when it was destroyed during the Roman conquest, although by this period—due to the disruption of Temple service by the Seleucid Greek empire (169–166 BCE) and politicizing of the priestly service under imperial rule—ritual practice and tradition had already begun transitioning away from the Temple and toward the home.

5. Translated as "Release." Commonly referred to as the sabbatical year. In the Torah, this time period is also called Shvi'it ("the Seventh") and Shabbat Ha'Aretz ("the Sabbath of the Land").

6. The Mishnah (literally, "to repeat" or "to study") is a codification of Jewish oral teachings, known as the Torah She'Ba'al Peh, which supplement and clarify the Torah She'Bichtav, the Written Torah (the Jewish Bible). The Mishnah was compiled by numerous early rabbis (*tann'aim*) in Israel during the first and second centuries and was completed around 200 CE, after the destruction of the Second Temple. The content of the Mishnah focuses on sets of rituals and legal and social practices that form the basis for rabbinic Judaism and Jewish law (*halacha*).

7. Shavuot is celebrated fifty days after the holiday of Pesach (Passover), which recalls the exodus from Egypt. These holidays are linked through the Sefirat Ha'Omer ("the Counting of the Grain Sheaf"), a forty-nine-day counting period in anticipation of the wheat harvest (Leviticus 23:15–16). The fiftieth day, after seven weeks of counting, marks Shavuot and the start of the ingathering of the wheat. This arc is a bridge from slavery to liberation; it takes us from the agriculture of empire (Egypt) to the agriculture of sacred avodah. This counting is also a fractal of the seven shmita cycles (forty-nine years) and the jubilee (fiftieth year).

11

Religion, Local Community, and Sustainable Agriculture

Anna Peterson

Most discussions of sustainable agriculture focus on its practices and scale. However, religious and cultural dimensions often play just as great a role in a farmer's ability to achieve social, environmental, and economic goals. This chapter examines two different types of rural communities, both of which have used their religious commitments as the foundation for innovative efforts to farm in environmentally and socially sustainable ways. The practices of the Old Order Amish are among the longest-running and most successful examples of sustainable agriculture in North America. Amish achievements are rooted not only in their farming methods but also in their social organization, which in turn rests on their religious values and worldview. The same is true of the communities created by displaced peasants in postwar El Salvador. A comparative analysis of the two groups reveals both the promise and the demands of religiously based sustainable farming practices and communities.

The Old Order Amish

The Amish are Anabaptists, part of a minority group of Christians (also called the Radical Reformation) that emerged in the 1520s in Switzerland and Holland. Anabaptists rejected infant baptism, believing that baptism entailed a serious commitment on the part of the believer that should be entered into only by willing and knowledgeable adults. They created voluntary communities that sought to live by biblical values of mutual aid, peaceableness, and isolation from the corrupting influences of the world. Despite their rejection of violence, Anabaptists frequently came into conflict with civil authorities because of their refusal to swear oaths or serve in the military.

Anabaptist nonconformity also led to persecution by Catholics, Lutherans, and Calvinists. By 1535, ten years into the movement, about fifty thousand Anabaptists had been killed, probably more than all the Christians killed by Romans in the first three centuries of the common era. Thousands more were killed in the succeeding decades. Nonetheless, the movement grew rapidly, confounding its persecutors. As one German official lamented, "What shall I do? The more I cause to be executed, the more they increase."[1] Many Swiss, German, and Dutch Anabaptists sought safety by migrating, first to eastern Europe and then, in the nineteenth century, to the United States and Canada.

Most Anabaptists today belong to liberal or acculturated denominations, mainly Mennonites. The smaller but more visible "old orders" include the Hutterites, Brethren, Old Order Mennonites, and Amish. While the old order groups vary, most reject certain modern technologies and practices, including electrical utilities, home telephones, and automobiles. They seek separation from mainstream society and depend upon the extended family and religious community for material and moral support. The largest old order group by far is the Amish, with 180,000 baptized adults and children under baptismal age. Most live in Pennsylvania, Ohio, and Indiana, although there are Amish settlements in many other states as well as in Canada and Latin America. The Amish population has grown rapidly in recent decades, fueled by birth rates averaging six children per family and retention rates of over 80 percent for most communities.

Along with this growth, Amish life has changed significantly since World War II. Although most communities remain rooted in agriculture, the proportion of families that earn their living by farming has fallen, mostly due to the decreasing availability and rising cost of farmland. A growing number of Amish adults work in small businesses making items for the local community or the tourist trade or even for Internet sales.[2] A smaller proportion have gone to work in factories, which leaders fear threatens their ability to continue to live according to their religious values.

Still, agriculture remains not only a major way of making a living but also a key element of collective identity in most Amish settlements in North America. This continues a tradition begun in Europe when persecuted Anabaptists fled to remote rural areas in the seventeenth and eighteenth centuries. The environmental pressures of farming on hilly, infertile land encouraged them to try novel methods of raising livestock, fertilizing, clearing land, and rotating and diversifying crops.[3] The main goal of Amish agriculture is self-sufficiency, so that the religious community can live without depending on the outside world (this is one reason the Amish reject electricity from public lines, which they describe as an "umbilical cord" to secular society).

The Amish have largely defied national and global trends toward larger farm size, mechanization, and industrialization. A 1990 study of Amish farms in Iowa showed a dramatic contrast between Amish and non-Amish farms. In Buchanan County, center of the second-largest Amish community in Iowa, only two Amish farms out of about 180 total in the county were sold during the farm crisis of the 1980s. This represents "an incredible survival rate" in contrast to the numerous foreclosures of non-Amish farms in the same county.[4]

The economic resiliency of Amish farmers rests on their rejection of most aspects of mainstream agriculture. First, their farms are small, averaging around sixty acres, in contrast to over four hundred acres for conventional farms. In addition to this quantitative difference, Amish farms are distinguished by their rejection of government subsidies and many technologies as well as of most forms of nonrenewable energy. Their unconventional approach has led to average higher profits per acre and per animal than conventional farmers achieve, along with lower debts and expenses.[5] As sustainable agriculture advocate Gene Logsdon summarizes: "The fact is that Amish farms have mostly survived, if not thrived, right on through the current economic malaise despite the fact that, by the standards of the technocrats, small farms are supposed to be on the way out."[6]

This success has encouraged sustainable agriculture advocates to study Amish farming, with an eye to what mainstream society can learn from their practices. Amish sustainability begins with an extremely efficient use of energy. A 1977 study comparing the energy budgets of Amish and non-Amish farms in three states concluded that despite regional variation, overall the Amish farms had much better energy ratios of inputs to outputs.[7] Even the most efficient modern farms, "according to this research, 'require 65% more energy per kg of milk' than the Amish ones."[8] The greater energy efficiency of Amish dairy farms results not only from the use of draft animals and the lack of electricity but also from specific agricultural practices such as crop rotation, greater use of manure, and heavier reliance on hand cultivation and human labor.

In addition to efficient use of nonrenewable energy, other Amish farming practices minimize their environmental impact. While few Amish farmers are completely organic, most use commercial fertilizers and chemical herbicides in much lower quantities than conventional farmers. This stems from their traditional reliance on animal and green manures and on controlling weeds by tilling, crop rotation, and crop diversity. One study found that when Ohio Amish farmers did use petrochemical fertilizer, they applied about half the recommended dosage, and they rarely used soil insecticides.[9]

Amish farms also excel at conserving topsoil, one of the most urgent challenges facing farmers in the United States and elsewhere. Most contemporary experts advocate "no-till" methods, which reduce soil erosion but require large quantities of herbicide. In theory, the intensive plowing on Amish farms should cause serious loss of topsoil year after year. In the mid-1980s, the Soil Conservation Service in Holmes County, Ohio, estimated that Amish farms ought to be losing seven to fifteen tons of topsoil per acre each year. Since many of the Amish farms have been under tillage for as many as 150 years, concluded one researcher, "these farms should have no topsoil left. This is obviously not the case." Soil analyses found that not only did Amish farms have sufficient quantity of topsoil but the soil was of excellent quality.[10] This is due to the use of horsepower and crop rotation, which enable Amish farmers to improve the soil rather than degrade it.

The soil erosion studies point to the inadequacy of conventional measures to evaluate Amish agriculture and perhaps other alternatives to the mainstream. They underline, further, the error of assuming that Amish farmers blindly continue practices that modern ways have rendered obsolete. Traditional Amish horse plowing, it turns out, is better for the soil than new no-till methods. Other studies have pointed to the contemporary relevance of other practices common among Amish farmers since the eighteenth century, such as crop rotation, the use of animal and green manures, and mixing production among diverse plant and animal species. The Amish combine new ideas and technologies with time-tested practices, creating horse-drawn mechanisms as sophisticated as those that conventional farmers pull with tractors. At the same time, skepticism about new technology means that innovations are not adopted simply because they are new but only when they have been shown to work better than traditional models. Some Amish farmers use wind or solar power or employ diesel generators to power refrigerators for milk they sell commercially. The Amish, in sum, "are not opposed to incorporating new ways of farming which have stood the test of time and increase the economic self sufficiency of their families and communities without jeopardizing their religion."[11] Religious priorities and values shape both social and agricultural choices, replacing mainstream goals such as profit or efficiency.

Amish farms provide a unique opportunity to study the results of sustainable practices over time. "Because the Amish have such a long history of experience as designers and practitioners of low-input sustainable agriculture," an Ohio study concluded, "researchers and non-Amish practitioners of sustainable agriculture could learn a great deal from Amish farmers; not only in terms of determining what types of sustainable systems are economically and environmentally viable, but also in terms of understanding why they work."[12] The "why" is related to religion, which makes possible the supportive community, local focus, and reflective decision making that have enabled Amish farms to thrive despite the transformation of conventional agriculture.

Repopulated Communities in El Salvador

Just as the Amish demonstrate the possibility of more sustainable agriculture in contemporary North America, a very different kind of community shows an alternative to conventional farming in a Global South setting. El Salvador is not only one of the smallest but also among the poorest and most violent nations in the Americas. Small farmers have long been at the bottom of the country's highly unequal social hierarchy. By 1975, three-quarters of Salvadoran rural families owned no land or less than one hectare, often in the least fertile and accessible regions. Tens of thousands of peasants migrated in search of work, and many small farmers fell heavily into debt. Economic deprivation was exacerbated by a lack of democratic institutions and a military that responded to criticisms with rapid and harsh repression.

For most of the country's history, the dominant Catholic Church sided with the military, ruling class, and government to keep peasants in their place. This changed as a result of the Second Vatican Council (1962–1965) and the subsequent flowering of progressive ideas and practices. The best known is liberation theology, which combined a participatory approach to religious community with a commitment to human rights and social justice. Progressive pastoral agents in El Salvador began working with peasant farmers in the late 1960s and early 1970s, building communities, teaching literacy and other skills, and imparting new theological ideas. These experiences were radicalizing for both the pastoral leaders and the laypeople involved, many of whom went on to participate in reform efforts such as agricultural cooperatives, land reform, and peasant organizations. Their guiding question, as one early participant remembers, was "How does God want us to organize?"[13] How, in other words, could they translate their religious values into concrete practices and new ways of living together?

The peasant movements faced harsh repression, including torture and assassination of many leaders. The violence in rural areas intensified after civil war began in early 1981 between the military-dominated and (US-supported) government and the guerrillas of the Farabundo Martí Front for National Liberation (FMLN). The war led to the depopulation of rural areas, especially in northern provinces (*departamentos*) like Chalatenango, Morazán, and Cabañas, all FMLN strongholds that suffered aerial bombings, massacres, and the destruction of their homes and crops. Tens of thousands of peasants fled to refugee camps in Honduras, mostly run by the United Nations, where they waited for the war to end.

By the mid-1980s, with no end to the war in sight, refugees began to plan their return. The Salvadoran government had wanted them to go to areas under government control, but refugees insisted on returning to their places of origin in the war zones.[14] In October 1987, the first repatriation occurred when four thousand residents of the Mesa Grande refugee camp returned to several villages in and around Chalatenango province, the heart of FMLN territory. The first repopulators (*repobladores*) encountered daunting conditions as a result of bombings and subsequent years of neglect. Most buildings were destroyed, fields were overtaken by shrubs and weeds, and wells and water lines had been obliterated.

The repopulators faced not only the uphill work of reclaiming farmland and building shelter but also constant harassment from the government military, including aerial attacks, ground incursions, and selective assassinations of teachers, health workers, and other community leaders. Skirmishes between government and guerrillas also took place frequently in the villages and surrounding hills. Despite the dangers and hardships, the repopulations quickly became known as a success story because of their ability to put their social vision into practice. As refugees still in Honduras learned about the experiences of the first returnees, thousands more returned over the next few years.

During this period, the military conflict between the government and the FMLN remained stalemated, broken finally by a major guerrilla offensive in November 1989. In the midst of the fighting, government soldiers assassinated six Jesuit priests, among the country's most prominent intellectual and moral leaders. Their deaths brought the number of progressive Catholic pastoral agents killed by the government military and right-wing death squads to over two dozen.[15]

The aftermath of this offensive led to serious discussions about a negotiated end to the war, which had claimed over eighty thousand lives, mostly of civilians, and displaced about a third of the population. The two parties reached agreement on New Year's Eve, 1991, and signed the treaty in Mexico City on January 16, 1992. The peace accords called for the demobilization of the FMLN, which would become a legal political party. The accords also mandated the transformation of the government army, including the removal of several units and of over a hundred officers associated with human rights violations. In addition, the accord called for greater political openness, judicial autonomy, and most important for rural areas, the transfer of thousands of acres to peasants and former combatants, transforming the pattern of rural land ownership.

After twenty years of peace, El Salvador is still marked by severe economic inequities and widespread poverty as well as numerous social and environmental problems, including severe air pollution, soil erosion, and water contamination, all intensified by deforestation and urban sprawl. The ability to address these issues is key to the success of the repopulations and their religiously grounded vision of a different kind of society. Environmental sustainability, in this context, cannot be separated from efforts to achieve social justice, economic security, and democratic political participation. Many forms of environmental destruction, including overlogging and the use of chemical pesticides and fertilizers, are directly related to poverty and economic insecurity. Without economic security, people will continue to cut down trees for firewood and plant on exhausted soils. Most peasants realize that environmental problems make their struggles for economic subsistence, health, and equitable development more difficult. However, structural reforms will be required to change these practices at the local level.

Of the various environmental issues facing the repopulated communities, the most important are related to agriculture, especially the loss of topsoil. To control erosion, many communities in war zones have begun reforesting and terracing. In addition, most farmers have stopped burning fields and now leave old crops and leaves for mulch. Some have also begun rotating crops, to the extent possible on their small plots. They have made less progress in reducing their reliance upon chemical fertilizers and pesticides. Most farmers realize that chemicals can harm the environment and human health but continue using them to make the poor local soils produce. There have been some successful initiatives to encourage sustainable practices, including land-use plans to diversify crops and encourage reforestation. Instead of only corn and beans, most families now have some land in fruit trees, some in perennial herbs, some in vegetable gardens, and some in trees for shade and firewood. Some of the most successful experiments with sustainable agriculture have taken place in repopulations in the Lower Lempa River region, where the war created an advantage: the abandonment of fields and orchards for years has facilitated the process of organic certification, which enables local producers to take advantage of international markets.

In addition to developing more sustainable ways to farm, communities have tried to set some land apart from intensive human use and create better habitat through reforestation efforts in both rural and urban areas. In Guarjila, one of the first repopulations, the local Comité

Ecológico and other organizations have provided fruit and shade trees to all the families in the village. The hills surrounding Guarjila are now covered by a green canopy, where before there was only dust. Residents also regularly see many formerly rare or absent bird species, including parrots, which have returned due largely to reforestation efforts. Older people take pleasure in seeing species they remember from their youth and sharing their knowledge with younger people, who interpret this in relation to concepts such as biodiversity, endangered species, and even wildlife corridors, with which they are increasingly familiar, thanks to the work of Salvadoran and international NGOs and education efforts.

Salvadoran farmers know that if they exhaust the land, they have no source of survival. As a peasant leader put it, "We live from the land, so we have to take care of it" (Vivimos de la tierra, tenemos que cuidarla).[16] This balancing act characterizes all efforts at sustainable agriculture: how to make a living while caring for the land. In El Salvador, as elsewhere, governments, community organizations, and individual farmers must find resources to develop alternative energy and organic fertilizers and pest controls and to restore forests and watersheds, and they must do so in a political and economic context that is acceptable to local residents and with methods that match their vision of their communities' future. In El Salvador, this vision is strongly shaped by religious values and traditions.

Religion and Sustainable Agriculture

Agriculture is not always seen as part of the solution to environmental problems. Some environmentalists see the domestication of plants and animals and the cultivation of land for agriculture as the beginning of the end of sustainable lifeways. By enabling people to live beyond the limits of their local ecosystem, in this view, agriculture fosters both unsustainable ways of life and instrumental views of nature.[17] This is what Wes Jackson calls the problem *of* agriculture, larger than problems *in* specific techniques or practices.[18] However, there are no clear alternatives to agriculture to support the current human population, at least in the short or medium term. Other critics suggest that the problem is not necessarily the existence of agriculture per se but rather the more recent emergence of particular agricultural methods, scales, and technologies.

Traditional family farms throughout the world were small in size, used simple technology, relied primarily on human and animal labor, and produced a variety of plant and animal products for home consumption, selling any surplus in local markets. This model of agriculture was relatively benign environmentally. Most such farms allowed room for wildlife at the margins of cultivated land and in woodlots, preserved topsoil, and consumed few fossil fuels and chemical inputs. Small farms also supported diverse local economies, widespread ownership of land, and neighborly cooperation.

Of course, not all traditional farmers are or have been consistently attentive to ecological and social values. Even before the advent of industrial agriculture, agriculture contributed to deforestation, species extinctions, and water contamination in many parts of the world. Farming, further, has often been an insecure and dangerous way to make a living, both because of its dependence on unpredictable natural systems and because of its entanglement in social, economic, and political structures. These problems have intensified many times with the rise of industrial agriculture since the Second World War. These issues have given rise to countless discussions and initiatives to create a more environmentally and socially sustainable way to make a living off the land.

Amid great diversity, some common principles for sustainable agriculture have become evident. One is a return to a smaller scale of production. Smaller farms are more likely to be diverse in their crops, to use fewer pesticides and artificial fertilizers, and to be tied to local economies and cultures.[19] They are also more productive, when productivity is measured by yield per unit of land or per unit of energy used, producing as much as triple the output per unit area as industrial forms of farming.[20] Although small farms have a potential for high productivity and more positive environmental and social contributions, economic trends and government policies have made it increasing difficult to farm on a small scale in the United States and elsewhere. This makes the experience of Amish farmers especially important. Their success stems from a variety of factors: reliance on inexpensive human and animal labor, modest standards of living, and probably most important, a supportive social context. The latter, in particular, depends on shared religious values and community structures. The same is true in the repopulated communities in El Salvador, although some of their limited scale has been based less on deliberate choices than on poverty and lack of access to land.

A second important feature of sustainable agriculture is a focus on local markets and an intent to meet the needs of local consumers. Production for distant markets is destructive not only because it requires energy for transportation and storage but also because it discourages diversity in production and reduces local autonomy. Despite recent changes, Amish communities remain largely self-sufficient and locally oriented in food and other necessities. In El Salvador, repopulated villages have sought to produce more for local markets, recognizing that they cannot compete nationally with imported products. The Salvadoran government's policy of supporting export-oriented crops and importing basic foods makes it harder for subsistence farmers to sell their small surpluses locally and make an adequate cash income. The stable, relatively self-sufficient local economies of Amish settlements might serve as a model for other areas, but government policies and poverty make realization of this goal unlikely in the short term. Larger structural changes, in other words, must complement local efforts to achieve economic security.

A third characteristic of sustainable agriculture is that rather than working against the ecosystem in which it is placed, the farm uses its existing strengths and possibilities, asking what is possible in a particular place. Ecologically sound farms include diverse plant and animal species, do not rely on chemical fertilizers and pesticides, and engage in intensive recycling. In this regard Anabaptists have been pioneers. Amish farms have a relatively closed cycle: most food needed by animals and humans is grown on the farm, most crop fertilizer comes from animal and green manures produced on-site, most labor comes from resident animals and humans, and most energy is provided by renewable sources such as wind (to pump water), wood (for heating), and the sun (to produce food and fertilizers). Many Amish farms use some fossil fuels, for example, for diesel generators or tractors used around the barn, but their consumption of nonrenewable resources is vastly lower than that of conventional agriculture. Fossil fuel consumption is also low in repopulated communities in El Salvador, but again this is due primarily to poverty and lack of access to machinery and fuel. Most work is done by people and domestic animals, although under conditions of hardship rather than deliberate design. The use of some sustainable practices is increasing in the repopulations, particularly when national and international organizations, including ecumenical and church-based groups in both North America and Europe, provide technical and financial assistance. Some small farmers have experimented with waste recycling, compost and organic fertilizers, and alternative methods of pest and weed control. It remains to be seen, of course, what will happen to per capita consumption if and when overall prosperity increases in the communities.

Sustainable agriculture also leaves room for wild nature, for example, in hedgerows, woodlots, and wild areas on the edges of cultivated land. Aldo Leopold defined a good farm as one where the wild flora and fauna have lost ground without losing their existence.[21] Amish farmer David Kline cites this as a guiding principle for his own farming and for traditional Amish practices that create better wildlife habitat.[22] In El Salvador, many native species no longer survive due to habitat loss and hunting, although ironically the war helped some species recover, since it caused human populations to decline sharply in many rural areas. Intensive restoration work will be required to develop additional habitat in the repopulations and other rural areas. Some successes have already come as a result of reforestation projects that have expanded habitat for bird species.

The Role of Religion

Sustainable agriculture as defined above requires cohesive and stable communities. Religion plays central roles in both the Amish and Salvadoran cases studied. It is important both ideologically, insofar as it provides reasons to value nonhuman nature, and socially, because it undergirds the institutions and relationships that make it possible for people to put their values into practice.

In theological terms, both Catholic and Anabaptist traditions are theocentric, or God-centered, rather than anthropocentric, or human-centered. In this framework, animals, plants, wild places, and other aspects of the natural world have special worth because they are created and valued by God. Respectful treatment of nature thus reflects respect for God; mistreatment of nature is a sin against the creator. Amish farmer David Kline points to a traditional prayer that includes the line: "Help us be gentle with your creatures and handiwork so that we may abide in your eternal salvation and continue to be held in the hollow of your hand." This, Kline asserts, is the Anabaptist "theology for living."[23] It is also a good summary of a theocentric ecological ethic: recognizing that other creatures, and nature in general, are God's creations, and that humans' religious duty and hope lie in the effort to treat this creation properly. This perspective differs from some radical ecological views in its assertion that creation is given for human use; it does not have to be left alone as wilderness, or at least not all of it does.

Similar themes emerge in thinking about social justice. It is God's intention for people to live in particular ways, both Anabaptist and Roman Catholics argue, and failure to do so represents not just a social but a religious problem. Catholics and Anabaptists share a conception of humans as social beings who can live full and proper lives only as part of a community and who can be saved only collectively. The social community is necessarily a natural community as well. Humans depend on each other, on God, and on nature. This mutual interdependence is part of an overarching natural and divine order, and what happens between humans has real and lasting effects on what happens in the natural world and vice versa. Common principles guide both natural and social realms, and human behavior in both areas ought to be consistent. For both Catholics and Anabaptists, religious ethics rest on a model of the reign of God as an ideal community in which all members adhere to the same high standards.

This religious community is also one that should allow no member to "fall through the cracks." This principle is expressed in the Anabaptist tradition of mutual aid and the long-standing Catholic insistence on a preferential option for the least well off, both of which are as relevant to farming practices as to any other social issue. The US Conference of Catholic Bishops, for example, extends special concern to agricultural workers, noting, "While some are doing well, others are vulnerable or struggling and poor. Those who farm, work in the fields or on ranches, and process our food must have decent wages and a decent life. . . . An important moral measure of the global agricultural system is how its weakest participants are treated."[24] Equally important is solidarity with those who are hungry and do not benefit from current agricultural production or distribution systems.

Another shared value is a conviction that social, economic, and environmental problems are connected in both their causes and their potential solutions. This differs from ecocentric perspectives that seek to solve environmental problems by keeping humans out of wild areas. Many Third World environmentalists believe this perspective reflects the self-interest of wealthy North Americans who want to enjoy nature for its aesthetic and recreational possibilities. In contrast, they argue, poor people who have to make a living from the land cannot afford to set aside parcels of wilderness. If the poor majorities in countries like El Salvador are to survive, then environmental protection must take place in the context of human inhabitation and use of nature. Critics also point out that the high-consumption practices of wealthy elites, in both developing and industrial societies, create more environmental damage than subsistence farmers.

This political critique agrees with Catholic and Anabaptist arguments that God intended creation for the use of all and Christians should use it in respectful and conservative ways. The good life does not require excessive consumption. Some human use of natural resources is necessary, but overexploitation causes both environmental destruction and deprivation for other humans. The Amish keep consumption low because they do not want consumerism to penetrate their culture and displace more important values such as community well-being and faithful discipleship. In El Salvador, consumption is also low, due in large part to extreme poverty. Few people have cars or telephones because few can afford them. Still, the differences are not only a result of economic pressure. Salvadorans in the repopulations are working explicitly to create communities that offer an alternative to individualist and consumerist values that dominate in both Salvadoran and US society.

Sustainable farms, and the communities in which they are embedded, are much more complex than idealized and abstract images allow. Further, they are intimately involved in and shaped by what happens at larger scales. Even the most apparently self-sufficient communities, such as the Amish, are enmeshed in national and international policies, structures, histories, and social relations. The promise of sustainable agriculture, including its religious dimensions, cannot be understood in isolation from economic and political processes, institutions, and forces. These interactions may be more intense now than ever, especially in the face of increasing globalization, climate change, and other contemporary complications. Thus the biggest challenge facing sustainable agriculture, at least in the communities profiled here, is how to negotiate these conditions and forces that are beyond the control of individual farmers and local communities. Production and distribution structures as well as government subsidies (especially in the United States) for industrial-style agriculture all make it hard for small farmers to compete with the prices offered by corporate farmers.

On the other hand, both the Amish and Salvadoran communities demonstrate that it is possible to survive and even to thrive while farming in environmentally and socially sustainable ways. Their success is possible in large part because of their religious commitments, which hold together their communities and provide grounds for making collective and individual decisions on the basis of principles rather than profit or expediency. However, religious identity is not something that can be created or held together spontaneously. Its benefits, for individuals and communities, often come only as the result of shared history as well as shared values. Comparing these religious communities to other successful experiences of sustainable agriculture may provide grounds for identifying the distinctive contributions of religion to alternative ways of farming as well as point to aspects that might be replicated in secular settings.

Study Questions

1. What common principles underlie the Catholic and Anabaptist approaches to sustainable agriculture?
2. When might these communities' religious principles conflict with ecological principles for sustainability?
3. What is the connection between social and ecological sustainability in these communities?
4. What differentiates these communities and their approach to sustainable agriculture from secular approaches?

Notes

Portions of this chapter are taken from Anna Peterson, *Seeds of the Kingdom: Utopian Communities in the Americas* (New York: Oxford University Press, 2005).

1. Thieleman J. van Braght, *Martyrs Mirror* (Scottdale, Pa.: Herald, 1938), 437.

2. Some Amish families raise puppies for pet store and Internet sales, often under inhumane "puppy mill" conditions.

3. Jean Séguy, "Religion and Agricultural Success: The Vocational Life of the French Mennonites from the Seventeenth to the Nineteenth Centuries," *Mennonite Quarterly Review* 47 (1973): 187.

4. Rhonda Lou Yoder, *Amish Agriculture in Iowa: Indigenous Knowledge for Sustainable Small-Farm Systems* (Madison: University of Wisconsin, 1990), 43.

5. Steven Stoll, "Postmodern Farming, Quietly Flourishing," *Chronicle of Higher Education*, June 21, 2002, B7–B9; Gene Logsdon, "Amish Economy," *Orion Nature Quarterly* 7, no. 2 (1988): 22–33.

6. Gene Logsdon, *Living at Nature's Pace: Farming and the American Dream* (White River Junction, Vt.: Chelsea Green, 2000), 140.

7. Warren Johnson, Victor Stoltzfus, and Peter Craumer, "Energy Conservation in Amish Agriculture," *Science*, October 28, 1977, 373–78.

8. Peter Craumer, "Farm Productivity and Energy Efficiency in Amish and Modern Dairy Farming," *Agriculture and Environment* 4 (1979): 292.

9. Deborah H. Stinner, M. G. Paoletti, and B. R. Stinner, "In Search of Traditional Farm Wisdom for a More Sustainable Agriculture: A Study of Amish Farming and Society," *Agriculture, Ecosystems, and Environment* 27 (1989): 83.

10. Mary Jackson, "Amish Agriculture and No-Till: The Hazards of Applying the USLE to Unusual Farms," *Journal of Soil and Water Conservation* 43 (November–December 1988): 483–84.

11. Stinner, Paoletti, and Stinner, "In Search of Traditional Farm Wisdom," 86.

12. Ibid., 86–87.

13. Juan Fernando Ascoli, *Tiempo de guerra y tiempo de paz: Organizacion y lucha de las comunidades del nor-oriente de Chalatenango (1974–1994)*, adapted by Miguel Cavada Diez (San Salvador: Equipo Maiz, n.d.), 21.

14. While most refugees did return to their regions of origin, many did not go back to their particular home villages, at least not initially. Since the repatriations took place in stages, people wishing to return to El Salvador at a given time could not choose their exact destination.

15. The victims included Archbishop Oscar Romero, twenty-two Catholic priests and nuns, a seminary student, and a Lutheran minister.

16. Carlos Quintanilla, president of *junta directiva*, interview by author, Guarjila, Chalatenango, El Salvador, January 5, 2002.

17. Paul Shepard, *Coming Home to the Pleistocene* (San Francisco: Island, 1998).

18. Wes Jackson uses this phrase frequently in his writings and lectures. See, for example, Wes Jackson, *Altars of Unhewn Stone: Science and the Earth* (San Francisco: North Point, 1987).

19. Wendell Berry, *The Gift of Good Land: Further Essays Cultural and Agricultural* (San Francisco: North Point, 1981), xi.

20. Jack Manno, "Commoditization: Consumption Efficiency and an Economy of Care and Connection," in *Confronting Consumption,* ed. Thomas Princen, Michael Maniates, and Ken Conca (Cambridge, Mass.: MIT Press, 2002), 84–85.

21. Aldo Leopold, cited in David Kline, "An Amish Perspective," in *Rooted in the Land: Essays on Community and Place,* ed. William Vitek and Wes Jackson (New Haven, Conn.: Yale University Press, 1996), 39.

22. Kline notes that not all Amish farmers observe these practices, especially in communities with high land prices, where the temptation is to "farm to the road" in order to maximize tillable acreage. David Kline, *Great Possessions: An Amish Farmer's Journal* (San Francisco: North Point, 1990), xx.

23. David Kline, "God's Spirit and a Theology for Living," in *Creation and the Environment: An Anabaptist Perspective on a Sustainable World,* ed. Calvin Redekop (Baltimore, Md.: Johns Hopkins University Press, 2000), 61–62.

24. United States Conference of Catholic Bishops, "For I Was Hungry & You Gave Me Food: Catholic Social Teaching and Agriculture," 2003, http://www.usccb.org/issues-and-action/human-life-and-dignity/agriculture-nutrition-rural-issues/for-i-was-hungry-cst-and-agriculture.cfm (accessed November 28, 2012).

12

Heideggerian Reflections on Three Mennonite Cookbooks and a Mennonite Farm in Northwest Ohio

Raymond F. Person Jr. and Mark H. Dixon

Stephanie Nelson posits, "Farming... implies a particular understanding of nature, of the cosmos, and of the divine, and of our own relation, as human beings, to the physical world, to other human beings, and to God."[1] This connection between a philosophical, spiritual, and rustic life—with its farms, gardens, and animals—is an ancient one. In the *Georgics* Virgil reflects on agriculture, viniculture, human nature, and historical events. Virgil's poem is a response to the Hesiod's *Works and Days,* an older poem in which Hesiod advises Perses on the interconnection between matters moral, religious, and agricultural. The Confucian philosopher Mengzi builds the ideal political system around a specific agricultural model—the well-field system. Zen gardens mirror certain cosmological principles and facilitate philosophical reflection. The Cistercians sought to combine the spiritual with the rustic through a grange-based agricultural system, uniting solitude and an intimate connection with the natural environment.

These illustrations raise a question that is as crucial as it is pressing: How do we, as human beings, interpret our connection to the earth as embodied and emplaced beings, and live our lives accordingly? Below we will strive to reflect on this question from a Christian, specifically an Anabaptist/Mennonite, perspective. Too often Christian theology and ethics have been based upon a dualistic understanding of spiritual versus material and a marked anthropocentric view of the world. Mennonites are not immune to this theological problem, which admittedly has contributed to the ecological crises we face today.² However, some Mennonites have recently insisted that the Anabaptist theological tradition, which is "strongly identified with the values of simplicity, nonresistance and nonviolence, and community living," can provide resources for theological solutions to these problems.³

Mennonites have a long history in agriculture, starting in Europe and continuing into North America and other places where they sought refuge from the persecution they experienced in Europe.⁴ Despite their long experience of working the land, current Mennonite practice too often reflects the theological problems within the tradition, thereby bringing current practice into conflict with Anabaptist values. Thus, Heather Bean concludes: "In North America, Anabaptists seem to become less interested in environmental issues and land ethics as their ties to their traditions decrease and they depart from the simple life of agricultural communities to assimilate into the materialistic mainstream culture. . . . The resulting prosperity and consumption is in direct conflict with Anabaptist values of simplicity and nonviolence." She notes that most North American Mennonite farmers farm no differently than their non-Mennonite neighbors.⁵

We agree that the Anabaptist values of simplicity, nonviolence, and community are excellent resources for developing an environmental ethic.⁶ Below we attempt to develop "simple living" as an embodied theology that provides an answer to the question concerning how we humans should interpret our connection to the earth and live accordingly. In order to understand simple living more thoroughly, we delve deep into what it means to be human beings, mammals, earthlings (in the sense of the Hebrew of Genesis 2:7: an *adam* [earthling] made from *adamah* [earth]).⁷ We do this, first, by exploring the philosophical thought of Martin Heidegger, who provides an invaluable lens through which to understand the relationship between humans and the natural world. This abstract exploration helps us more clearly explicate the more mundane and concrete expressions of the theology of simple living as expressed in three cookbooks commissioned by Mennonite Central Committee and embodied (at least partially) in the community that farms one of the historical Swiss Mennonite homesteads in northwest Ohio.

To Live, to Dwell, to Cultivate

In "Building Dwelling Thinking" Heidegger formulates the concept of "dwelling"—"To be a human being means to be on the earth as a mortal. It means to dwell." Heidegger then argues that *to dwell* is *to build*, which results in the equation *To be is to dwell is to build*. "To be" is, in the most basic sense, to exist, but it is more than existence simpliciter. "To be" also concerns the mode in which we exist or live. As human beings, this means our inevitable thrownness—*Dasein*—which, in Heidegger's conception, is never abstract but rather a radical factical existence. "To dwell" is the basic mode that underlies or defines human existence—"The manner in which we humans *are* on the earth is *Buan,* dwelling." "To build"—*bauen*—implies more than architectural construction; it also means to "to till the soil, to cultivate the vine." "To build," then, encompasses both architecture and agriculture. "To build" implies more than the obvious realization that to be human is to live on the earth. It also establishes a fundamental connection between builders, what is built—whether cathedral or vegetable garden—and the environment in which it is built and which provides the resources and materials for its building. This connection requires human beings to cherish, to protect, to preserve, to care about the environment.[8] Thus Heidegger writes, "To dwell, to be set at peace, means to remain at peace within the free, the preserve, the free sphere that safeguards each thing in its nature. *The fundamental character of dwelling is this sparing and preserving.* It pervades dwelling in its whole range."[9]

This much is exegesis of Heidegger's essay. What follows is an attempt to elaborate on Heidegger's concepts to articulate an explicit ethical foundation (a foundation that may be absent in Heidegger). The language Heidegger uses to describe the relationship—"to cherish," "to protect," "to preserve"—suggests a definite ethical dimension to the connection between human beings and the natural environment. Thus, as dwellers/builders/cultivators, human beings *ought* to cherish, protect, and preserve the land—this is dwelling's *fundamental character.* The ethical dimension, then, is neither epiphenomenal nor emergent but foundational, built into the necessary connection between human beings and the natural environment.

Hence, in order to realize their fundamental nature as dwellers, human beings must nurture a lifestyle that incorporates a close connection to the earth. Our contention is that a paradigmatic lifestyle is the farm, with its gardens, animals, and orchards. Moreover, the ethical dimension of this lifestyle operates on multiple levels, among and within the personal, social, and ecological communities that intersect on the farm. Thus, there is unequivocally an ethical dimension to cultivation. The question remains, though, whether there is also a spiritual dimension. As with the ethical dimension, the spiritual dimension emerges as we explore what it means to dwell in more depth.

Christine Swanton notes that what distinguishes Heidegger's perspective is that it focuses on human beings as beings *in the world*—dwellers, builders, cultivators, conservers, and protectors—and also as mortals. Heidegger even argues that human reason is bound to the essence of human beings as dwellers. To ignore or dismiss what it means to dwell makes it impossible to appreciate either human reason or the human virtues. Then what does it mean to dwell? Swanton writes that, in order to appreciate what it means to dwell, we must grasp: (1) Heidegger's ideas that human beings are *beings-there-in-the-world* (*Dasein*), and (2) the orientation to the world that is appropriate to *dwelling*.[10] The second point is most essential to our current purposes.

In essence, the appropriate orientation to dwelling is that we be open to the earth's mysteriousness as a whole Being and as a dwelling place, and that we be open to the earth as a place that cares about human beings and also a place that human beings must care about in return. These are impossible goals, Swanton argues, unless we are able to overcome our incessant need to calculate the natural environment's value and instead see through to the environment's *holiness*.[11] Thus, the fundamental orientation that is appropriate to dwelling is one that "opens up" the earth as sacred.[12]

This "holiness" or "sacredness" is where the spiritual dimension enters into the connection between human beings and the natural environment. Moreover, the spiritual dimension and the ethical dimensions are inseparable, since both presuppose that what motivates our attitude to (and interactions with) the earth is a fundamental sense of benevolence or compassion. Thus, to see the earth's holiness requires that human beings cultivate certain virtues, such as caring, benevolence, and love. Swanton also notes that the wonder and awe that define the sacred orientation underlying dwelling operate on a universal as well as a local level.[13] As individual dwellers we are connected to the specific place where we dwell. Moreover, in dwelling we find the recognition that the entire earth deserves our respect and concern. When we are able to dwell in this sense, then we will "be at peace . . . be brought to peace."[14]

As we noted above, a farm encompasses more than the relations between the individual and the farmland and between the farmland and the natural environment; there is also a social dimension—that is, in the same manner that farms exist within the larger natural environments, farms also exist within larger social (religious, political, economic) environments.

The farm-social environment relation raises certain questions. Does Heidegger's equation *To be is to dwell is to cultivate* have direct implications on one's lifestyle, over and above the obligation to cherish, protect, and preserve—to see the world in its holiness? In particular, does it advocate a simple and sustainable lifestyle? Does his equation require that we also place similar constraints on relations between humans and humans (that is, does it have a social component) and on humans and nonhumans (that is, does it have a species component)? Does it require that we act to cherish, protect, and preserve other human beings and nonhuman animals?

In one sense the answer to these questions is simple. When we realize that human beings, their cities, and their farms are themselves components with the natural environment, the inevitable conclusion is that whatever obligation we have to cherish, protect, and preserve the environment must also include both human beings and nonhuman animals. The details, however, are rather more complex.

The Simple Life

The constraint on lifestyle seems the more direct implication. To act in a manner that cherishes, protects, and preserves the earth and therefore recognizes and acknowledges the earth's holiness would necessitate certain constraints on how we live. In Heidegger's discussions about *Dasein* in *Being and Time*, he implies that dwelling, among other things, is the realization that not all manipulations and utilizations are permissible.[15] "To save the earth is more than to exploit it or even wear it out. Saving the earth does not master the earth and does not subjugate it."[16] What is essential to avoid and what "dwelling in holiness" allows us to overcome is the mindset that sees the natural environment as containing mere resources to pursue human ends. What further inculcates and reinforces this mindset are the technologies upon which much human life relies.[17]

The obvious problem, though, is that since human beings (indeed, all animate creatures) must manipulate and utilize environmental resources in order to survive, are there some guidelines here? How do we determine when manipulations or utilizations are inappropriate? An answer to these questions lies in the Anabaptist value of simple living. Do Heidegger's "to dwell" and "to cultivate" require simple living? We think so. There can be no doubt that a genuine commitment to cherish, protect, and preserve the earth is incompatible with overconsumption and wanton environmental destruction. Moreover, as this concerns agricultural practices, the evidence is unequivocal that modern commercial, monocultural agriculture is destructive and unsustainable. In contrast, an agriculture that focuses on smaller-scale practices, with appropriate technologies, promotes both human and environmental health.

Dasein *and the Social Life*

Heidegger's analysis seems to emphasize the individual. In particular, this appears to be true in *Being and Time*, where, as Todd Lavin writes, Heidegger's analysis foregoes a "real political or collective dimension and remains private and personal."[18] Moreover, this is the principal characteristic that separates Heidegger's phenomenological analysis from traditional Western philosophical analysis: the elevation of the individual's subjective experiences to legitimate philosophical status. This assumption, though, is rather too precipitate.

Our being in the world and the manner in which we relate to Being (*Dasein*) can be authentic or inauthentic. Lavin argues that implicit in Heidegger's analysis is the recognition that an authentic existence is impossible unless there is also a commitment to social activism.[19] Though

Lavin's main interest is in political commitments, it seems obvious that these commitments must include ecological ones. In fact, we can translate Lavin's conclusions to insist that Heidegger is implicitly advocating values closely associated with the Anabaptist tradition—simple living, nonviolence, and the importance of community.

This analysis suggests that human relations to the natural environment through agriculture operate over at least three levels—the relation between farmers and the environment, the relation between farmers and the other animals within the farm environment, and the relation between farmers and their larger social communities.

We Are What We Eat: Toward a Mennonite Food Ethics

Below we use the connections drawn above between human beings, nature, farms, and spiritual and religious realization to examine one specific ethical approach to cultivation and consumption—a simple-living approach as advocated in three cookbooks that were commissioned by Mennonite Central Committee (MCC): *More-with-Less, Extending the Table,* and *Simply in Season.*[20] "MCC is a relief, community development, and peace organization of the Mennonite and Brethren in Christ churches in Canada and the United States."[21] That such a church-related organization would commission cookbooks is unusual enough, but these are not mere collections of recipes. We will begin with a brief discussion of Mennonite cookbooks as expressions of the Christian gospel before looking specifically at these three cookbooks. We will close this section by returning to our reflections on Heidegger as a lens through which to better understand the cookbooks' theology of food.

In his review of various Mennonite cookbooks—including *More-with-Less* and *Extending the Table*—Matthew Bailey-Dick concludes: "Moving . . . toward a more explicitly missiological understanding of Mennonite foodways, we can both acknowledge how values and convictions become visible and enacted in particular ways, including the way a community conveys, enacts and remembers its cuisine (as visible contextualization), and see the cookbook as one place in which the Gospel is being translated into 'understandable terms' for other cultures. The cookbook, therefore, stands as a witness to the Gospel and a mission partner for God's work in the world."[22] From his analysis,

Bailey-Dick identified the following eight themes found in Mennonite cookbooks: simple living, globalization of Mennonites, remembering the past, Mennonite migration patterns, gender roles, Anabaptist history, acculturation, and inter-Mennonite cooperation. For our purposes, the most important of these themes is the first—simple living.

MCC commissioned each of these cookbooks for reasons directly related to its Christian vision: "MCC envisions communities worldwide in right relationship with God, one another and creation."[23] According to their title pages, *More-with-Less* was commissioned "in response to world food needs"; *Extending the Table* "to promote global understanding and celebrate the variety of world cultures"; and *Simply in Season* "to promote the understanding of how the food choices we make affect our lives and the lives of those who produce the food."

More-with-Less also established the process by which these cookbooks were compiled. Notices were published in Mennonite and Brethren in Christ publications, soliciting recipes and anecdotal contributions. From the thousands of submissions, the authors chose recipes that were then sent to various "home economists," who tried the recipes in their homes and provided feedback. The authors then selected the final recipes for inclusion. Thus, the process itself was truly a communal project of the churches participating in MCC. The recipes and anecdotes were published with the names of their contributors immediately following the entry, and the acknowledgments in each volume include a long list of those involved in the evaluation and selection process.

The three cookbooks are certainly more than mere collections of recipes. *More-with-Less* includes a lengthy introduction by editor Doris Janzen Longacre. First, she describes how North Americans currently live "less with more": "We are overspending money. We are overeating calories, protein, fats, sugar, superprocessed foods. We are overcomplicating our lives." She concludes, "Getting more with less means eating joyfully. We can release resources for the hungry. We can gain more nutrients for ourselves. We can get more for our food dollars, and eat with more creative expression and good taste."[24] Similar thoughts are expressed in the foreword by Mary Emma Showalter Eby originally published in the first edition of the book in 1976: "More joy, more peace, less guilt; more physical stamina, less overweight and obesity; more to share and less to hoard for ourselves."[25] As well as theoretical discussions, Longacre provides practical advice, including a chart outlining complementary meatless proteins. Thus, she addresses two interrelated concerns—world hunger and the unhealthy diets of North Americans.

Although the other two cookbooks' introductions are not as extensive, they nevertheless provide information other than recipes. *Extending the Table* includes stories of the generosity that missionaries experienced from the poor in various parts of the world as well as practical advice about where to locate some of the unfamiliar ingredients. *Simply in Season* is organized according to the seasons of the year to guide readers in buying seasonal local produce. At the bottom of some of the pages are remembrances of times past ("Nothing sweeter than grandma's peas"), discussion of current trends ("Globesity: too much, too little"), and other bits of wisdom. Each section ends with "Invitations to Action"—such as "Support locally-owned grocery stores, restaurants, and cooperatives."[26]

Thus, all three cookbooks call for resisting the current corporate models of food production and supporting new paradigms of production and consumption that are more consistent with simple living. Moreover, all of this is connected to an understanding of the Christian gospel. In the words of Mary Beth Lind, in her "Foreword to the Anniversary Edition" of *More-with-Less:* "When we make food an integral part of our lives and our homes, it becomes part of our theology. We are connected to our food—cultivating it, preserving it, and preparing it. We are nurturers instead of consumers. This shift affects our relationship to the Giver of our daily bread. We become co-creators with God and stewards of God's garden."[27] Lind's words hold true for all three cookbooks.

In "Mennonite Cookbooks and the Pleasure of Habit," Rebekah Trollinger analyzes these three cookbooks, concluding that they are concerned with Americans' spiritual and physical addiction to unhealthy superprocessed foods. She argues that the cookbooks are interested in encouraging their readers to consciously decide to resist such food addictions in order to create good habits that not only address the concerns of world hunger and North American obesity but also bring more pleasure. Trollinger describes how each cookbook "builds on and transforms the previous": "First, *More-with-Less* develops the ethic of habits. Then, *Extending the Table* introduces new tastes, rejecting the idea that habit must be automatic. Finally, *Simply in Season* celebrates the pleasure of promoting local foods. In its simplest form, this progression moves away from habit and toward pleasure."[28] However, she also notes that this schema may give the false impression that the habits created through *More-with-Less* need to be abandoned in order to enjoy the pleasure found in *Simply in Season*. "The three cookbooks never regard habit as static, but rather flexible, often providing room for pleasure"—for example, all three cookbooks encourage readers to alter the recipes as ways of exploring the joys of cooking. Thus, Trollinger concludes,

> In a society in which free will is continually lauded as the path to pleasure, and in which that free will is simultaneously dismembered by the specter of compulsion, opting for habit is surely radical. But more radical than that is to suggest that pleasure is inherent in and necessary to habit. Taken together, the three M.C.C. cookbooks approach this radicalness: *More-with-Less* reveals the social justice of moving outside the paradigm of free will and compulsion and into a practice of habit; *Extending the Table* revives sensory experience; and *Simply in Season* shows the moral imperative of allowing habit, each day, to unfurl the pleasure of the familiar.[29]

As Trollinger's analysis shows, simple living can bring joy and pleasure to those who have developed the necessary good habits. However, just as simple living is a radical rejection of the current hypersubjectivist and individualist notions of free will as well as of the societal addiction to overconsumption, simple living should also be understood as a radical rejection of the easy life or the lifestyle of the rich. The easy life glorifies individual choices to consume excessively those unnecessary "goods" whose production depends on the unjust and violent exploitation of the poor and natural resources. The simple life strives to minimize this exploitation. As such, living the simple life is not effortless, and these cookbooks recognize that. In her section entitled "Building a Simpler Diet," Longacre argues that a simpler diet saves money and reduces waste but also requires more time spent acquiring food (through gardening and shopping), preserving food (through home canning, freezing, dehydrating), and preparing food (by using nonprocessed foods). Such work is not easy. "Home food preservation is hard work, but it can provide meaningful opportunities for family members to work together."[30] Of course, as we can see in this comment, hard work is not necessarily drudgery, for the time spent doing such work can also be time spent building community.

If simple living is not easy, then what is "simple" about it? The "simple" in simple living refers to the whole system of production—for example, in contrast to the food production system that provides "convenience" foods for the easy life made from ingredients culled from a wide geographical range, a simpler diet greatly reduces "food miles"—the input of transportation. This is illustrated well by Cathleen Hockman-Wert's remarks at the bottom of the page for cornmeal wheat pancakes in *Simply in Season:*

Three good reasons to support local farmers:

1. Good food: We all need to eat. Fresh food tastes better and retains more of its nutrients.
2. Good for the environment: When farming is spread out on small diversified farms, rather than concentrated in huge industrial facilities, it tends to have a lower impact on the environment. Less packaging may be needed, so less waste ends up in landfills. Foods sold fresh soon after they're harvested require less processing and refrigeration and thus use less energy. Eating locally-produced foods also reduces the use of transportation powered by fossil fuels; those emissions release carbon dioxide into the atmosphere which contributes to global warming.
3. Good for communities: In areas with large corporate farms, small towns tend to die off. Skilled jobs are reduced, machines do more work, and low-wage/high turnover positions increase while corporate profits are pocketed by distant shareholders. Local farmers, in contrast, reinvest in local businesses, hire local labor and contribute to the civic strength of their communities. The results: higher employment and a more vibrant social fabric.[31]

Hockman-Wert contrasts the easy life, with its "huge industrial facilities" that disproportionately benefit "distant shareholders," with simple living and its more ecologically sustainable means of food production that support "a more vibrant social fabric." Therefore, even though the simple life may involve hard work, it creates healthier, more livable communities in which pleasure for a greater number of individuals may be found.

We have argued that inasmuch as Heidegger's dwelling implies building and cultivating, it has direct ethical implications, most prominently obligations to protect and preserve the natural environment. Moreover, it also encourages us to inculcate certain virtues, such as love and compassion. Thus, Heidegger's analysis has direct implications on how we ought to use the generous and bounteous returns that nature provides conscientious cultivators. In other words, it would seem farcical to cultivate the land with tenderness and benevolence and then waste the result in conspicuous overconsumption. Such waste violates Heidegger's concept in both spirit and letter.

The Mennonite food ethics that underlie and motivate the three MCC cookbooks push Heidegger's ideas to their logical conclusion. To dwell as human beings who approach the natural environment with awe and reverence encourages a lifestyle that is simple in the footprint that we leave through building and cultivation as well as in our consumption patterns. Thus we realize our ethical and spiritual connections to the environment in both its cultivation and our consumption. Over and above the basic virtues of benevolence and love, we ought to cultivate other virtues as well—frugality, generosity, and humility. Perhaps even more essential is to inculcate gratitude. In other words, Heidegger's notion of dwelling requires the Anabaptist virtues of simplicity and nonviolence as practiced within a community in which gratitude and thanksgiving are expressed to God in worship, in service to others, and in care of creation.

More-with-Less, Extending the Table, and *Simply in Season* provide the means to appreciate the harvest that our cultivation provides in a manner that respects the land's offerings. What is even more special is the collaboration and inclusiveness that these cookbooks represent. In numerous religious traditions meals are seen as a time to reflect upon life as well as to commune with those with whom we share our meals. These cookbooks encourage intentional awareness about the meal's source, production, collection, and preparation. It is this intentional awareness that (at least in part) comprises dwelling.

Dwelling is not a given. We might exist, we might survive, and still fail to dwell in Heidegger's sense, where true dwelling implies bringing forth a truthfulness in our existence. We must learn to dwell, and the initial stage in this process is, as Trollinger suggests, to begin to inculcate certain habits. Although we have an inclination to see habits as negative character traits, habit is the essential prerequisite to virtue as well as vice. And the habits of simple living that Trollinger identifies in the cookbooks place us on the path to a simple, pleasurable, wholesome, sustainable—that is, virtuous—lifestyle.

The Farm as Spiritual Canvas

We have argued above that for human beings *to dwell* we must nurture a lifestyle that incorporates a close connection to the earth and that recognizes "natural" things—both animate (humans, nonhuman animals, and plants) and inanimate (land, water, air, and minerals)—as well as "built" things, such as houses, barns, pastures, bird nests, and spiderwebs. In the ecological sense, a fundamental connection exists between the animate and inanimate as well as the "natural" and the "built." Hence, we are fundamentally related to all other beings in and of the earth—that is, a fundamental connection exists between builders, what is built, and the environment in which it is built and which provides the resources and materials for its building. When we are aware of this interconnectedness we can dwell with awe and wonder. Therefore, what is essential to avoid and what "dwelling in holiness" allows us to overcome is the perverted mindset that views the natural environment as the distant "other" and therefore as mere resources to manipulate and use for human ends.

However, as is true of all animate beings, we must take from inanimate matter; as is the case with all other animals, we must take life (animal and/or plant) in order to survive. This fact of life creates a conundrum. How do we determine whether manipulations or utilizations are appropriate or inappropriate? Our answer includes simple living as is demonstrated, although imperfectly, on the farm discussed below. Although any type of building—whether agriculture or architecture—requires some manipulation and utilization of the environment and its resources, simple living provides a framework for reflecting on the relationships between human beings and the inanimate beings in the environment, between human be-

Aerial view of the farm. (Photo by Raymond F. Person Jr.)

ings and other animate beings, and between human beings and the larger social communities in which they live.

In June 2002, Ray and Elizabeth bought one of the historic Swiss Mennonite homesteads located between Bluffton and Pandora, Ohio.[32] The twenty-acre farm is a portion of a larger farm originally homesteaded by John Amstutz in the 1850s. The original portions of the current house date to the 1860s, and the current barn was built from hand-hewn timber from the property in 1868.

Before Ray and Elizabeth bought the property, they and four other families shared the vision of running the farm as a cooperative. This group now includes nineteen families in their twelfth growing season. Three main goals guided their collective vision: (1) to produce quality food using organic practices, (2) to build human community, and (3) to (re)build nonhuman community through the improvement of wildlife habitat. These goals grew out of their common Christian faith that living more simply leads to being better stewards of God's creation.

The farm comprises a large garden, an orchard, a pasture, a hayfield, beehives, Ray and Elizabeth's house, the barn, a workshop, and the garden tool shed. In the garden grow over fifty varieties of vegetables plus a large

A closer view of the farm. (Photo by Raymond F. Person Jr.)

strawberry patch. The orchard has raspberries, cherries, pears, grapes, hickory, black walnut, and other fruit and nut trees that have not yet matured. The barn, pasture, orchard, and hayfield are home to domesticated livestock—goats, sheep, chickens, ducks, and pigs as well as two dogs and a cat—and wildlife—mostly insects, birds, amphibians, reptiles, and rodents as well as various larger mammals that wander through.

Even while he was working on his dissertation in the late 1980s, Ray had a vegetable garden in his yard. His passion for gardening was nurtured by the small gardens his parents and grandparents had during his childhood and while he was a college and seminary student at Phillips University in Enid, Oklahoma. As a student, he became actively involved in peace and justice issues, especially opposing nuclear weapons and promoting the sanctuary movement for Guatemalan and Salvadoran refugees. Thus, he interacted with a variety of people with shared values, including Mennonites, Quakers, Roman Catholics (especially Benedictine sisters), and Disciples of Christ. They often discussed simple living as an expression of the Christian gospel, and these discussions sometimes occurred during potluck meals. Some of the dishes shared were made from *More-with-Less* and Frances Lappe's *Diet for a Small Planet*. When he met his future wife,

Elizabeth, she shared these values, so the two have continued the struggle to embrace simple living. Buying a farm and starting a cooperative to run it were consistent with their long-held Christian values, and they easily found support for their endeavors within their faith community.

Although most of the families involved in the cooperative are members of First Mennonite Church, Bluffton, from the very beginning families from other congregations have also been involved. The current group includes families from three different Mennonite congregations as well as from a Roman Catholic parish, a United Church of Christ congregation, and a Quaker meeting. Many of the families include someone who has served overseas in missions work, including some who had full-time positions with MCC and/or have participated in Christian Peacemaker Teams. Many of the families regularly use recipes from the three MCC cookbooks, and the farm potlucks often feature dishes from these cookbooks.

Below we will reflect on how simple living has guided decisions on the farm, taking into account the relationship of the farmers to (1) the water, air, and soil; (2) the plants; (3) the nonhuman animals; and (4) the broader human community, beginning with the Bluffton-Pandora area and extending to the world at large.

The Water, Air, and Soil

Pesticide and fertilizer runoff from conventional agriculture remains a serious source of water pollution. The organic practices on the farm do not eliminate this problem, especially in relationship to manure; however, they certainly reduce the problem significantly. Thus, the organic practices minimize the agricultural contribution to water pollution. Furthermore, any rain runoff from the garden must first pass through large grassy areas that act as living filters, further reducing the possibility of runoff from the garden contributing to pollution in the Maumee River watershed. Although heavily dependent on rainfall, the farm includes three wells that provide water for the house, garden, and livestock.

The farm families remain too dependent on gas-powered vehicles for transportation and their farming still depends on them too—for example, a pickup, tillers, and a large lawnmower with attachments. Thus, they contribute to America's burgeoning carbon footprint. However, this impact is also lessened by the fact that some members of the cooperative ride their bicycles on the three-mile trip from town, and others have been known to walk or jog. For example, Steve and Monica with their three sons, Lucas, Julian, and Christopher, recently rode out on their bikes together. Also, relative to many other farms, many of the tasks are done with human labor rather than large machinery—such as cleaning the manure out of the barn using shovels and other hand tools rather than using, for example, a tractor with a front-end loader. Although the farm uses considerable electricity—for example, to keep the livestock's water from freezing in the winter—this is offset by the use of a hybrid solar/wind system that generates 33 percent to 50 percent of the electricity needed.

The use of synthetic fertilizers and pesticides negatively affects living microorganisms in soil. The farm's use of organic fertilizers—manure, blood meal, bonemeal, greensand—not only adds the necessary nitrogen, potash, potassium, and minerals but also encourages the living microorganisms in the soil. Thus, the soil in the garden is richer and healthier than it was ten years ago, as evidenced by a significant increase in the number of earthworms, more organic material in the soil, and larger harvests.

Plants

Plants on farms are generally divided into two groups, those whose growth is encouraged and those whose growth is inhibited, and this is certainly the case on this farm. Much time and energy is put into creating a good environment in which vegetables, herbs, fruits, nuts, pasture grasses, and hay can thrive. However, what might be considered a "weed" in the garden can also be understood as beneficial elsewhere on the farm. Thistles remain a thorny issue in the garden, but the goats prefer them in the pasture; therefore, while we are weeding the garden, we are also harvesting food for the goats. After garden plants have provided human food, the plants then provide winter forage for the livestock—for example, sweet potato vines, sweet corn stalks, and bean stalks are all put up in the hayloft only to return to the garden in the spring in the form of manure. Certain "weeds" that are pulled in the garden flourish elsewhere for the benefit of wildlife, which in turn benefits the garden. Red-wing blackbirds nest in the thistles outside the pasture fence, and praying mantises and other beneficial insects find their home among Queen Anne's lace, goldenrod, and other wildflowers along the garden fence.

Nonhuman Animals

As is true of plants, nonhuman animals on farms are generally divided into two groups: those that are beneficial, whose presence is encouraged, and those that are problematic, whose presence is inhibited, and this is the case on this farm as well. Much time and energy is put into creating a good environment in which the livestock can thrive in a way that more closely mimics their original natural environments—for example, the sheep and goats are pastured without grain-based feeds, and the chickens are free range. Some time and energy is also put into creating an environment to attract beneficial wildlife—for example, a bat box was hung on the barn wall. However, the farmers have learned that one of the easiest ways to attract beneficial wildlife is simply to create a variety of microenvironments—the garden, the pasture, and the orchard—that provide a welcoming place. Thus, the creation of such microenvironments has meant that species that did not thrive on the farm earlier now flourish, including toads, leopard frogs, barn swallows, red-wing blackbirds, kingbirds, and various insects. The barn houses not only the livestock but also bats, barn swallows and, for a fifteen-month period, a screech owl. The bats and barn swallows are welcomed to help reduce the fly, moth, and mosquito populations. The screech owl was an important part of rodent control.

Similarly, time and energy is put into discouraging creatures that are destructive to the agricultural processes on the farm, especially predators—rats, hawks, foxes, mink, raccoons, and opossums—and the "bugs" and other "pests"—rabbits, groundhogs, and deer—that damage the fruits and vegetables. Fencing helps significantly, but other measures are sometimes needed. The rodent-control plan includes old-fashioned spring-loaded traps, homemade poisons that target only the rodents (for example, those that use plaster of paris or baking soda as the active ingredients), and Earl, the barn cat. Skillet, the experienced red-bone hound, stays on the job protecting the livestock from predators. He has even learned the rooster's crow warning the hens of a distant hawk approaching and will run in the rooster's direction until he finds the hawk and chases it away. At times, the predator problem must be resolved by the teamwork of Skillet and Ray with a small rifle. Moreover, the free-range chickens are effective hunters of Japanese beetles, grasshoppers, crickets, and even small mice and voles that can damage the garden. Not only are the domesticated animals part of the team to keep destructive animals in check, but some of the wildlife also are as well—for example, Ray will always stop the tiller to move a toad or frog out of harm's way, since it can eat fifty to a hundred insects daily.

Because the domesticated animals have important jobs on the farm—providing meat, eggs, and/or pest control—the farm families strive to give them a good quality of life, even if their eventual fate is to provide meat. However, it must be noted that some farm families who are vegetarian or vegan choose not to participate in the livestock projects.[33] Therefore, some farm families, in the words of Wendy, choose to "get closer to [their] meat," while others choose to distance themselves from all meat. Ironically, similar values underlie both decisions, especially opposition to meat produced in concentrated animal-feeding operations.

The Broader Human Community

The farm is a place where friendships have been made and strengthened among the farm families as they work, eat, and play together, but the farm community extends far beyond nineteen families and twenty acres. Locally, the farm community includes all the other people whose work directly contributes to the food production—for example, nearby farmers who produce the livestock feed and breeding stock; the local merchants from whom they purchase minerals and bedding plants; the local butcher; and contractors and volunteers who have contributed to larger projects, such as the renovation of the house and barn or the building of fences. It includes those who purchase eggs and meat and who have enjoyed the excess produce and eggs donated to charities. It includes the various groups that have visited the farm to learn more about organic food production—for example, Girl Scouts, 4-H, church youth groups, and university classes—as well as those groups to which Ray and others have been guest speakers—for example, Sunday school classes and the 2011 Intercollegiate Peace Fellowship conference.[34]

The farm community extends even further, beyond the local community into the global community. Of course, organic practices and efforts to reduce our dependence on fossil fuels have global consequences, however insignificant such individual projects may appear. But the global farm community has a more personal relationship as well. For example,

Jo's family from New Zealand and Andy's family from Canada have visited the farm and helped with various chores. Furthermore, connections exist between the farm and Menno Village, an organic CSA (community-supported agriculture enterprise) in Naganuma, Hokkaido, Japan, founded by Aki and Ray Epp. Wendy and Andy and their daughters, Hannah and Sara, lived and worked at Menno Village for a year, and Ray and Elizabeth have visited there.[35] Thus, their experiences on an organic farm in Japan have had some influence on practices on the farm in northwest Ohio. Furthermore, the farm has hosted international visitors, such as peace ambassadors from the World Friendship Center in Hiroshima, Japan; a laywoman from the Lutheran church in Tanzania; and Tibetan monks from the Gaden Shartse monastery in southern India, all of whom share similar values. Furthermore, the global farm community consists of more than human members—some wildlife migrates from distant lands, for example, red-wing blackbirds, kingbirds, barn swallows, great blue herons, and monarch butterflies.[36]

The Interconnections among the Various Members of the Farm Community

Although the above discussion already hints at the interconnections among all of these relationships—water, air, and soil; plants; nonhuman animals; and the broader human community—there is a chance that these relationships could be misread as somewhat disconnected. Therefore, we will close with a description of two complex relationships that illustrate their interconnectedness.

Living in the country often includes accepting more insects in one's life than one might have in town (or at least different kinds of insects), and this farm is no exception. The flies were bad enough in the farm's first year, and in the second year when he purchased the first goats, that Ray feared that the problem would worsen. Fortunately, the opposite was the case. Despite the goats producing manure for the flies to breed in, there were fewer flies that year and every year since. The explanation is actually quite simple. When Ray opened the barn for the goats, the first barn swallows moved in. The barn swallows, as well as taking advantage of the safe nesting sites in the barn, were attracted to the open fields, especially the pasture, over which they constantly hunted insects on the wing. Quite by accident, then, the preparation of the pasture and barn for the goats was also a welcoming environment for the barn swallows, which gladly helped out with the fly problem. The chickens also contributed by scratching in the manure to feast on the fly larvae. Thus, the creation of a more holistic microenvironment helped remedy the fly problem. Of course, there are still flies on the farm, as there should be—they have their place in the manure management system of the farm—but the barn swallows and chickens help keep them in check.

Because of the farmers' commitment to organic practices, they rarely use pesticides, even organically approved ones. Because of this, the diversity of insects in the garden and elsewhere has increased significantly from what it was earlier with monoculture crops and synthetic pesticides. Recently, when Ray, Judy, and Jonah were digging potatoes, Jonah found a large hornworm and said, "Oh, wow! Look at this disgusting thing!" His outburst clearly communicated both fascination and revulsion. Judy and Ray stopped their digging and walked over to see what he had found. This led to a discussion about hornworms: they are destructive to potato and tomato plants, but also they can carry the next generation of the parasitic braconid wasp, a beneficial insect that preys on soft-bodied insects like the hornworm. After carefully examining the hornworm to make sure that it was not carrying braconid wasp larvae—because if it had been, he would have gently put it in the tomato patch to nurture the braconid wasps—Ray fed it to a nearby chicken, which enjoyed the snack. Once upon a time Ray would have had a reaction very similar to Jonah's. However, because over twenty years ago Bill, a friend and veteran organic farmer in North Carolina, taught him about the relationship between the hornworm and the braconid wasp, Ray had knowledge about that relationship to share with others.

Concluding Reflections

This episode also illustrates one of the ways in which simple living as a movement develops and spreads—not through commercial processes but through the experiences of those who have dedicated themselves to the same vision. Simple living represents a "living history" in which ideas and methods are shared, often through anecdotes. We have an obligation to share our knowledge and blessings. No one has an exclusive claim to the good; indeed, its nature requires that we teach and share it with others. Thus, just as Ray learned from Bill in North Carolina and Wendy and Andy learned from the Epps in Japan, the Ohio farm families understand the farm's mission to include reaching out to others, including groups visiting the farm and cooperative members speaking to groups.

What we hope we have conveyed is that to live in the fullness of our own being in the world is impossible in isolation. Being is a nexus that encompasses communities within communities within communities, each with inseparable connections to the others—connections ecological, biological, social, philosophical, theological, and spiritual. To separate these connections leads to violence and alienation. To lose our connection to the earth, to each other, and to other creatures is to injure our own being in the world. Although we ought to celebrate our differences, we nevertheless experience a core commonality as participants in the great and grand design of creation. This places certain obligations on us—to learn from the past, to protect the present, and to conserve for the future. Heidegger provides a philosophical explanation of the foundation of our very being, but Heidegger's notion requires that we live according to the Anabaptist virtues of simple living and nonviolence within faith communities that express gratitude and thanksgiving in acts of worship, in service to others, and in dwelling in creation with awe and wonder. Although as creatures we must take from creation in order to live, we must do so in a way that nevertheless cherishes and preserves the earth. The model of simple living, as expressed in the three MCC cookbooks and as embodied in the community that farms a historical Swiss Mennonite homestead in northwest Ohio, cherishes the good creation God has given us to preserve so that it can continue to be a blessing to us and others (both human and nonhuman) for many generations.

Study Questions

1. What is dwelling, and why is this important in evaluating human-nature interactions via the intersection of farming/agriculture?
2. The authors claim, "To act in a manner that cherishes, protects, and preserves the earth and therefore recognizes and acknowledges the earth's holiness would necessitate certain constraints on how we live." Do you agree with this? Why or why not? If their claim is

valid, what changes in lifestyle, especially in regard to food habits, might we as a society or as individuals need to make?

3. What are implications, for the authors, and for you personally, of shifting our concepts of self from being consumers to being nurturers? What food-related practices might emerge from undertaking such a shift?

4. How does "dwelling in holiness" influence ethics and spirituality in regard to the earth, to nonhuman animals, and to other humans in our communities, especially via the practice of agriculture?

Notes

We wish to acknowledge Forrest Clingerman and Elizabeth Kelly for their valuable comments and editorial assistance during this chapter's preparation.

1. Stephanie Nelson, *God and the Land: The Metaphysics of Farming in Hesiod and Vergil* (Oxford: Oxford University Press, 1998), v.

2. Heather Ann Ackley Bean, "Toward an Anabaptist/Mennonite Environmental Ethic," in *Creation and the Environment: An Anabaptist Perspective on the Sustainable World,* ed. Calvin Redekop (Baltimore, Md.: Johns Hopkins University Press, 2000), 183–86.

3. Ibid., 194. See also the other chapters in Redekop, *Creation and the Environment.*

4. Michael Yoder, "Mennonites, Economics, and the Care of Creation," in Redekop, *Creation and the Environment,* 71–77.

5. Bean, "Toward an Anabaptist/Mennonite Environmental Ethic," 187–88. See also Yoder, "Mennonites, Economics, and the Care of Creation," 75–76.

6. Our description of these values as Anabaptist is not intended to claim that they are exclusively Anabaptist, only that the Anabaptist tradition is often characterized as placing these values centrally in its understanding of faith and practice.

7. For an excellent discussion of how the Old Testament is best interpreted from an agrarian perspective, which also aids in the development of an environmental ethic, see Ellen Davis, *Scripture, Culture, and Agriculture: An Agrarian Reading of the Bible* (Cambridge: Cambridge University Press, 2009).

8. Martin Heidegger, *Poetry, Language, Thought,* trans. Albert Hofstadter (New York: Perennial Classics, 1971), 145.

9. Ibid., 147.

10. Christine Swanton, "Heideggerian Environmental Virtue Ethics," *Journal of Agricultural and Environmental Ethics* 23 (2010): 148, 149.

11. It is essential to note here that while holiness has distinctive religious overtones, in the sense in which Heidegger uses the concept it implies a myste-

riousness, awesomeness, and wonderousness (ibid., 160). In this respect it shares certain similarities with Rudolf Otto's *The Idea of the Holy.*

12. Ibid., 150.

13. Ibid., 161.

14. Heidegger, *Poetry, Language, Thought,* 149. In Heidegger, "to dwell" has a close connection to "the fourfold," a concept that further complicates the relations between human beings and nature. Since the aim here is to use Heidegger as a springboard, rather than explicate Heidegger's own ideas, we shall leave the fourfold to a future discussion.

15. Martin Heidegger, *Being and Time,* trans. John MacQuarrie and Edward Robinson (New York: Harper Perennial, 1962), 89.

16. Heidegger, *Poetry, Language, Thought,* 148.

17. Heidegger explores this idea in depth in "The Question concerning Technology," in *Technology and Values: Essential Readings,* ed. Graig Hanks (Chichester, U.K.: Wiley-Blackwell, 2010), 99–113.

18. Todd Lavin, "The Politics of Authentic Experience," in *Existentialist Thinkers and Ethics,* ed. Christine Daigle (Montreal: McGill-Queens University Press, 2006), 54.

19. Ibid., 62.

20. Doris Janzen Longacre, *More-with-Less,* 25th anniversary ed. (Scottdale, Pa.: Herald, 2000); Joetta Handrich Schlabach, *Extending the Table* (Scottdale, Pa.: Herald, 1991); Mary Beth Lind and Cathleen Hockman-Wert, *Simply in Season,* rev. ed. (Scottdale, Pa.: Herald, 2009).

21. Schlabach, *Extending the Table,* title page; and Lind and Hockman-Wert, *Simply in Season,* title page. A similar but fuller description of MCC is found in the author's preface in Longacre, *More-with-Less,* 4.

22. Matthew Bailey-Dick, "The Kitchenhood of All Believers: A Journey into the Discourse of Mennonite Cookbooks," *Mennonite Quarterly Review* 79 (2005): 163. *Simply in Season* was not available when Bailey-Dick undertook his analysis, even though he was aware that it was forthcoming (177n79).

23. http://www.mcc.org/purpose-vision-statements (accessed July 3, 2011).

24. Longacre, *More-with-Less,* 13, 48.

25. Mary Emma Showalter Eby, foreword to Longacre, *More-with-Less,* 7.

26. Lind and Hockman-Wert, *Simply in Season,* 225, 128, 278.

27. Mary Beth Lind, "Foreword to the Anniversary Edition," in Longacre, *More-with-Less,* viii.

28. Rebekah Trollinger, "Mennonite Cookbooks and the Pleasure of Habit," *Mennonite Quarterly Review* 81 (2007): 546.

29. Ibid., 547.

30. Longacre, *More-with-Less,* 43.

31. Lind and Hockman-Wert, *Simply in Season,* 296.

32. Ray is one of the present authors, Raymond F. Person Jr., who will hereafter be referred to in the third person. When discussing the other members of the

farm cooperative as well as other community members, we will consistently refer to them simply by their first names, avoiding the more formal academic use of last names. Furthermore, we will continue to use "we" for the collective author of this essay, even when we are (self-)describing Ray.

33. Mark, the other present author, is not a member of the cooperative due to distance; however, he purchases eggs from the farm because, even though he is a self-described vegan, he has met the chickens and respects the conscientious treatment they are given, which ensures them a high quality of life.

34. The Mennonite world sometimes can be small. For example, Bailey-Dick's brother, Andy, and his family are founding members of the cooperative. Moreover, Trollinger grew up in Bluffton and helped with the house renovation. Thus, in some sense they too are members of the extended community associated with the farm.

35. See Ray Epp, "Building a Farm with a Future," 2002, http://newfarm.rodaleinstitute.org/features/0802/japan1/print.html (accessed July 3, 2011).

36. For an excellent discussion of emplacement and migration as well as reflections on herons and butterflies, see Forrest Clingerman, "The Intimate Distance of Herons: Theological Travels through Nature, Place, and Migration," *Ethics, Place & Environment* 11, no. 3 (2008): 313–25; and Forrest Clingerman, "Butterflies Dwell betwixt and Between: Non-human Animals, Theology, and Dwelling in Place," in *Animals as Religious Subjects: Transdisciplinary Perspectives,* ed. Celia Deane-Drummond, Rebecca Artinian-Kaiser, and David Clough (London: Bloomsbury, 2013), 169–190. Forrest, Gail, Sabina, and Asa are also members of the farm.

13

Steward or Priest?
The Possibilities of a Christian Chicken Farmer

Ragan Sutterfield

My task here is to articulate what difference Christian faith might make in how we practice and understand eating and agriculture. This is a broad undertaking since Christianity is broad, with a two-thousand-year history of thinking that has not been unified. I will attempt here to describe some of the different approaches to the creation taken within the faith and what effects those have on agriculture. I articulate these ideas as a Christian practitioner and will frequently use the personal "we," though I hope to make myself clear to those of all orientations of faith.

At the close of Michael Pollan's *The Omnivore's Dilemma*, he and a group of friends gather for a feast. It is a meal of pure goodness, reflecting all the values Pollan has discovered along his journey through the American food system. Although a secular person, Pollan remarks that sitting there with friends and family around this meal, he found himself "reaching for . . . the words of grace."[1] This reach for a blessing is something a good meal demands. A blessing is even what a good meal becomes, for as Pollan goes on to say, "The meal itself had become . . . a wordless way of saying grace"—the sacred intersection of world and body.[2] That we so often don't say grace and bless a meal as a sacred thing might mean that we have lost our ability to recognize the sacred in a world of speed and achievement.

A few years ago the *Washington Post* did an experiment: the newspaper had Joshua Bell, one of the world's greatest violinists, play in a Washington, D.C., metro station during rush hour. Here was a virtuoso playing incredible music, and yet hundreds of people passed by with barely a pause—a display of pearls before swine. We might argue that this same rush past the sublime happens every day at our mealtimes.

Christianity, along with other faith traditions, offers a possible corrective to this desacralized view of a meal. Holding onto the sacredness of our food, there is a call within Christianity to bless a meal—to say "grace." This work is meant to recognize that we are a part of the whole of the creation, dependent upon the creation, and participants within it. As the early Christian theologian Clement of Alexandria (circa 200 CE) wrote, "It is befitting, before partaking of food, that we should bless the Creator of all; so also in drinking it is suitable to praise Him on partaking of His creatures."[3]

Such a blessing may seem simple at first, but in a time when our food systems are not always good, when our meals represent sins against the creation, how can Christians bless a meal that represents a desecration as much as the blessings of God? Would Michael Pollan have reached for the words of grace sitting down to a Big Mac, Coke, and fries?

One of the most memorable lines from the comedy movie *Talladega Nights* comes when Ricky Bobby, a fictional NASCAR driver, sits with his family at a table laden with fast food. He begins the blessing: "Dear lord baby Jesus . . . we thank you so much for this bountiful harvest of Domino's, KFC, and the always delicious Taco Bell."[4] We find the idea of blessing such a meal absurd, particularly with the phrase "bountiful harvest" thrown in to draw the maximum comedic contrast. Does a blessing at such a table even make sense given that it rests on so much exploitation of people and earth?

The difference between Pollan's meal and Ricky Bobby's starts well before the food reaches the table. Even if we should offer a word of grace at both, the meaning and nature of that prayer will change according to what comes earlier. The question that precedes grace is then a question of agriculture, which is more fundamentally a question of the human role within creation. We want to offer here two possibilities within the Christian tradition for how we might see our relationship with the land. One is the idea of "stewardship," which orients our farming and use of creation toward a management of creation. Creation here becomes a series of "resources" whose primary value is determined according to human use and interaction. This is the dominant view of creation within current

Christian practice as well as national understanding, even among many environmentalist organizations such as the Natural Resources Defense Council and the Christian college-based organization Earthkeepers.[5]

On the other side we have the idea of the human role as one of "priesthood." This view maintains a human exceptionalism within creation, but places that emphasis not on a separate managerial role or ontological status, but rather on a special "priestly role" for humankind. Stanley Hauerwas and John Berkman have articulated this idea in reference to the traditional Christian/Jewish view of the role of the Jewish people in the world—that all nations will be "saved" through the Jewish people not because of a special ontological status, but because they are the carriers of the world's story: "To put it most simply, the only significant theological difference between humans and animals lies in God's giving humans a unique purpose. Herein lies what it means for God to create humans in God's image. A part of this unique purpose is God's charge to humans to tell animals who they are, and humans continue to do this by the very way they relate to other animals. We think there is an analogous relationship here; animals need humans to tell them their story, just as gentiles need Jews to tell them their story."[6]

In the priestly view, the role of humankind is not merely to be a passive storyteller but to actively call the creation to its fullness within God's very being. The role of a priest is to perform sacramental rites that make the holy visible. A priest is one who prays the prayer that turns common bread and wine into the sacrament of the Eucharist, a sacrament many Christians take to involve the actual flesh and blood of Christ. The human role of priest in creation is to name, bless, and call forth the sacred nature of all created things. As Wendell Berry once wrote in a poem, "There are no unsacred places; / there are only sacred places / and desecrated places."[7] It is the naming of this reality that priests are called to speak. A priest does not view the creation as a resource to be managed, but rather as a subject in relationship with the divine.

Two Kinds of Chicken Farming: Two Views of Creation

Within agriculture there are correlates to these two views of creation. On the one side there is agrarianism, or agriculture on a human scale, a form of agriculture that sees the soil as a gift that has intrinsic value that we must respect. This is a kind of priestly agriculture. On the other side there is industrialism, or a form of agriculture that sees value purely on the side of human economy—the soil made meaningful only in the context of human interactions with it. Industrialism is agriculture as resource management, the agriculture of the steward, managing for increased yields to benefit the greatest number of humans and to increase economic profit at the same time.

As a way to explore these two forms of agriculture and the possible Christian views of human relationship to creation, we will use as a heuristic two chicken farmers, both well known as Christians. Though both these farmers proclaim allegiance to Christianity, their views of farming could not be more different. Their views are so divergent, in fact, that one might question whether they really do worship the same God, and this is a question well worth entertaining. In other words, their two ways of farming reveal two different theologies, two different ways of understanding the human role in creation that cannot both end up being right. Are we priests or stewards? What difference does it make to our mealtime blessings?

Type 1: John Tyson and the Industrial Mind

John Tyson is the current chairman of Tyson Foods, the world's largest meat producer.[8] Tyson Foods, founded by John Tyson's grandfather, is a pioneer in a vertical model of farming and food processing, with every step from the birth of a chick to the packaged chicken tenders in a grocery store freezer aisle controlled by the company. The success of this vertical model has made Tyson Foods one of the five largest food companies in the world and the world's largest "protein" company, as the business magazines like to refer to the business of producing and selling animal flesh. The company's revolutionary approach was developed in large part by rethinking what a farm is—no longer a place of husbandry, a way of raising animals whose standards are built around the household economy, but rather a factory using "animal science" for the production of a protein product.

The entire system of Tyson's food production is built around the maximization of profit, which means growing chickens as cheaply as possible, as quickly as possible, and producing as much meat as possible from each bird. This starts with selectively breeding broiler chickens for rapid growth. Tyson has been able to nearly halve the time it takes for a chicken to reach butcher weight, from eighty-five days in 1950 to forty-five days now. This rapid growth means that costs are cut significantly—farmers can now raise nearly twice the number of chickens with the same amount of feed, water, and electricity.

In addition to raising faster-growing chickens, Tyson pioneered methods of growing large numbers of chickens in confinement. Using long row houses, Tyson contract growers raise as many as twenty thousand birds in one house.[9] These birds are given only enough room to eat, drink, and grow. Because of their rapid growth, most of these broilers develop leg problems early and don't move around as actively as more traditional chicken breeds. Their lives are engineered around one purpose—to grow meaty breasts as quickly and cheaply as possible.

The life of a Tyson chicken starts in a hatchery, where eggs from breeding stock are hatched in large incubators. Since chickens are precocial (animals that are able to live relatively independently of their parents from birth) the chicks are able to eat and drink without the care of an adult. The birds are then sorted according to sex, a difficult task often performed by Hmong refugees living in northwest Arkansas. The male and female birds are distributed differently, based on the particular end product in mind—a big-breasted broiler, a smaller fryer, or a bird for a processed chicken dinner. Once the chicks are sorted, they are packed in plastic crates that are stacked on the back of a truck and driven to chicken farms.

Once the chicks arrive at the farm, they are released into a long row house filled with heavy wood-shaving bedding (a whole industry of bedding production is built around the forests of Arkansas). Low hanging lamps provide necessary warmth, and the chicks are given as much food and water as they can consume. They eat and eat, nonstop, with artificial lighting creating the illusion of a never-ending day. The birds do sleep, but their circadian rhythms are disrupted for the purpose of encouraging them to eat as much as possible. The food they are given is a mix of corn and soybeans, mixed with vitamins and arsenic or other antimicrobial agents that are technically not "antibiotics," allowing Tyson to make the claim that their meat is antibiotic free (though this has been challenged).

The birds eat and eat until they reach butcher weight, but many of them, an acceptable percentage on the accounting sheets, die. This method of mass production involves a great susceptibility to disease and death, sometimes on a large scale. Growers have freezer units outside of their chicken houses where they collect dead chickens on a daily basis, but when major disease outbreaks happen, large numbers of birds are buried, often illegally. The work of collecting dead chickens is often performed by immigrant workers, who must bear the heavy ammonia stench of chicken manure as well as its dust, which can cause respiratory problems. The manure itself is a major management concern. It is frequently used as fertilizer on nearby fields, but because of the arsenic used as an antimicrobial in many chicken feeds, the fertilizer is toxic, leading to heavy levels of arsenic pollution in rivers and streams where the excess manure runs off.

Once the birds are ready for slaughter, they are packed into cages so tightly that they can barely move. The birds are then driven to a slaughter facility, often many miles away, where they are killed either with a beheading machine or a dip in an electrified bath. The skin is scalded, the feathers plucked, and then the birds are eviscerated by hand by workers, often illegal immigrants with few legal recourses in the case of abuse by managers. This is how most chicken dinners get their start.

This way of raising chickens represents a major shift in agricultural practice, but more so a shift in thinking about land and animals. In writing about the demise of family-scale farming in America, the agrarian Wendell Berry explains that the "principal reason for this failure is the universal adoption, by our people and our leaders alike, of industrial values." These values, argues Berry, are based on three assumptions:

a) That value equals price—that the value of a farm, for example, is whatever it would bring on sale, because both a place and its price are "assets." There is no essential difference between farming and selling a farm.

b) That all relations are mechanical. That a farm, for example, can be used like a factory, because there is no essential difference between a farm and a factory.

c) That the sufficient and definitive human motive is competitiveness—that a community, for example, can be treated like a resource or a market, because there is no difference between a community and a resource or a market.[10]

These three assumptions are clearly at play in Tyson Foods' approach to farming. The value of a farm is purely locked up in its ability to produce meat as cheaply as possible. The farm is obviously a factory, and the chickens raised on a farm are treated, insofar as possible, like any industrially produced widget. The fact that they create manure is an inconvenient reminder of the biological beings they are. As Matthew Scully wrote in the *American Conservative*, "All of factory farming proceeds from a massive denial of reality—the reality that pigs and other animals are not just production units to be endlessly exploited but living creatures with natures and needs."[11]

These natures and needs are, according to Christian theology, given by a creator. Tyson did not create the chicken as Ford created the automobile. Engaging with a chicken, raising it, is a matter of taking account of and responsibility for its divinely appointed givens. Farming is what Aristotle would have called a stochastic art, one that requires dynamic work, discernment in relationship to the object (the Greek refers to divination). A stochastic relationship with a chicken would be one in which one observes the needs of the chicken and answers them with no preset conditions other than an end toward flourishing.

But this is the exact opposite of the industrial mind. In the industrial way of farming, the animal becomes subject to a system to which it must be molded. Its suffering is a result of nonconformity to the factory, as are its diseases, solved (but never permanently, of course) by engineering and drugs. There are even researchers attempting to create animals without the cognitive capacity for stress so that the conditions of the factory will no longer be a source of pain and struggle—animals will be "happy" whatever their conditions.[12]

It is critical to understand that John Tyson, the man who holds in his mind this industrial approach to farming, is a Christian. It is common, of course, for one's religious beliefs to be relegated to the world of the private—that has been the American way since John F. Kennedy told the nation that his faith was private and not a factor in his role as president, and the privatization of religion in America goes back much further. But Tyson, a former drug addict who had a dramatic conversion to faith, does not hold his faith and his business as separate things. In fact, he is one of the leading advocates of bringing faith into the workplace, and Tyson employs the largest number of business-based chaplains in the world.

Tyson has dedicated a great deal of his resources to founding the Tyson Center for Faith and Spirituality in the Workplace at the Sam M. Walton College of Business at the University of Arkansas. He also speaks frequently on how to integrate faith into a multicultural workplace (Tyson's chaplains come from many faith backgrounds, though Tyson himself is a Christian). John Tyson has also left his mark on Tyson Food's "core values," which because of him include "We strive to be a faith-friendly company" and "We strive to honor God and be respectful of each other, our customers, and other stakeholders." Among these core values we also find this: "We serve as stewards of the animals, land, and environment entrusted to us."[13]

How do we make sense of Tyson's clear interest in explicitly living his faith in and through his business, and the reality of the factory farming upon which his business is founded? We might argue that it begins with Tyson's theology of "stewardship" and how that concept mixes with a view of creation that has served as the basis of the American economy since its founding.

The understanding of creation most influential on the American mind might be linked to John Locke, one of the Enlightenment's foundational thinkers, whose philosophy profoundly influenced the founding fathers of the United States. Locke assumed, in step with the dominant Christian theology of his time, that the creation was given to humankind. But the creation that was given was not valuable in itself; rather, it was raw material that could be made meaningful only through human management and enterprise: "God gave the world to men in common; but since he gave it for their benefit, and the greatest conveniences of life they were capable to draw from it, it cannot be supposed he meant it should always remain common and uncultivated. He gave it to the use of the industrious and rational, (and labour as to be his title to it); not to the fancy or covetousness of the quarrelsome and contentious."[14] The creation does not, then, belong to all, but only to those who will make the best use of it—"the industrious and rational." To put this in modern terms and to broadly stereotype, God gave the creation for entrepreneurial businesspeople, not lazy hippies.

A critical point to draw from this is that in Locke's view, the goods of creation are not value laden. It is our "stewardship" of those goods that gives the raw materials value and makes them useful to human ends. In such a view, a cheap package of chicken breasts can only be a good thing because a chicken has no value in itself. It is interesting to note here that some of the same logic was used by Christians to justify slavery: Africans and their American descendants, judged to not be particularly "industrious" in their tribal settings, might be made useful through labor in American plantations and homes. In 1773 Richard Nisbet, a West Indian plantation owner living in Philadelphia, argued that the slave trade was responsible for "rescuing many millions of Africans, as brands from a fire, and even compelling them to the enjoyment of a more refined state of happiness, than the partiality of fate has assigned them in their native state."[15] Africans in their native state, then, were the raw stuff of creation. It was only through the intervention of "civilization" by the slave owner that the slave could be made valuable. This same logic is at work, it seems, in John Tyson's "stewardship."

Stewardship as a theological concept has been used to good effect in articulating the Christian call to care for the creation by writers such as Calvin Dewitt, but too often the model exhibits a mistaken theology. As Kelly Johnson argues in her wonderful book *The Fear of Beggars: Stewardship and Poverty in Christian Ethics,* stewardship language, relatively new on the theological scene, "shifts the center of gravity within economic ethics, the daily disciplines and language of preaching, toward the presumptions that the key moral agent, the model disciple, has disposable wealth and that existing property rights are underwritten by God."[16] In other words, the concept of stewardship shifts the ideal of the Christian life away from one in which a person is completely dependent upon the gifts and grace of God to one in which a person carefully manages the "assets" of creation in order to gain wealth. The aphorism "God helps those who help themselves," a phrase so common that many Americans mistakenly think it is in the Bible, captures the stewardship ethic.[17] In the context of Tyson, the ethic of stewardship shifts meaning and responsibility from givens within the creation to the sphere of human use.

It was without irony, then, that Tyson Foods released a booklet titled *Giving Thanks at Mealtime* in 2005. The booklet contains thanksgivings from a variety of major faith traditions, mostly Christian but also some Buddhist, Muslim, Hindu, and Jewish blessings. As John Tyson wrote in the introduction, "This Giving Thanks at Mealtime booklet is designed to help you discover (or rediscover!) the joy and power of saying a word of thanks before mealtime."[18] The blessings all center on thanking God for the gifts of food and family, and the many other blessings God bestows. Some of the prayers are quite touching, such as this one from Mount St. Mary's Abbey: "Lord bless our shared meal, a sacrament to our shared unity." But in praying these thanks to a God who has delivered these gifts, should we not as Christians be concerned with how they came to us?

The shift to stewardship enabled Christian ethicists to move the act of blessing from the inevitable questions of right livelihood to an abstracted use of wealth for good. As Johnson notes, stewardship "requires no disruption of social structures . . . leaves the order of property rights largely uncontested, and yet makes room within that order for a critique of selfishness, for a stern demand for attention to the needs of others, for an obligation for generosity."[19] Tyson Foods can thus proudly tout its work to feed the hungry while at the same time building its profits from the abuse of creation and the exploitation of low-wage migrant labor, all while maintaining its role as "stewards of the animals, land, and environment entrusted to us."

While Tyson represents a common thread in Christianity practiced in and synchronistic with Western capitalism, we are right to feel uneasy with this theology and skeptical of its facile blessings. The scriptures and the best of the Christian theological tradition would seem, on the whole, to indicate that the kind of food system Tyson Foods fosters is more of an abomination than a "sacred trust." How can we know the fullness of a Tyson chicken's life and say with the psalmist that all creation looks to God "to give them their food in due season" (Psalm 104:27)? Nowhere in the mealtime blessings is there a confession for the food. That might be the most appropriate prayer.

Type 2: Farming as if "the Earth is the Lord's"

The most critical mistake in the view of property put forward by Locke is that the world is made valuable only through human interactions with it. This makes stewardship a matter not only of the proper use of gifts but of making value from them, but this is not the biblical witness. In the Christian and Jewish scriptures we find frequent references to a creation whose value is found only in relation to the creator, not to the human "steward" within the creation. Psalm 104 gives us a beautiful portrait of a creation that finds its life in the breath and love of God. Even predators that pose a threat to human economic well-being are given sustenance through God: "The young lions roar for their prey, seeking their food from God" (21). At the close of this psalm we are told, "Let sinners be consumed from the earth and let the wicked be no more" (35a). Walter Brueggemann has suggested that the sinners and wicked here are those who abuse the land, seeking their own ends rather than respecting the whole of the land.[20] This psalm, which is best read as an interpretation of Genesis 1, where human beings are told "to fill the earth and subdue it" (1:28), should disabuse us of any notions that human dominion means that we are taking the place of God in this creation that God sustains with God's own breath (29–30).

In the book of Job, as God explains to Job that Job is not the center of the universe, God uses the example of the wild donkey and the wild ox, both nondomesticated versions of beasts of burden that are, as Wendell Berry points out, "said to be 'free,' precisely in the sense that they are not subject or serviceable to human purposes":[21]

> Who has let the wild ass go free? Who has loosed the bonds of the swift ass, to which I have given the steppe for its home, the salt land for its dwelling place? It scorns the tumult of the city; it does not hear the shouts of the driver. It ranges the mountains as its pasture, and it searches after every green thing. Is the wild ox willing to serve you? Will it spend the night at your crib? Can you tie it in the furrow with ropes, or will it harrow the valleys after you? Will you depend on it because its strength is great, and will you hand over your labor to it? Do you have faith in it that it will return, and bring your grain to your threshing floor? (Job 39:5–12)

Even when land is given and promised to the Israelites, God makes clear that at best the people are tenants: "The land shall not be sold in perpetuity, for the land is mine; with me you are but aliens and tenants" (Leviticus 25:23). We could be better described as shepherds on the commons, raising our lambs on land we do not own, than as "stewards" of the land, with the loftiness that position has come to represent. Nevertheless, Wendell Berry uses the term *stewardship,* clearly employing it differently than Tyson, asking, "If 'the earth is the Lord's' and we are His stewards, then obviously some livelihoods are 'right' and some are not. Is there, for instance, any such thing as a Christian strip mine? . . . Is there not, in Christian ethics, an implied requirement of practical separation from a destructive or wasteful economy?"[22] If we are honestly reflecting on the biblical witness, we must answer yes, and we must also ask, "Is there such a thing as a Christian factory farm?"

If John Tyson's system of farming is based on a misunderstanding of the nature of the creation and our role within it, then what would be a proper view of creation? Who would be a proper role model of the Christian farmer? We might look to another chicken farmer, Joel Salatin.

Joel Salatin's farm in the rolling hills of north-central Virginia is a masterpiece of integrated systems. On Polyface Farm, so named to reflect the many faces present—pigs, chickens, cattle, rabbits, goats—Salatin raises his animals in mixed systems meant to maximize natural patterns. Birds once followed buffalo, eating the insects the larger animals would stir up. Salatin imitates this pattern by bringing his laying hens to scavenge where his cattle recently grazed. Everything in this system is meant to work toward the long-term development of the farm's soil and in turn its pasture, which provides most of the food for the farm's animals.

The chickens Salatin raises are purchased from a nearby hatchery. They come to his farm in boxes and are then turned out into a large brooding area with bedding and heat lamps to keep the chicks warm. After a few weeks, when the chicks have developed enough feathers to offer some protection from the elements, they are moved out into the pasture. There they are raised in mobile chicken pens that are moved daily, sometimes multiple times a day, to fresh grass. The chickens eat the grass as well as a mix of grains and protein Salatin makes specifically for his farm.

The manure from the chickens helps to fertilize the grass that cattle and chickens will come back to eat once it has grown back fully, meaning that Salatin's farm is cultivating the future rather than mining resources. Because the chickens are constantly moving at a pace that the surrounding grassland can properly incorporate, Salatin has none of the runoff or water pollution problems that Tyson growers generate. Manure is not a problem for Salatin, it is an asset—an indication that his farm is working in the pattern of nature rather than against it.

When the birds have reached slaughter weight, they are packed in crates and moved a short distance across the field to the slaughter shed, which is open air. The birds live and die in an area that can be measured by yards rather than miles, which means, among other things, that the birds do not experience the great stress of travel before their death. The birds are killed individually with a cut to the jugular vein by a skilled butcher paid a living wage, following in the pattern of humane killing that has long been a part of the Kosher and Halal traditions.

Salatin is a Christian and he speaks in clearly moral terms about what and how he farms: "We should at least be asking, Is there a righteous way to farm and an unrighteous way to farm? . . . What is a moral farm? What is a moral way to raise a chicken?"[23] Salatin's answer is farming in a way that allows the animals he raises to express their given nature—a cow her cowness, a chicken her chickenness, a pig her pigness. This means that Salatin receives the animals he has on his farm as gifts that already have a value that he did not determine, a reality he did not make. While he still raises them as animals and ends their life to feed people just as Tyson does, Salatin also believes that he has a responsibility to those animals by which he will be judged.

A *New York Times* profile of Salatin dubbed him the "High Priest of Pasture."[24] Such a title is apt, not only because it names Salatin's role within the community of sustainable agriculture but because it most properly names the role of human beings within creation that Salatin represents. If we are not to be value-making stewards of a creation without intrinsic worth, then our role might be better called priests of creation, members of the community of creation with a special role—to call creation to its life within God and to work toward its fullness.

Orthodox theologian John Zizioulas describes the priestly stance within creation in this way: "The priest is the one who freely and, as himself an organic part of it, takes the world in his hands to refer it to God, and who, in return, brings God's blessing to what he refers to God. Through this act, creation is brought into communion with God himself. This is the essence of priesthood, and it is only the human being who can do it, namely, unite the world in his hands in order to refer it to God, so that it can be united with God and thus saved and fulfilled." As Zizioulas goes on to argue, this priestly role transcends the instrumentalism of stewardship:

> The human being is related to nature not functionally, as the idea of stewardship would suggest, but ontologically: by being the steward of creation the human being relates to nature by what he does, whereas by being the priest of creation he relates to nature by what he is. The implications of this distinction are very significant. In the case of stewardship our attitude to nature is determined by ethics and morality: if we destroy nature we disobey and transgress a certain law, we become immoral and unethical. In the case of priesthood, in destroying nature we simply cease to be, the consequences of ecological sin are not moral but existential.[25]

In restoring the creation to communion with God, the priestly role of the person does not leave nature alone. Instead, and this is a key aspect of the agrarian vision, the human mediation in creation is meant to call creation to its full flourishing. What is at stake isn't simply our moral status, but our very life. That is what "sustainable agriculture" means—a way of farming that continues and draws forth the being of creation. This results in a way of farming that is full of grace rather than a competition for ever-scarcer resources.

As Norman Wirzba describes Salatin's "forgiveness farming" in contrast to industrial agriculture, this priestly role becomes clear: "Industrial food production has no room for kindness or mercy. . . . But in the dance that is Polyface Farm, the fields and the animals play off each other's strengths. There is room for failure and acceptance here because each member of the dance can be itself, and each is in harmony with the activity of the other. There is time for delight."[26]

In the industrial vision all improvements are oriented to maximizing profit. The priest, on the other hand, is oriented to the fulfillment and flourishing of the creation, which it is unable to achieve on its own: "In a priestly approach to nature we develop it not in order to satisfy our needs as human beings, but because nature itself stands in need of development through us in order to fulfill its own being and acquire a meaning which it would not otherwise have."[27] This is the difference between a farm and the wild, a garden and a wilderness.

Gary Nabhan, an ethnobiologist who works with Native people in the Sonoran Desert, has written about the desert oasis Quitobaquito. This oasis was a traditional source of irrigation for the Tohono O'odham people and as such was cultivated and cared for by them. But when the oasis was made a "preserve" as a part of the Organ Pipe National Monument, the Tohono O'odham stopped using the oasis so that it could return to its "natural" ecosystem. However, just across the US-Mexico border from the national monument, the Tohono O'odham still use a very similar oasis that they actively cultivate. The amazing thing is that the biodiversity of the oasis at the Organ Pipe National Monument has significantly declined, while the oasis still used by the Tohono O'odham thrives with a greater variety of native plants and wildlife.[28] This is a view of the role of priesthood at work developing the flourishing of the creation. No factory farm could fulfill such a vocation because it is an exploitative monoculture—*to flourish* is an impossible verb for a Tyson chicken house.

When we look at Joel Salatin's farm, or the farms of many good farmers, we see the flourishing of creation unleashed. There is an abundance not only of food for human use but also of wildlife, native plants, and at the base of it all, healthier soil. Joel Salatin's work is the work of a priest—blessing the creation through his work, which must always be work in concert with his congregation.[29] This is the kind of farming and the food grown from it that Christians are called to bless in their prayers. This is the intersection that is to come together at the table as every meal becomes a Eucharist—a sacramental thanksgiving that shows forth God's grace in the world. Over such a meal, every human person is called to be a priest before the holy elements of creation.

Such flourishing cannot come from a system of agriculture that treats the members of creation as purely utilitarian elements—that denies their life as creatures of intrinsic and independent value. John Tyson's sin is not that he has played the steward poorly, but that he has failed to name and call forth the sacredness of the life of the creation. He has failed to "honor the chickenness of a chicken," as Joel Salatin would say. The vision of stewardship is inadequate in describing the Christian's view of the human vocation within creation—it calls for management when we should be blessing. Tyson, it would seem, has failed to honor the creatures God declared as "good" and has instead entered into an economy of abstract value where goodness is made rather than accepted or recognized. Such stewardship results in difficult prayers, thanksgivings that fail to give thanks for what is truly good and confess the ways in which we have violated those givens.

The task of "saying grace" begins well before mealtime. Without such grace calling forth the flourishing of the creation, we cannot thank God for the "bountiful harvest" but only confess our abusive theft. Inevitably, many of our meals will contain within them mixed blessings—some of the sacred and some of the desecrated. The Christian tradition is familiar with such a condition—it is the nature of our life in a fallen world. The Christian task is to confess and return again to the work of blessing the flourishing of the creation.

Study Questions

1. How does the way we grow food change the way we say thanks for it?
2. Do practices of penance and confession have a role in how we eat?
3. How does the role of steward differ from the role of priest?

Notes

1. Michael Pollan, *The Omnivore's Dilemma: A Natural History of Four Meals* (New York: Penguin, 2006), 407.

2. Ibid.

3. Clement of Alexandria, "The Instructor," in *Anti-Nicene Fathers: The Writings of the Fathers Down to A.D. 325,* vol. 2, ed. Alexander Roberts et al. (Peabody, Mass.: Hendrickson, 1994), 249.

4. *Talladega Nights: The Ballad of Ricky Bobby,* dir. Adam McKay (Sony Pictures, 2006).

5. "We seek to establish sustainability and good stewardship of the Earth as central ethical imperatives of human society." Natural Resources Defense Council, "Mission Statement," http://www.nrdc.org/about/mission.asp (accessed

May 20, 2012). See also the Earthkeepers website, http://www.messiah.edu/offices/publications/the_bridge/fa1109/earthkeepers/index.html (accessed May 20, 2012).

6. Stanley Hauerwas and John Berkman, "The Chief End of All Flesh," *Theology Today* 49, no. 2 (1992): 199–200.

7. Wendell Berry, "How to Be a Poet," http://www.poetryfoundation.org/poetrymagazine/poems/detail/41087 (accessed April 24, 2016).

8. http://www.tysonfoods.com/ (accessed June 3, 2012).

9. Tyson and other companies work with independent farmers who actually raise their chickens for them. Tyson provides feed and chicks and stipulates the infrastructure needs, but the company does not pay for them. This is a way for Tyson to remove itself from the actual risks of farming.

10. Wendell Berry, *Home Economics: Fourteen Essays* (New York: North Point, 1987), 168.

11. Matthew Scully, "Fear Factories," *American Conservative*, May 23, 2005, 12.

12. Armelle Casau, "When Pigs Stress Out," *New York Times*, October 7, 2003, http://www.nytimes.com/2003/10/07/science/when-pigs-stress-out.html (accessed May 28, 2012).

13. http://www.tysonfoods.com/About-Tyson/Company-Information/Core-Values.aspx (accessed May 27, 2012).

14. John Locke, *Two Treatises on Government* (London: Whitmore and Fenn, and C. Brown, 1821), 214.

15. Quoted in Kelly Johnson, *The Fear of Beggars: Stewardship and Poverty in Christian Ethics* (Grand Rapids, Mich.: Eerdmans, 2007), 95.

16. Ibid., 98.

17. Barna Research Online, "Americans' Bible Knowledge Is in the Ballpark, but Often off Base," July 12, 2000 (website no longer available).

18. http://web.archive.org/web/20061111233246/http://www.tyson.com/Recipes/GivingThanks/Tyson_Prayer.pdf (accessed October 23, 2012).

19. Johnson, *The Fear of Beggars*, 99.

20. "Walter Brueggemann's Guide to the Bible," Christianity Podcast, November 10, 2012, http://homebrewedchristianity.com/2012/11/10/brueggemanns-guide-to-the-bible/ (accessed November 20, 2012).

21. Wendell Berry, *The Gift of Good Land: Further Essays Cultural and Agricultural* (New York: North Point, 1982), 273.

22. Ibid., 275.

23. David Grant, "Joel Salatin Advocates a Better Way to Raise Food," *Christian Science Monitor*, November 24, 2009, http://www.csmonitor.com/Environment/Living-Green/2009/1124/joel-salatin-advocates-a-better-way-to-raise-food (accessed October 23, 2012).

24. "High Priest of Pasture," *New York Times*, http://query.nytimes.com/gst/fullpage.html?res=9D0CE7DF173EF932A35756C0A9639C8B63&smid=go-share (accessed April 24, 2016).

25. John Zizioulas, "Proprietors or Priests of Creation?" (address delivered to the Religion, Science, and the Environment Symposium, June 2, 2003), 5, 7–8, http://www.rsesymposia.org/themedia/File/1151679350-Pergamon.pdf (accessed October 23, 2012).

26. Norman Wirzba, "Barnyard Dance," *Christian Century,* January 23, 2007, 9.

27. Zizioulas, "Proprietors or Priests of Creation?" 8.

28. Gary Nabhan, *The Desert Smells Like Rain: A Naturalist in O'odham Country* (Tucson: University of Arizona Press, 2002).

29. Gene Logsdon has suggested that farmers should follow a priestly model of income as well. He notes that if growing food for profit doesn't work, then perhaps "farmers should take a cue from religious orders, form a service organization and take vows of poverty in return for producing free food. The laity contribute generously for free spiritual services; they'd surely do the same for free food." "What if Farming for Profit Really Isn't Possible," n.d. (website no longer available).

Religion and Agriculture

How Islam Forms the Moral Core of SEKEM's Holistic Development Approach in Egypt

Maximilian Abouleish-Boes

In order to understand SEKEM (Egyptian for "vitality from the sun") and our approach to sustainable agriculture and holistic development, it is necessary to first understand the history, and thus the vision, of SEKEM's founder, Dr. Ibrahim Abouleish. Born in Egypt in 1937, Dr. Abouleish studied chemistry and medicine at the University of Graz, Austria, from which he received his Ph.D. in 1969. In 1977 he decided, together with his Austrian wife and two children, to return to Egypt and establish a sustainable development initiative called SEKEM, which aims to set out a blueprint for the healthy corporation of the twenty-first century.

SEKEM was the first entity to apply Rudolph Steiner–inspired biodynamic farming methods in Egypt. Its strong commitment to innovative development led to the nationwide application of biodynamic methods to control pests and improve crop yields. Since its founding in 1977, SEKEM has grown exponentially into a nationally renowned enterprise and market leader of organic products and phyto-pharmaceuticals, which are now also exported to Europe and elsewhere. The inspiration of this growth is seen in Dr. Abouleish's summary of his founding vision:

> We are often asked: "What is SEKEM?" When I say SEKEM is a holistic development initiative, people think that it is about a few social institutions that live on donations. When I say that SEKEM consists of some

successful enterprises, this is not wrong either, but again it is not sufficient explanation. I could also say: SEKEM is a political initiative, a research initiative, an education initiative. All of this is right, but still, it does not fully explain what SEKEM really is. SEKEM is a consonance; the interaction between all these different areas, similar to a symphony, where everyone contributes to the whole, knowing that the composer, conductor and the rest of the players are all together serving a divinely inspired idea.[1]

The economic branch of the SEKEM initiative includes five main companies: SEKEM for Land Reclamation, for farming and organic seedlings, fertilization, and pest control; Isis, for fresh fruits and vegetables as well as for organic foods and beverages (such as juice, dairy products, oils, spices, and tea); Lotus, for herbs and spices; NatureTex, for organic cotton and textiles (children's clothing and home wear); and Atos, for phyto-pharmaceutical products.

The societal challenges of the future—such as climate change, resource scarcity, population growth, extreme poverty, and food security—will all need innovative, problem-solving solutions. Since the beginning, the SEKEM Development Foundation and later also the Heliopolis University for Sustainable Development have contributed to solving these challenges. This is born out of the holistic model for sustainable development that SEKEM is founded on, which includes sustainably balancing the dimensions of human economic, cultural, and societal life and embedding these within the regenerative limits of local ecosystems.

The Egyptian Context and Challenges Related to Agriculture

Egypt's landmass comprises 1 million square kilometers (386.6 square miles), with a population of about 80 million people, who live on 40,000 square kilometers. Only 3 percent of the Egyptian land is arable. Agriculture takes up 3.3 million hectares (8.1 million acres), out of which 825,000 hectares (about 2 million acres) come from desert reclamation.[2] Although the government is investing in land reclamation, the area cultivated has not changed a lot, due to urban and industrial expansion at the expense of agricultural land.[3] The result is that agricultural land per person has shrunk from 923 m²/person in 1960 to 456 m²/person in 2005.[4] This implies that existing agricultural land must be preserved and protected and new desert land needs to be reclaimed in order to have the basis for an adequate food supply for the Egyptian population, which is growing annually at a rate of around 1.5–2 percent.

According to the Egyptian National Competitiveness Council's (ENCC) Food Security and Safety Sub-council, increasing food prices coupled with reduced effective earnings have increased the percentage of Egyptians living below the poverty line (extreme poverty) from 16.7 percent in 1999 to 25.2 percent in 2010.[5] The development of the agricultural sector is recognized as a prime contributor to inclusive growth and poverty reduction as well as to Egypt's food safety, given that it relies on local goods and services and as such indirectly stimulates the economy in rural areas. The urgency of the need to improve the productivity of Egypt's agricultural sector has become greater since the January 25 revolution in 2011, when the people expressed a few basic, simple demands: for bread, freedom, dignity, and social justice. The demand for bread is a reflection of the pressing need of the people to be able to provide decent food for their families, to be assured, at minimum, of guaranteed food security or income. While the January 25 uprising was a rather peaceful revolt of the people, political analysts warn of a second "Revolution of the Hungry."

The uprising materialized when the military overthrew former president Mohamed El-Morsi, who was the first democratically elected president in Egypt representing the Muslim Brotherhood. The disturbance was inevitable, in the view of many, because politicians did not take immediate concrete and significant steps to reform Egypt's economy, in particular the reformation of agricultural policies that impede Egypt's ability to ensure its own national food security. At the heart of this food security is the well-being of the smaller farmers (those owning 4.8 acres or less) who cultivate 90 percent of Egypt's agricultural land and therefore form the foundation of this sector. Hence, moving the Egyptian small farmer from subsistence levels to sustainability and then abundance is critical socially, economically, and politically.[6]

It is important to recognize that the energy-water-food nexus represents a huge challenge for sustainable development in Egypt, and agriculture is strongly related to this nexus. Sustainable desert reclamation plays a key role in addressing these issues and therefore contributing to political stability and the related transition to an authentic form of democracy. This is relevant not only for Egypt but for the whole region. Within this context of food insecurity and social and environmental challenges, SEKEM represents a viable alternative, one that builds upon a praxis of sustainable agriculture that resonates strongly with Muslim insights and teachings.

SEKEM's Holistic Development Approach

Introduction to SEKEM

To gain an appreciation for SEKEM, it is best to hear from its founder, so his voice figures prominently in this chapter. Dr. Abouleish states:

> The joy that I feel every morning, the enthusiasm for my work, the boundless love which fills my heart for all around me, brings forth a vision of the community: a community in which people of all nations and cultures work and learn in peace, and resonate together in harmony as a symphony; a community in which vocations from all walks of life, from all age groups, from all levels of consciousness, acknowledge, nurture and love the divine world and strive towards noble ideals; a living, ever regenerating community maintaining its dynamism by reaching towards the science of the spirit; a community pursuing truth and tolerance, generously offering its understanding in service of earth and man; a people where modesty and diligence prevails over vanity and comfort, and all endeavors are blessed.[7]

After living for twenty-one years in Austria, Dr. Ibrahim Abouleish decided to return to Egypt. Abouleish could not stop thinking about images and events of his previous visits to his home country. "During my trip I had become aware of the excessive use of artificial fertilizer, and I discovered far worse things about Egypt's economy, education and health situation and agriculture and trade relations than I had already learned through my discussions with Egyptians."[8] In an effort to help eradicate poverty and to contribute to the advancement of the individual, society, and the earth, he established the SEKEM initiative. The initiative began on seventy hectares of virgin, arid soil located near Belbes, around sixty kilometers northeast of Cairo.[9]

SEKEM's mission is to achieve sustainable development while assuming corporate social responsibility for human improvement. SEKEM does this by capitalizing on information technology, research and development, and active management of supply chains, quality, customer relationships, and the natural environment.[10]

Company and Practice

SEKEM was the first organization to promote pesticide-free farming techniques in Egypt, influenced by the idea that organic—specifically, biodynamic—farming enhances soil fertility, develops biodiversity, and eliminates waste.[11] Biodynamic agriculture entails a self-containing and self-sustaining ecosystem without any unnatural additions. Soil, plants, animals, and humans together create a holistic living organism. The concept developed out of eight lectures on agriculture given in 1924 by Rudolf Steiner (1861–1925), an Austrian scientist and philosopher and the founder of anthroposophy, to a group of farmers near Breslau (then located in the eastern part of Germany, now Wroclaw in Poland). Most people will agree that this is one method of sustainable agriculture. But is sustainable agriculture grasping all aspects of SEKEM's way of cultivation and application of biodynamic agricultural methods?

Most agricultural experts understand "sustainable" as "self-sustaining." The verb *to sustain* itself means only "to endure" or "to last," and *sustainable* therefore became an expression for anything that has the capacity to endure. A plastic bag that is buried in soil and does not decay is thus also "sustainable," at least linguistically speaking. SEKEM's approach to sustainable agriculture includes the regenerative powers of agriculture. In regenerative agriculture lost ecological systems can ultimately begin "regenerating" back into existence.[12]

The desert is such a lost ecosystem, at least from an agricultural perspective. The soil has lost most of its nutrients and completely lacks any organic matter and water-holding capacity. Under these conditions, agriculture is not possible unless organic matter can serve as both a foundation and a catalyst for growing plants and accommodating animals. The very fact that SEKEM's approach turns desert into living soils through the application of compost and biodynamic methods shows that desert land can be reclaimed, and thus regenerated. For more than thirty-five years, SEKEM has been building up living soils in desert land and implementing closed nutrient cycles with livestock integration and a diverse range of crops, plants, and trees. By farming without chemicals, the health of the farmers and the consumers who eat organic products regenerates. The returning wildlife also benefits, which in turn gives back to the farm by helping to keep down insect pests.

SEKEM's approach to agriculture stands in direct contrast to business-as-usual industrial agriculture. The latter relies heavily on external inputs, spreads vast areas of monocultures over the planet, and even changes plants' genetic source codes to increase resistance to pests and aid adaptation to climate change. Numerous scientific studies have shown, however, that industrial agriculture and the application of genetically modified organisms (GMOs) affects the ecosystems negatively and in fact degrades rather than regenerates them.

In 1994, SEKEM helped form the Egyptian Biodynamic Association (EBDA), which conducts research on sophisticated biodynamic cultivation practices and increases awareness of biodynamic agriculture through collaboration with other institutions. To date, the EBDA has supported the transition of over 400 farms with more than 8,000 acres to organic farming practices, including some 4,500 acres on 120 farms that were reclaimed from arid land. The EDBA was also a global pioneer in growing and producing biodynamic cotton. The presence of the EBDA helped in solving many challenges SEKEM was facing. The training that EBDA provides to farmers in international methods of biodynamic agriculture raised awareness of this method of agriculture and facilitated the acquiring of organic products certification in order to open up new market channels.[13]

Impact on Society

SEKEM has succeeded in informing public decision makers about the advantages of its agricultural approach—with remarkable results. In cotton production SEKEM has not only reduced synthetic pesticides in its own operations but, in cooperation with the Egyptian Ministry of Agriculture, cut chemical use by more than 90 percent on Egyptian cotton farms in the 1990s.[14] More recently, the leadership of SEKEM actively engaged in policy advocacy at the ENCC, with the result that sustainable agriculture was integrated into the Egypt National Competitiveness Strategy 2020, which is currently being taken forward by the United Nation's Environment Programme's Green Economy Initiative, launched in early 2013.[15]

Despite these achievements, the overall implementation of sustainable agricultural practice remains marginal. Of the total cultivated land in Egypt, only 1 percent is currently cultivated organically.[16] This can be explained by strong energy subsidies and the lack of adequate water-pricing mechanisms, conditions that do not incentivize resource efficiency in farming operations. On the contrary, huge amounts of social and environmental externalities and costs are currently outsourced to nature or future generations through unsustainable agricultural practices. Other reasons for the slow upscaling of organic and/or biodynamic agriculture is the lack of capacity of people and a missing long-term perspective of farming businesses.

It can be concluded that in Egypt, SEKEM has a highly unconventional business model that incorporates social and environmental externalities and considers these the bases for increasing competitiveness in the future. While it is a profit-making enterprise, it does not aim for profit maximization. Through a profit-sharing methodology, it shares its returns with the smallholder farmers in its network. The SEKEM Development Foundation (SDF) has launched many beneficial community development initiatives such as establishing schools and a medical center, celebrating culture and diversity, and promoting peace, cooperation, and understanding among human beings.[17]

How Islam Informs the Concept of Sustainable Development and the Moral Core of SEKEM

The Context and Role of SEKEM

For Ibrahim Abouleish, the development of SEKEM had to begin with the development of himself. During his time in Austria he interacted with European cultures and philosophies. He found a way to integrate this influence with his own Islamic faith and Egyptian cultural background. As he writes, "My spiritual inspiration came out of very different cultures: a synthesis between the Islamic world and European spirituality. I moved around freely in these different areas as if I was in a great garden, picking the fruits of the different trees. I would have felt I was restricting myself if I had had to limit myself to one way of thinking. But I felt there was enough inner space for everything in me."[18] This passage reflects very well the transcultural values that characterize SEKEM, also evident in the subtitle of Abouleish's book, *The Sekem Vision—The Encounter of Orient and Occident Changed Egypt*.

Recent demographic studies suggest that out of Egypt's population of approximately 84 million, 90 percent are Muslim (mostly Sunni), 9 percent are Coptic Christian, and the remaining 1 percent mostly includes other forms of Christianity. Overall, Islam plays a major role in Egyptian society, and most people are very religious, observing times of prayer and visiting mosques regularly. Islam means "submission to God," and this attitude is deeply embedded in the Islamic-Muslim tradition. But often the deeply felt religion is limited to the personal, private realm and does not radiate into practical life.

Like everyone else in the world, most Egyptians are in general working for financial and/or other egoistic reasons, not out of a deep connection to religion or God. Hence, SEKEM employees do not always understand the whole vision of the initiative or even fully appreciate its multifaceted activities. When SEKEM was smaller, it was easier for Dr. Ibrahim and other members of the core community to connect to each coworker and convey the bigger picture and idea behind the organization. Nowadays, with more than fifteen hundred employees, this is much more difficult. This lack of understanding can translate into a loss of operational control and quality. In an effort to convey SEKEM's overarching vision, the founder and his son, Helmy, meet weekly with different groups of employees to talk about the initiative's mission. They hope to overcome the gap between religion and practical life and implement in the workplace Islamic values and principles, while still being open to other spiritual sources. Given this challenge, in the following discussion, the reader will explore how some of the main principles of Islam are connected to SEKEM's practices, and how those at SEKEM try to bridge the gap between a personal commitment to Allah and application of religious ethics to society and the environment.

Living Examples of Islamic Principles

For Ibrahim Abouleish, the qualities of the ninety-nine names of Allah were very helpful in difficult situations he faced and provided much space for meditation. Some of the names will be presented and elaborated below with regard to key concepts and principles of Islam, taken from an inspiring book, *Islam and Sustainable Development,* by Odeh Al-Jayyousi, former vice president of science and research at the Jordanian Royal Scientific Society.[19]

He Is the One and Only: The Concept of Tawhid (Unity) According to Islamic belief, the universe was created by a supreme being who is one and unique, whom Muslims call Allah. Everything created by Allah has a purpose, which gives meaning and significance to the existence of the universe, of which humans are a part. Allah is aware of and deeply concerned with even the minutest details. The Qur'an states, "He is the First and the Last, the Evident and the Immanent: and He has full knowledge of all things" (057:003).

The concept of unity has different permutations, which are reflected in different ways at SEKEM.

Unity of individual and community. True individual value and freedom are found in doing what is good and beautiful for the community. This unity is expressed by the people who continually build up the SEKEM community. Here, the efforts of individuals gain in value as they add up in synergy, all aligned to the same vision of sustainable community building.

Unity of tradition and modernity. When Ibrahim Abouleish started to build a garden in the desert, he was working with local peasants and Bedouins who came from a very rural context and who had mastered the art of living on the desert over thousands of years. But Bedouin knowledge is based on their tribal culture; they have traditionally moved from place to place according to the availability of food for their livestock, mainly sheep and goats. This nomadic lifestyle of the old tribes did not require any systematic long-term planning for cropping and caring for soil. These ideas were introduced through settlements and traditional farming systems. So the knowledge of the Bedouins had to be adapted and transferred to this new context. Agriculture based on well water and artificial irrigation was completely new to them, as were the ways of production and some crops that SEKEM cultivated and exported. The first export crop was a local plant called *khella* (*Ammi visnaga*), also called bishop's weed, which was originally cultivated by the ancient Egyptians, who used it to treat many ailments, including urinary tract diseases and vitiligo (loss of skin pigments). Traditionally, the plant has also been used to treat respiratory system diseases such as asthma and bronchitis, and this knowledge is retained among the elder people who live around the SEKEM community, facilitating successful cultivation.

Ways of cultivating crops thus became a combination of old wisdom about local crops and farmers' time-honored manual techniques with an increasing degree of mechanization and sophisticated research on organic ways of fertilization and pest control. The melding of tradition and modernity is often realized in business cooperation, like that exemplified by MIZAN (part of SEKEM for Land Reclamation), founded in 2006 as a fifty-fifty joint venture between Grow Group Holland and SEKEM Group Egypt that offers grafting and plant cultivation services for fruit and vegetable plants.

Unity of localism and globalism. SEKEM would never be what it is today without its global network. Many friends and partners helped SEKEM to build its infrastructure and capacities. Special emphasis must be placed on the associations of SEKEM's allies in Germany, Austria, the Netherlands, Switzerland, and Scandinavia, which mainly support the cultural work of the initiative. Biodynamic agriculture itself has its origin in central Europe. The member organizations of Demeter-International work together in the spirit of an international confederation with democratic principles.[20] Likewise, SEKEM shares its knowledge and strongly advocates its approach to sustainable agriculture in other global networks, such as IFOAM, Global Compact, and the World Future Council.[21] In this way SEKEM acts locally but creates value globally by being in contact with the arts, civil society, science, the business world, and even the public sphere. The idea is also to create more favorable or at least fair conditions for organic agriculture in gen-

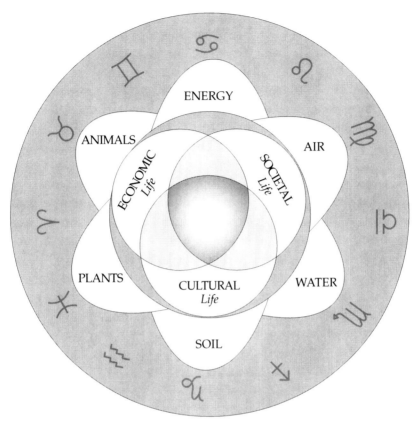

SEKEM's sustainable development flower. (Courtesy of SEKEM.)

eral, efforts often hampered by direct or indirect subsidies for conventional agriculture.

Unity of economic, societal, cultural and ecological life. SEKEM's model for sustainable development integrates different spheres of life into a holistic whole of which all parts are at the same time independent and interconnected.²²

He Is the Giver of Life: The Concept of Fitra (Natural State) Allah's creation, or the natural state (*fitra*), can be referred to as the natural equilibrium or balance (*mizan*) in which there exists full harmony of nature, people, and the built environment. In the agricultural context, it is obvious that there is a divine force that creates life. Growing plants or raising animals is more than just transforming inputs into outputs. Ibrahim Abouleish has often explained this approach to religious people who at first distrusted SEKEM's approach to agriculture, quoting the Qur'an: "It is Allah Who causeth the seed-grain and the date-stone to split and sprout. He causeth the living to issue from the dead, and He is the one to cause the dead to issue from the living. That is Allah: then how are ye deluded away from the truth?" (006:095).[23]

Ibrahim Abouleish explained about the millions of microorganisms and their work in the earth and told people that the living earth is connected to the heavens. He asked, "How can we assist this connection to the heavens? What is the essence of a plant? Is it just a seed we place in the earth, or does this seed receive life from Allah, so that out of it all the different types of plants can grow? Because Allah says: it is not you who cultivates, but Allah cultivates. He lets the plants grow!"[24]

Biodynamic farming, with its composting process and its preparations adding vitality to the soil, acknowledges its context in the larger picture, and hence is related to a divine idea. Moreover, waiting for specific star constellations before planting can be interpreted as being inspired by Allah to act correctly: "He has made subject to you the Night and the Day; the sun and the moon; and the stars are in subjection by His Command: verily in this are Signs for men who are wise" (016:012).[25] For those at SEKEM, to read and understand Allah's signs in nature, people need wisdom and knowledge.

He Is the All-Knowing One: The Concept of Ilm (Knowledge) The first surah that the Prophet Muhammad received from Allah started with the following imperative to all humans: "Proclaim! (or read!) in the name of thy Lord and Cherisher, Who created" (096:001). This underlines the value of knowledge in Islam. But according to Al-Jayyousi, the English term *knowledge* falls short of expressing all the aspects of ilm. Knowledge in the Western world means processed information, while ilm is an all-embracing term covering theory, action, and education.[26] Another surah also illustrates that God wants people to gain knowledge: "He granteth wisdom to whom He pleaseth; and he to whom wisdom is granted receiveth indeed a benefit overflowing; but none will grasp the Message but men of understanding" (002:269).

One example for a deeper meaning behind things we see in our daily life is given in Ibrahim Abouleish's book where he explains the relation between the glucose cycle of plants, the course of the sun, and prayer times:

> What had the Prophet seen?—A tree illuminated with light. He saw the living form of the tree, the living processes occurring inside the tree. These processes are reflected physiologically as follows in the plant: when dawn nears, the plant produces glucose until the sun reaches its zenith; this process then ends once the plant's shadow is the same length as the actual plant. This is in the afternoon. The exact time changes depending on the time of the year. Between the afternoon and sunset the plant transports the glucose it has previously produced to all its organs. After sunset a third process begins, the plant starts transforming the glucose into active substances. This process ends when darkness sets in and during the night the plant starts to grow. These processes correspond to the five positions of the sun, and allow the internal life of the plant to appear. Prophet Mohammed recommended we think about Allah and turn to the supersensory at these five times throughout the day. Because of the relationship of the five prayers with the sun's course and the rhythms of the plant, around the world the praying human connects to the cosmic processes.[27]

Harmonizing human action with cosmic processes and recognizing that dealing with nature has a divine context are embedded in SEKEM's approach to agriculture. To grasp this demands a high consciousness and deep human understanding. This is the reason why SEKEM puts so much emphasis on education and cultural life in general. Moving arts or painting does not directly result in the transfer of specific knowledge, but it makes the human soul much more receptive and hence more open to nature and knowledge (ilm). The aim of conducting artistic sessions with farmers and agricultural engineers and workers is—besides the positive effect on teamwork and group dynamics—to lead them to experience something new and to develop their sense of openness and curiosity. The objective is to transfer this experience to their work context, to view their world in a different light. Islam refers to the process of consciousness development as a form of beauty and excellence (*ihsan*).

Employees at SEKEM's farm in Minya engaged in artistic activities. (Courtesy of SEKEM.)

He Is the Beautiful One: The Concept of Ihsan (Beauty and Excellence)
Ihsan refers to excellence and inner beauty in the form of the conscious evolution of individuals, organizations, and society (*ummah*). It also entails continual development as well as value and knowledge creation for all humanity.[28]

This realization of beauty is embedded in the approach of SEKEM, which aims to develop individuals, organizations, and societies at large. Nature provides great inspiration for that. According to SEKEM's vision statement, "In nature, every organism is independent and at the same time systemically inter-connected to other organisms. Inspired by ecological principles, representing the wisdom of nature and the universe, we continuously strive to gain and sustain a harmonious balance between [this polarity] and to integrate [it] into our development."[29]

The development model of SEKEM is holistic and rooted in sustainable agriculture. Many challenges need to be overcome on the way to excellence, and much time and many generations will be required to see the completion of this work. Most people, on first hearing Ibrahim Abouleish's ideas before they bore fruit, thought it was an impossible mission, that organic desert greening could never succeed. However, it did, and it still does—but only because of the people who believe in it. It is that inner transformation that must take place before the outer transformation can succeed—the ability to see a vision and the will to bring a holistic and divine idea "down to earth," into reality. The importance lies in the way toward that vision, not on the final achievement.

He Who Feels Responsibility: The Concept of Khalifa (Stewardship of the Earth) In Islam, broadly speaking, the human is viewed as a trustee or steward to ensure that all resources, physical and human, are utilized in a reasonable, equitable, and sustainable manner in order to uphold and develop the natural state. This is seen in the Qur'an: "We did indeed offer the Trust to the Heavens and the Earth and the Mountains; but they refused to undertake it, being afraid thereof: but man undertook it; He was indeed unjust and foolish" (033:072).

The application of chemical fertilizers and pesticides is harming the environment and destroying the fragile balance of the ecosystem. It represents a mechanistic worldview that does not realize the bigger context in which agriculture takes place. The perverse large-scale and destructive forms of "modern" agricultural systems do not consider ecological and ethical values. That is one of the reasons why SEKEM is convinced of the importance of its sustainable approach to organic and biodynamic agriculture, which excludes chemical fertilizers and pesticides and the use of GMOs.

Stewardship (khalifa), moreover, implies the social equality and dignity of all human beings—regardless of skin color, gender, or social status—a cardinal element of the Islamic faith. Within such a worldview, the right attitude toward others is not that "might is right," the struggle to serve one's own self-interest, or "the struggle to survive," but rather mutual cooperation, in order to develop the entire human potential. Second, resources are a trust (*amanah*), provided by Allah, whereby the human being is not the primary owner but is just a trustee (*amin*). So resources are for the benefit of all, for the well-being not just of oneself and one's family but of the community at large.[30] This community encompasses the whole web of life, from soil organisms to insects to plants to wildlife. This is explained in the Qu'ran: "There is not an animal [that lives] on the earth, nor a being that flies on its wings, but [forms part of] communities like you. Nothing have We omitted from the Book, and they [all] shall be gathered to their Lord in the end" (006:038).

Acting responsibly in the interest of communities implies also that resources should not be wasted. This is also made clear several times in the Qu'ran, such as here: "O children of Adam! Wear your beautiful apparel At every time and place Of prayer: eat and drink: But waste not by excess, For God loveth not the wasters" (007:031). Waste recycling is the underlying principle of composting in general. At SEKEM organic waste produced by all SEKEM firms and some additional green waste from the surrounding farms is used to produce high-quality compost, some that is applied on the organization's own fields and some that is sold to external customers. This and many other practices of organic and biodynamic agriculture represent responsible human action and justice toward nature and society. Because justice is such an important cornerstone in Islamic belief, it is presented separately below even though it is strongly connected to the principle of khalifa.

He Is the One Who Brings Justice: The Concept of Adl (Justice) Justice (adl) corresponds to cosmic, ecological, and human justice as well as harmony with the universe. Therefore, ethical governance is the cornerstone for attaining and sustaining progress and thus a good life (*hayat tayebah*) or, in other words, sustainable development.

Human brother- and sisterhood, regardless of faith, would be a hollow concept without socioeconomic justice. There are no fewer than a hundred different expressions in the Qur'an embodying the notion of justice, which is placed nearest to piety. For example, "O ye who believe! Stand out firmly for justice, as witnesses To God, even as against Yourselves, or your parents, Or your kin, and whether It be [against] rich or poor: For God can best protect both. Follow not the lusts [of your hearts], lest ye Swerve, and if ye Distort [justice] or decline To do justice, verily God is well-acquainted with all that ye do" (004:135).

The concept of adl has different dimensions and can have different meanings. As well as social justice, which basically means upholding the dignity and freedom of the individual, adl stands also for economic justice, such as one can find in the Fairtrade practice in the context of agriculture. All SEKEM operations are conducted according to Fairtrade principles, even though the official logo does not appear on every product; because accreditation is so expensive, it is sought only based on customer demand. All SEKEM's supplying farmers get a premium price for their organic or biodynamic products and have long-term contracts with regulated pricing mechanisms, constituting the basis for a fair relationship. In the Qur'an there are passages describing this: "O ye who believe! Eat not up your property among yourselves in vanities: But let there be amongst you Traffic and trade by mutual goodwill: Nor kill [or destroy] yourselves: for verily Allah hath been to you Most Merciful!" (004:049).

Then there is the principle of ecological justice, which refers back to the responsibility of humans toward the earth and nature. Islam teaches that species including plants and wildlife are in a state of prayer, referring to the intrinsic value of nature and the ecosystem. Prayer in its wider meaning is about fulfilling a task in the bigger context of God's creation. The harm of any species means that we are disrupting the symphony of life and silencing worshipers.[31] According to the Qu'ran, "Seest thou not that it is Allah Whose praises all beings in the heavens and on earth do celebrate, and the birds [of the air] with wings outspread? Each one knows its own [mode of] prayer and praise. And Allah knows well all that they do" (024:041).

The reader was introduced to SEKEM as a holistic development initiative based on the principles of organic and biodynamic agriculture. Islam is a main pillar of the cultural context of SEKEM and its spiritual source of inspiration. Another pillar is European culture and philosophy, which were authentically integrated in the personality of Ibrahim Abouleish and SEKEM in general.

Islam offers a story of the origin of the universe and human beings. Humans are trustees (*khulafa*) who must ensure that all resources are used in a sustainable manner. Reading the Qur'an informs the mind and soul that our natural and social capital are interconnected and interdependent. SEKEM builds on that insight and provides a practical example of Islamic values in agriculture, at the same time building communities, developing humans, and conducting economic value creation.

SEKEM offers a role model for addressing burning societal issues, such as the energy-water-food nexus, in an authentic way born and rooted in its cultural context. It shows that a compatibility with religious values does not mean following the stereotype of closed-mindedness and conservatism—rather, embracing the ethics of Islam is to be future oriented and sustainably competitive.

Study Questions

1. This chapter advocates "harmonizing human action with cosmic processes and recognizing that dealing with nature has a divine

context." What does this mean, and how does it relate to other chapters in this book?
2. What is biodynamic agriculture?
3. What are the six living examples of Allah, and how do these inform the sustainable agriculture undertaken by SEKEM?
4. Why might SEKEM be considered a highly unconventional business model when compared to other businesses, including agricultural ones, in Egypt?

Notes

1. Ibrahim Abouleish, *Sekem: A Sustainable Community in the Egyptian Desert* (Edinburgh: Floris Books, 2005).
2. Lowell Lewis, "Egypt's Future Depends on Agriculture and Wisdom," 2011, http://www.egyptianagriculture.com/ (accessed June 8, 2013).
3. Ibid.
4. FAO, Egypt Country Profile, http://www.fao.org/countryprofiles/index/en/?is03=EGY (accessed June 14, 2013).
5. World Bank, Egypt Country Profile, http://data.worldbank.org/country/egypt-arab-republic (accessed June 8, 2013).
6. "ENCC (2013): Improving Egypt's Competitiveness through Policy Reform: A National Campaign for an Integrated Solution for Food Security," Egyptian National Competitiveness Council, March 2013.
7. Abouleish, *Sekem*.
8. Ibid.
9. Tarek Hatem, "Sekem: A Holistic Egyptian Initiative," UNDP, 2008, http://growinginclusivemarkets.com/media/cases/Egypt_SEKEM_2008.pdf (accessed June 8, 2013).
10. Ibid.
11. Ibid.
12. Regenerative agriculture dates back to the late 1970s and 1980s, when it was introduced by the Rodale Institute in the United States.
13. Hatem, "Sekem."
14. World Economic Forum, "Redefining the Future of Growth: The New Sustainability Champions," 2011, http://www3.weforum.org/docs/WEF_GGC_SustainabilityChampions_Report_2011.pdf (accessed June 5, 2013).
15. H. M. Abaza and P. M. Boes, "Sustainable and Green Growth for Egypt," in *The 8th Egyptian Competitiveness Report: A Sustainable Competitiveness Strategy for Egypt* (Cairo: Egyptian National Competitiveness Council, 2012).
16. International Trade Center, Egypt Country Profile, http://www.intracen

.org/trade-support/organic-products/country-focus/country-profile-egypt/ (accessed June 20, 2013).

17. More information on all of these initiatives can be found on the websites of SEKEM (www.sekem.com) and Heliopolis University (www.hu.edu.eg).

18. Abouleish, *Sekem*.

19. Odeh Al-Jayyousi, *Islam and Sustainable Development: New Worldviews* (Farnham, U.K.: Ashgate/Gower, 2012).

20. www.demeter.net.

21. www.ifoam.org; www.unglobalcompact.org; www.worldfuturecouncil.org.

22. For a detailed discussion and explanation of these spheres, please refer to the annual report on sustainable development from SEKEM at www.sekem.com/rsd.

23. Abouleish, *Sekem*.

24. Ibid.

25. Maria Thun and Matthias Thun, *The Biodynamic Sowing and Planting Calendar, 2013* (Edinburgh: Floris Books, 2013).

26. Al-Jayyousi, *Islam and Sustainable Development*.

27. Abouleish, *Sekem*.

28. Al-Jayyousi, *Islam and Sustainable Development*.

29. SEKEM Report on Sustainable Development, 2012, www.sekem.com/rsd (accessed June 17, 2013); SEKEM Report on Sustainable Development, 2011, www.sekem.com/rsd (accessed June 17, 2013).

30. Al-Jayyousi, *Islam and Sustainable Development*.

31. Ibid.

Tohono O'odham *Himdag* and Agri/Culture

Tristan Reader and Terrol Dew Johnson

Prelude: O'odham New Year, 2014

The soft swishing sound of a gourd rattle breaks the silence before dawn in the Sonoran Desert. Resolving itself into a gentle beat, it is joined by the powerful voice of Michael Enis as he sings:

Wañ ge s-wepegim o babahi	They are ripening red
Wañ me:k o s-ma:masim o kekiwa	Standing so visible in the distance
Ia ñ-da:m g ce:wagî da'iwuñe	Cloud rises over me
Heg o g jewed wa'usida	And moistens the earth.

This traditional saguaro harvest song wakes—and blesses—more than one hundred people camped at the Alexander Pancho Memorial Farm. Bleary-eyed, people load themselves and their *kukuipad*—the fifteen-foot-long harvesting sticks made from the inner ribs of the cactus itself—into the backs of pickup trucks. As they bump and roll their way down a dusty dirt road into the mountains, the dawn light creates a silhouette of Baboquivari Peak, home of I'itoi (Elder Brother). So begins a new year for the Tohono O'odham—the People of the Desert. So continues the *himdag*—their ways and lifeways.

Each summer—under the blistering sun of Arizona's Sonoran Desert—Tohono O'odham families harvest *bahidaj*—ruby-red fruit atop the towering saguaro cacti (*Carnegiea gigantean*). The fruit is cooked over mesquite wood fires into *sitol*, a mahogany-colored syrup used to make the sacred wine for the *jujkida* (rain ceremony). When families harvest fruit with kukuipad, they are "pulling down the clouds." When families sing, dance, and pray during the jujkida, they are calling the rains to return to their land. After everyone has become satiated with saguaro wine, they "vomit up the clouds." The rains begin the Tohono O'odham New Year and make possible the lives of the plants, animals, and People of the Desert.

> Come together!
> You shall see this thing, which we have always done
> And what must truly happen.
> Because we have planned it thus and thus have done.
> Right soon, indeed, it will happen.
> It will rain.
> The fields will be watered.
> Therein we shall drop the seed.
> Seed which bears corn of all colors;
> Seed, which grows big.
> Thus we shall do.
> Thereby we shall feed ourselves;
> Thereby our stomachs shall grow big;
> Thereby we shall live.
> —Tohono O'odham call to the jujkida[1]

Tohono O'odham culture is truly an *agri/culture*. In the very first sentence of her 1946 book *Papago Indian Religion*, anthropologist Ruth Underhill noted that the Tohono O'odham are "an agricultural people, thoroughly adapted to farming and food gathering in an arid country."[2] Songs, ceremonies, and legends all have their roots in the food system. So too the roots of economic exchange, health and wellness, and social relations are deeply embedded in that food system. Tohono O'odham—the People of the Desert—have a profound understanding that "you are what you eat." Food is health. Food is culture. Food is community. Food is the basis of a sacred economy of what you grow, what you harvest, and how you share. Understanding the traditional Tohono O'odham food system is essential to understanding the O'odham Himdag—the Desert People's lifeways. Moreover, there is an inextricable link between the living practice of the himdag and the physical, spiritual, social, and economic well-being of the Tohono O'odham.

We begin with an exploration of the traditional Tohono O'odham food system. We then investigate how Tohono O'odham culture and spirituality emerge from the people's relationship with the food system. We describe the devastation to the traditional food system and the resulting damage to the cultural, spiritual, and physical well-being of the Tohono O'odham community. Ending on a hopeful note, we describe recent efforts to revitalize the himdag and Tohono O'odham foodways.

Economies of Abundance: The Traditional Tohono O'odham Food System

The Tohono O'odham have lived in the heart of the Sonoran Desert—North America's largest and hottest—for countless generations. As Castetter and Bell explain, "The Papago, living in a very arid area with meager agricultural possibilities, had imposed upon them a seminomadic life, planting and caring for their fields in the rainy period, and wandering about exploiting to the utmost the wild animals and vegetable resources of the land the rest of the year. For a period of at least three months, while hunting was done by men, wild plants were gathered by women."[3] Traditionally, they combined a series of well-adapted strategies for producing food in such an arid land. This food system supported a local economy, maintained the people's physical well-being, and provided the material foundation for Tohono O'odham culture—the O'odham Himdag. Four strategies drove the annual semi-migratory life of the Tohono O'odham: *ak chin* farming, harvesting wild foods, hunting, and sharing.

Ak chin farming. With the arrival of the summer monsoon rains, Tohono O'odham families would return to their "field" villages. "The run-off of the mountain areas collected in streams with well-defined channels, which, on reaching the large un-dissected alluvial basins, lost their momentum and spread out in broad sheets over the level land for a square mile or more. Places where this spreading occurred were known to the Papago as Akchin, 'arroyo mouth.'"[4] These *oidag* (field) villages were the summer homes of families and communities as they nurtured a set of traditional crops specifically adapted to the hot, arid conditions. Among these staples are tepary beans—the most heat- and drought-tolerant legume in the world. O'odham corn varieties included the fastest-maturing corn known—just sixty days from germination to maturity. Desert-adapted varieties of squash, amaranth, and fiber plants (for example, devil's claw) took advantage of the summer rains.

In 1919, the Tohono O'odham had more than sixteen thousand acres of ak chin fields under cultivation, yielding more than 1.8 million pounds of beans and three hundred thousand pounds of corn.[5] With a population at the time estimated at under six thousand people, indigenous agriculture provided important nourishment for the Tohono O'odham community, as it had for countless generations.

Harvesting wild foods. In 1933, Tohono O'odham elder Maria Chona reflected on her life harvesting wild foods: "All the year round we were watching where the wild things grew so we could pick them. Elder Brother [I'itoi] planted those things for us. He told us where they are and how to cook them. You would not know if it had not been Given. You would not know you could eat cactus stems and shake the seeds out of weeds. Elder Brother did not tell the Whites that. To them he gave peaches and grapes and wheat, but to us he gave the wild seeds and cactus. These are good foods."[6]

As Maria Chona and other Tohono O'odham women know, the desert provides a wide variety of wild foods to be collected, preserved, and eaten. Historically, the diversity of these foods provided for a varied and healthy diet. Foods from cacti, agaves, and succulents included saguaro cactus fruit, prickly pear fruit and pads, cholla flower buds, banana yucca fruit, and agave hearts. The seedpods of mesquite, paloverde, and ironwood trees were collected and ground into flours. Wild greens such as amaranth and lamb's quarter were an ephemeral but highly valued part of the diet. "Whether superabundant and storable, or as rare and perishable as manna, these plants helped shape and succor cultures within the Sonoran Desert. They served as calories, cures, and characters in tribal legends."[7] Even salt was harvested while on pilgrimage to the waters of the Sea of Cortez. Historically, these wild foods made up as much as two-thirds of the annual diet, especially in years when drought made ak chin farming even more difficult than normal.[8]

Hunting. The animals of the desert also provided an important source of nutrition. The hunting of rabbits, deer, packrats, quail, and many other desert dwellers was a significant supplement to the foods grown in O'odham fields and plants collected in the desert. From the communal rabbit hunts that often accompanied the jujkida to the multiday deer hunts in the mountains, hunting provided much-needed protein and variety to the traditional Tohono O'odham diet.

Sharing and "trade." The traditional food economy of the Tohono O'odham was primarily based upon a system of gift giving and reciprocity. "Poor as the Papago country was, its economics were those of abundance. Papago did not hoard property; they did not quarrel about land boundaries; they were constantly giving, as though from an inexhaustible supply."[9] Traditionally, there was nothing that capitalism would recognize as trade. Instead, there was a system of gift giving, ceremonial exchanges, gambling, and sharing. Through this process, "what available wealth there was [was distributed] fairly evenly throughout the entire group. Human beings invested in one another rather than in material goods. To do otherwise was to invite isolation, and there is no place in the desert for islands."[10]

Himdag as Agri/Culture: Communal, Individual, and Adaptable

Before the arrival of the Jesuit missionary Eusebio Francisco Kino in 1686—and for many more generations for most Tohono O'odham—there was no word for "religion" in the O'odham language. Instead, all aspects of collective life and identity are expressed with the term *himdag*. Himdag has no literal translation into English, nor does it have any official definition among the Tohono O'odham. At its most basic level, himdag encompasses the lifeways of the Tohono O'odham. It includes kinship and social relationships, ceremonial and cultural expressions, foodways, and values such as balance, industriousness, and obligation. "*Himdag* ultimately involved all things, all people, and all action. It concerned the rights, dignity, and propriety—as well as the potential power—of human being, of ceremonies, of living things, and of such life-giving entities as salt, the sun, and the ocean (from whence rains come)."[11]

As expressed through himdag, Tohono O'odham culture and religion are—at their roots—an agri/culture. The *symbolic* culture of the Tohono O'odham—ceremonies, legends, songs, dances, language—all grow out of the fertile soil of the *material* culture—the food system. "Food ceremonies form the more conspicuous part of Papago religious practices. They were carried on according to a yearly schedule for the making of rain, the growth of crops, the cleansing of food, the celebration of the harvest, and, once in four years, 'to keep the world in order.'"[12] Ruth Underhill described three related realms of Tohono O'odham religious expression—communal ceremonies, ceremonies for individual power, and the use of power (primarily for healing). Each of these communal ceremonies—and even some of those for individual power—is a fundamental expression of Tohono O'odham food culture.

Communal Ceremonies

As described above, perhaps the most sacred of Tohono O'odham cultural practices is the jujkida. Designed to "sing down the rain" that makes agriculture—and life itself—possible in the dry desert, the saguaro harvest and wine ceremony served as a cornerstone of O'odham ceremonial life, marking the beginning of the New Year. After four nights of singing sacred song cycles, communal dancing, and prayers by *mamakai* (medicine men), the people become satiated with the sacred saguaro wine and cleanse the body by "vomiting up the clouds"; thus are the monsoon rains summoned back to the parched land.

> Drinking the [saguaro cactus] wine was a sacramental communion bringing good and purification to the communicant. It brought a supernatural presence and an ability for every Indian to call for rain. Singing brought power and was a supplication to the supernatural for rain. . . . Papago wine-drinking filled the participants with liquid from the skyward reaches of the Saguaro which symbolically united with the earth when vomiting occurred. . . . Vomiting wine on the ground completed the cycle, returning moisture to the ground that had been taken up by the roots of the Saguaro, and stored in its body during the time of drought, incorporated into the fruit, and made into rain.[13]

Once the rain arrives, it is time for planting the ak chin fields with traditional crops and engaging in ceremonies to promote growth.

Blessing songs—"song magic"—were sung by individuals, families, and whole villages at virtually every stage of the farming process, including the Corn Woman songs:

> Songs begin.
> At the west they begin.
> Thereby the corn comes up.
> Upward its heart [the ear] is stretching.
> At the west they begin.
> *Hiro a' áro!*
> Songs are ending.
> At the east they are ending.
> Thereby squash comes up.
> In a row its heart [fruits] are standing
> At the east they are ending.
> *Hiro a' áro!*[14]

There were songs for planting corn: for when it sprouts, when it is one foot high, when it is knee high, when it begins to tassel, when the tassel is fully developed, and when the corn is ripe. There were songs for squash, beans, and other crops.[15] This song magic was a powerful and ubiquitous part of daily O'odham agricultural life.

Just as there were songs and rituals for each part of the farming process, so too were there hunting ceremonies—fasting and ritual oratory before the deer hunt, songs before butchering and eating the animal, for cleansing the hunter after the kill, and for the deer itself.

> What man is it
> who has killed the deer?
> Oh, it is the wind man, the deer he has killed!
> Yonder he struggles to carry it on his head.
>
> Do you see, yonder coming toward us
> The tail of the black deer?
> Poor thing brought here;
> Poor thing brought here;
> Poor thing brought here;

Poor thing, walking yonder on the mountains
It nearly fell.[16]

"The intervillage athletic meet which was one of the most important occasions of Papago life," notes Underhill, "ranks fundamentally as a food ceremony."[17] The event took place in the weeks following the harvest of ak chin fields. Runners summoned people and challenged villages far and wide. This celebration of the bounty of the new crop provided a mechanism for gift giving and gambling (sharing of the harvest), and brought luck (in the form of rain) to the winners. Athletic competitions such as *toka* (a women's game similar to field hockey) and *songiwul* (a men's long-distance kickball race) were accompanied by song cycles that told stories, including wind songs, ocean songs, cloud songs, and crane songs. At the end of the competition, as the people headed back home and into the dry winter, they sang:

Then we, having sung,
Go back whence we came.
Watch! It will not be long
Till the earth grows very green.[18]

The *wikida* ceremony renews and cleanses the world. Symbolic representations of each village's fields were blessed by corn dancers and singers:

Hí híanai hu!
Here on my field
Corn comes forth.
My child takes it and runs,
Happy.
Here on my field
Squash comes forth.
My wife takes it and runs,
Singing.[19]

A final communal cultural event—not mentioned by Underhill and not precisely a ceremony—is *storytelling,* including the sharing of legends, during the winter. Whether it be the tale of the Milky Way made up of white tepary beans scattered across the sky by coyote, the emergence of the first saguaro cactus, the hunter captured by a deer, or countless other stories, these legends serve many functions. They teach ethical behavior and cultural norms, reinforce ecological and agricultural practices, build group identity, and entertain.

Ceremonies for Individual Power

Expressions of himdag include several ceremonial practices that were less communal and more related to the powers—spiritual and physical—of the individual. Rituals and/or purification related to warfare, the pilgrimage to the Sea of Cortez to collect salt and seek the power of the ocean, the killing of an eagle, and the puberty dance for girls are among these central parts of himdag. Although they do not as directly affect foodways, the ways in which they manifest themselves nevertheless reflect their importance to concerns related to food. The sea—an incredible source of power for a desert people—not only bestowed strength and blessings upon the young men brave enough to seek its power; it also provided salt to season the food of the entire community. Fasting was an important part of puberty rites. Even the experience of warfare was related to that necessary ingredient for growing food, rain:

> Papago war ritual never mentions the glory of war. . . . War is conceived as an onerous duty. The descriptions are not of battles but of the visions which precede and follow them. Actual fighting is expressed by metaphor and disposed of in one hurried sentence. Emphasis is all on the obligation to the community, the steps taken to work the fighters up to action, and the final reward. This last is neither prestige, booty, nor satisfied vengeance, but that sine qua non of a desert people—rain. This is the attitude of busy agriculturalists who fought, as the Papagos say they did fight, only in self-defense.[20]

Thus, even the most private expressions of ceremonial life—the emergence of a girl into womanhood, healing from the trauma of war, finding one's power and place in the world—were expressed and experienced in the context of foodways.

Disease and the Use of Power for Healing

Singers, herbalists, and mamakai (medicine men) had distinct powers and roles in the diagnosis and healing of disease. Although the causes of disease were varied—ceremonial lapses, disrespect to an animal, intrusion of a foreign object into the body (often by sorcery)—each reflected a central premise: disease was the result of veering away from himdag. "The Piman Way [himdag] was reinforced among its followers by the threat of diseases which could befall one who transgressed against it—the principal role of the shaman [or medicine man] being to diagnose and direct cures for such illnesses."[21]

As we shall come to see, this concept of disease being the result of the move away from himdag returns with devastating consequences for the Tohono O'odham community.

Himdag and Christianity

Since the arrival of Padre Kino and other Catholic missionaries—and the later arrival of Protestant ones—Christianity has played a significant role in the life of the Tohono O'odham community. From the references to saints in the Spanish names for villages (for example, San Pedro, Santa Cruz, San Miguel, and Santa Rosa) to the roadside crosses that mark the location of fatal traffic accidents, from the Catholic chapels in almost every village to the shrines to the saints in the homes of elders, the Christian influence is evident. Each autumn, many Tohono O'odham make the pilgrimage to the shrine of Saint Francis Xavier in Magdalena, Mexico.

Yet for the majority of Tohono O'odham, there is no contradiction between Christianity and himdag. They are able—quite literally—to hold both. Alexander Pancho, the grandfather of coauthor Terrol Dew Johnson—was a makai, a medicine man. During his blessings, he would hold a crucifix in one hand and an eagle feather in the other. The blending is not a form of syncretism in which the two worldviews are merged; rather, many O'odham are simply able to embrace them as a duality with separate truths and meanings. This ability to simultaneously adhere to two distinct worldviews plays out in more than the religious realm. Today, young Tohono O'odham speak of walking in two worlds: the world of mainstream American culture—formal education, hip-hop music, shopping at Walmart—and traditional Tohono O'odham culture—legends as a form of pedagogy, singing for rain, harvesting saguaro fruit. The two worlds are not separate; they are simply peoples' reality.

This ability to walk in two worlds emerges from an approach to himdag that is nondogmatic. In contrast to the Roman Catholic faith, "there is no one authorized version of Papago culture. Each of the more than 20,000 Papagos is a carrier of it although some people certainly are better versed or more interested in the Native traditions than others. The culture thrives through variations. This is a strength for its survival for it is not dependent on a few individuals."[22]

In addition, there is a plasticity in himdag that responds to context. A legend told in one village today may emphasize certain themes, with the storyteller well aware of how it relates to the recent experiences of listeners; elsewhere, on a different day, the story may differ in focus and tone, responding to that social context. Himdag is more like a river—flowing and changing, yet maintaining its integrity and identity—than a desert pond—stagnant and eventually drying up in the heat.

In contrast to Christianity—particularly its Protestant forms—himdag is less about faith and belief and more about ways of living. In discussing traditional ecological knowledge (TEK), Deborah McGregor (Anishnabeg) identifies this approach as based on action rather than belief:

> TEK is a "way of life"; rather than being just about knowledge of how to live, it is the actual living of that life. One way of looking at the differences between the Aboriginal and non-Aboriginal views of TEK is to state that Aboriginal views of TEK are "verb-based"—that is, action-oriented. TEK is not limited, in the Aboriginal view, to a "body of knowledge." It is expressed as a "way of life"; it is conceived of as being something you do. Non-Aboriginal views of TEK are "noun-" or "product-based." . . . TEK is viewed as a thing rather than something you do.[23]

Similarly, himdag is something you *live* rather than something you *believe*. Himdag is enacted in daily life.

Damage to the Traditional Food System

According to a Tohono O'odham origin story, "The reason Elder Brother planned this was that some day in the future . . . the rains would not come down all over the earth very often, only once in a while, and the crops that the people raised wouldn't be irrigated anymore by rain water."[24] Eighty-year-old Dolores Lewis carried a bucket of water from his adobe home to the squash, tepary beans, and black-eyed peas growing in his field a few hundred meters away. In the traditional farming village Ge Oidag (Big Fields), the floodwaters no longer arrived in the fields. Years before, the US Army Corps of Engineers had built an earthen dam to prevent flooding in the village. Yet those floodwaters were the entire reason for the village's existence; without them, farming was impossible. In 1997, Mr. Lewis was the only person in the village—perhaps on the entire Tohono O'odham Nation—still growing traditional crops on family land. When he died a few years later, all such production ceased.

How could such devastation to the traditional Tohono O'odham food system have come to pass? After all, within Mr. Lewis's lifetime, the Tohono O'odham had been almost entirely food self-sufficient. As late as the 1920s, the community utilized traditional methods to cultivate over twenty thousand acres in the floodplain of the Sonoran lowlands. By 1949, that number had declined to twenty-five hundred acres. By 2000, that number was certainly less than five. In 1919, the Tohono O'odham produced 1.8 million pounds of tepary beans; in 2001, fewer than one hundred pounds were produced. At the same time the once common practice of collecting and storing wild foods declined in an equally dramatic way.[25]

The causes of this loss of food sovereignty are complex and multifaceted. They include (but are not limited to):

1. *Federal work and assimilation policies.* US governmental policies removed families from their own lands to work as migrant laborers in commercial cotton fields. Entire families would leave their communities for several months each year, during which it became impossible for many families to plant, tend, and maintain their own fields or to collect wild foods.

2. *World War II.* During the war, most young Tohono O'odham men—those responsible for much of the farming and many parts of ceremonial life—served in the military, leaving many fields empty of crops and ceremonies unperformed.
3. *Commodity food programs.* In order to keep market prices high, surplus commodity programs and price supports became federal policy. For reasons ranging from ignorance to charity, commodity foods were imported to Native communities. Most of these foods were processed. In fact, the prevalence of commodity lard and white flour across American Indian reservations resulted in the pan-Indian "food" known as fry bread. The introduction of and easy access to processed foods led many people to alter their diets and decrease the amount of traditional foods produced and consumed. These same social programs—although well intentioned—often created dependency relationships where self-sufficiency had previously existed. Over the course of a few short decades, the Tohono O'odham community went from being almost entirely food self-sufficient to being almost entirely food dependent.
4. *Termination and relocation programs.* In the mid-twentieth century, the American Indian relocation program, which moved Native people off reservations and into urban areas, had a devastating effect upon Tohono O'odham culture and food systems.
5. *Boarding schools.* Large numbers of Tohono O'odham children were forcibly placed in boarding schools, where they were prohibited from speaking their language and practicing their culture. This meant that they did not learn the skills necessary to farm in the desert or how to collect, process, and cook wild foods.
6. *Drought and climate change.* As the Tohono O'odham origin story warned, environmental factors (such as a lowering of the water table due to nearby development) combined with misguided governmental flood-control efforts to make water sources for traditional agriculture even more scarce than normal in the arid Sonoran Desert. Drought has only accelerated in this era of climate change.[26]

It must be noted that virtually none of these causes of the decline in Tohono O'odham agriculture were the result of malice. As Bernard Fontana has observed: "The Papago Reservation is in many ways a graveyard of good intentions, an enclosure of mutual misunderstandings. The all too pervasive habitual drunkenness, violence, suicide, incarceration, and economic dependency [to which we add diabetes and disease] among the reservation's Indian people are not there because anyone wanted it that way. They are partially the result of terrible mistakes made by people who have been well meaning: federal bureaucracies, school teachers, missionaries, and others whose job it has been to 'improve' the lot of Papagos."[27]

The problems are also the product of the Papagos' own miscalculations concerning their long-range interests. Time and again, Papagos have traded away the self-respect that comes with economic and social independence, sacrificing control of family and community affairs for short-term gains. Too frequently they have relinquished their right to make significant decisions on their own behalf. Rather than formulating their own plans and setting their own goals, Papagos often find themselves in the position of having to respond with a "yes" or "no" to plans formulated for them. These involve healthcare, housing, community food programs, educational programs, and economic development. When individuals or communities can no longer devise and carry out their own schemes, the collective psychological and sociological impacts are devastating.[28]

The Army Corps of Engineers did not set out to make it impossible for Dolores Lewis and others in Ge Oidag to continue farming as they and their ancestors had for countless generations. The government saw a village that was flooded on a regular basis and set out to help. Yet today, standing on top of the earthen dam the Army Corps of Engineers built on the eastern edge of the village looking upstream, one sees lush, wild greenery that has been watered and nourished by monsoon floodwaters. Turning around, one faces many hundreds of acres of abandoned fields—empty of beans, corn, and squash, empty of dancing feet, and empty of song.

Agricultural Decline, Damage to Himdag, and Disease

Himdag in Decline

Given the nature of Tohono O'odham agri/culture, it should come as no surprise that the near-total destruction of the Tohono O'odham food system led to dramatic damage to the vitality of himdag.[29] We have described the ways in which public ceremonies for making rain, growing crops, hunting, and celebrating the harvest were the central communal expressions of himdag. By the time Delores Lewis dropped the last seed in the ground, only a tiny portion of the O'odham community participated in any of these sacred rites. In 1994, the most central of these—the jujkida—was held in only two villages, and many elders complained that it had stopped being sacred and had simply become another reason to drink heavily. There were *no* communal ceremonies for the growth of plants or hunting. Although a few women played toka during the annual rodeo and fair, there was no athletic competition or harvest feast.

The reason for this decline is relatively simple: almost no Tohono O'odham produced their own food. Grocery stores and federal commodity programs—rather than the desert—had become the source of food. The endangerment of Tohono O'odham *symbolic* culture followed directly the decline in *material* culture. People did not stop planting the fields because the ceremonies were dying out; the ceremonies began to die out when people stopped planting their fields. After all, if you never plant crops, the importance of rain is diminished. And as we have seen, the bringing of rain, growth of crops, and relationships with animals were the heart of the himdag.

Given the connections between the Tohono O'odham food system and cultural traditions, it is virtually impossible to imagine a scenario in which the O'odham language, songs, ceremonies, stories, dances—the O'odham Himdag (Desert People's Way) itself—can remain a vital and living tradition without the rejuvenation of the food system. However, as cultural practices are reconnected with their material foundation, they once again take on their central role in the community. Putting ceremonies, songs, stories, and other cultural traditions back into their original context strengthens them.

Disease

"Wheat flour makes me sick! I think it has no strength. But when I am weak, when I am tired, my grandchildren make me gruel out of the wild seeds. That is food."[30] When ninety-year-old Maria Chona spoke these words seventy-five years ago, she provided an accurate assessment of the effects of nontraditional foods on the health of the Tohono O'odham. The near-total destruction of the Tohono O'odham food system has proven himdag's concept of disease to be accurate within the Tohono O'odham community. For centuries, traditional desert foods—and the physical effort it took to produce them—kept the Tohono O'odham healthy. The introduction of processed foods, however, changed all of that, leading to unprecedented rates of adult-onset diabetes. As recently as the early 1960s, type 2 diabetes was virtually unknown among the Tohono O'odham. Today, more than 60 percent of the adult population develops the disease, the highest rate in the world. Diabetes has even begun to appear in children as young as seven.

Several scientific studies have confirmed what Maria Chona already knew: traditional Tohono O'odham foods help regulate blood sugar and significantly reduce both the incidence and effects of diabetes. Two primary attributes of traditional Tohono O'odham foods are thought to contribute to reductions in both the incidence and severity of diabetes. First, the soluble fiber, tannins, and inulin in one group of traditional foods (which includes mesquite bean pods, acorns, and tepary beans) help reduce blood sugar levels, slow sugar absorption rates, and improve insulin production and sensitivity. Second, a complementary group of traditional foods (including prickly pear fruits and pads, cholla cactus buds, chia seeds, and mesquite bean pods) contains mucilaginous polysaccharides gums that slow the digestion and absorption of sugary foods. Combined, these two categories of desert foods provide a low-glycemic diet that can help prevent and reduce the effects of diabetes. When this diet is further supported by increased physical activity, the positive health outcomes can be quite dramatic.

The Tohono O'odham understanding that it is divergence from himdag that leads to disease has been proved correct on more than a metaphorical level. The move away from himdag—Tohono O'odham agri/culture—has indeed brought disease and devastated the health of the community.

For twenty years, there has been a growing movement within the community—supported by Tohono O'odham Community Action (TOCA)—to revitalize all aspects of himdag, including kinship and social relationships, ceremonial and cultural expressions, foodways, and values such as balance, industriousness, and obligation. It is only through this return to himdag that the Tohono O'odham community once again can achieve cultural vitality, abundance, and wellness.

Seeds of Change: Efforts to Revitalize Tohono O'odham Food Systems

According to Christine Johnson, an elder from Nolic Village, "Every year, I sang the songs that called down the summer rains. But this year, I had a garden filled with devil's claw and corn, melons and squash. This year, I sang for them. This year, I sang like I really meant it."[31]

Since 1996, Tohono O'odham Community Action has worked with hundreds of community members like Christine Johnson to revitalize himdag.[32] Taking a holistic approach, these efforts have been built on the premise that himdag is "wisdom from the past creating solutions for the future." The goal has been to revitalize Tohono O'odham symbolic culture (such as songs, ceremonies, and legends), values (like balance, industriousness, and mutual obligation), physical and spiritual wellness, and the food system in which they are rooted.

Working with community members, tribal programs, health providers, and others, TOCA has accomplished much, including (but hardly limited to):

- Since 1996, TOCA has sponsored over 150 events to harvest traditional wild foods (saguaro fruit, cholla buds, prickly pear, wild greens, and mesquite beans). The three-day Saguaro Harvest Camp has been held for eighteen years in a row, attracting over two hundred attendees annually.
- Starting with a half-acre community garden in 1996, TOCA's farming efforts have grown to include a 3-acre traditional ak chin farm and a 120-acre production farm. TOCA's farms now produce over one hundred thousand pounds of traditional crops annually.
- In 2009, TOCA opened Desert Rain Cafe to serve healthy, delicious, traditional Tohono O'odham foods to the community.
- TOCA has worked with local schools to introduce traditional foods to the school lunch program and to provide nutrition education to over one thousand students.
- The annual Harvest Feast and Traditional Games Festival brings over four hundred community members together to celebrate food traditions.
- Each winter, TOCA sponsors several Storytelling Nights in which the legends and oral history of the Tohono O'odham are once again

shared. Today young adults as well as elders are telling the people's stories.

TOCA's work, however, is about more than *what* is done; it is also about values and principles:

- O'odham Himdag: "Wisdom from our past creating solutions for our future"—The *O'odham Himdag* (Desert People's Way) guides TOCA as it seeks to develop culturally appropriate solutions to the challenges that confront our community. By drawing upon our heritage and cultural traditions we are able to create lasting solutions and a stronger community.
- Community Assets: "See our resources, not just our needs"—The Tohono O'odham community already possesses many of the assets that are necessary to create a healthy and sustainable community. TOCA encourages people to take stock of these community assets in order to develop indigenous solutions, rather than focus on the problems while importing "solutions" from the outside. The wisdom of elders, the enthusiasm of young people, the richness of the land, the centrality of extended families, and a desire to create a healthier community all contribute to the capacity to create solutions that will be culturally based and sustainable.
- Encourage Community Self-Sufficiency: Social programs on the Tohono O'odham Nation have too often created dependency relationships that destroy the sustainable structures that have previously supported the people. In response, TOCA attempts to re-empower the community to become increasingly self-sufficient, utilizing and promoting an empowerment model of organizing, rather than the service model so prevalent in most social service programs.
- Context is Crucial: Strengthening the material roots of O'odham culture—It is not enough to simply preserve cultural activities, such as ceremonies, songs, and stories. The material basis out of which these cultural practices grew must also be maintained. A ground blessing dance loses much of its power when only ever performed for an audience in an auditorium rather than in the fields where the O'odham have planted for generations. TOCA works to redevelop the material foundation of Tohono O'odham culture.[33]

Afterword: A Year at the Farm

The Saguaro Harvest Camp—the New Year—at TOCA's Alexander Pancho Memorial Farm is just the beginning of a cycle of events that mark the Tohono O'odham calendar, the life of the people, and himdag. Shortly after the harvest, the first rains arrive. A week later, as the monsoon rain clouds gather strength over Baboquivari Peak, a small group—farmers and singers, elders and youth—gathers to bless the ground, seeds, and people who will bring a bountiful crop to this land. In the coming weeks of toil under the desert sun, seeds will be planted, water will flow, and plants will be nurtured. Throughout the summer, the farm grows more than food; it grows a new generation of Tohono O'odham farmers in an agricultural training program. The produce provides local schools with traditional foods for their lunch programs.

In the fall, the farm hosts the O'odham Haicu Ha-Cicwidag, a harvest feast and traditional games festival. Before the dusty, rough-and-tumble toka competition, the women gather to sing the toka songs taught to them by elder Ena Lopez. Boys remove their shoes to run through the fields, kicking the wooden ball, in the songiwul race. *Gins* sticks—made of the inner wood of the saguaro cactus—fly through the air as men gamble on the outcome. The soft, rhythmic thumping of hundreds of feet accompanies song after song in a traditional round dance. And in the traditional "economics of abundance," food from the farm—more than one thousand pounds of squash, tepary beans, corn, and melons—is given as a gift to all in attendance.

Months later, on a winter's night, a fire crackles and sends sparks floating into the sky to join the countless stars. At first, the children are cold and look for warmth in those flames. Soon enough, however, they forget the cold, captivated by the legends, told as they have been since the beginning. They are frightened by Ho'ok, the child-eating witch who lived and died in a cave south of the farm. They laugh at the antics of Ban (Coyote). And they learn about how it was children who saved their people from a terrible flood. In the morning, they will return to school, eat tepary beans at lunch, and tend the school garden.

Himdag is not something you believe. Himdag is not something you know. Himdag is a life that you live, a life that continues for Tohono O'odham—the People of the Desert.

Study Questions

1. What do the authors mean when they write, "There is an inextricable link between the living practice of the himdag and the physical, spiritual, social, and economic well-being of the Tohono O'odham"?
2. What is himdag, and how are the material and "religious" interrelated for the Tohono O'odham? How does this influence their foodways, past and present?
3. What are key foodway ceremonies for the Tohono O'odham, and what ethics are encoded in ceremonial rituals, myths, and songs? How do these relate to contemporary goals of sustainable agriculture?
4. What traditional crops were grown where you currently live, by whom, and what rituals, ethics, and ceremonies were involved? Is it possible to return to this type of agriculture where you live? Why or why not?
5. What has caused the loss of traditional Tohono O'odham foodways, and what has been the impact physically, culturally, and environmentally? What movements have emerged to ameliorate these conditions, and what role does religion play in this?

Notes

1. Ruth Underhill, *Papago Indian Religion* (New York: Columbia University Press, 1948), 45–46. Here we present most songs and ritual oratory translated into English. Until the revival of the jujkida and other ceremonies in the late 1990s, many of these songs had fallen from regular use. Over the past twenty years, the coauthors have been active in Tohono O'odham Community Action, a community-based group that has been instrumental in reintegrating the Tohono O'odham Himdag into daily life.

2. Ibid. The name Papago was used by almost all non-Tohono O'odham until the 1980s. Most sources claim it to be a seventeenth-century Spanish name meaning "bean eaters." Any Spanish speaker, however, would be puzzled by this explanation. This linguistic mystery was resolved when Dr. Karen Smith Wyndham identified that the word was actually of Basque, not Spanish, origin. In Mark Kurlansky's *Basque History of the World* (New York: Penguin, 2001), he notes that the Basque pejorative *babazorra* means "bean eater." The Basques were actually the first Europeans to eat maize imported from the Americas; other Europeans fed corn to livestock. Perhaps these food alliances contributed to the naming of this part of the New World as Nueva Vizcaya, or

New Basquelands, on many maps of the time. In this chapter, we use Tohono O'odham; however, when referencing the writings of others, we have chosen to keep their use of Papago. This serves as a reminder of the ways in which the views of the community by "outsiders" may have differed from those of the people themselves.

3. Edward Franklin Castetter and Willis Harvey Bell, *Pima and Papago Indian Agriculture* (Albuquerque: University of New Mexico Press, 1942), 40.

4. Ibid., 168. Despite Castetter's recognition that the Tohono O'odham continued to farm in this way, he referred to the Papago in the past tense. Castetter, Underhill, and many other non-Native writers of that and subsequent generations based their writings on information provided by living people in living communities still engaging in these practices, yet they succumbed to the cultural bias of seeing Natives as a dead—or at least dying—people. As we shall see, in the case of the Tohono O'odham, this is far from the truth.

5. Ibid., 78.

6. Ruth Underhill, *Papago Woman* (Prospect Heights, Ill.: Waveland, 1979), 39.

7. Gary Paul Nabhan, *Gathering the Desert* (Tucson: University of Arizona Press, 1986), 5.

8. Castetter and Bell, *Pima and Papago Indian Agriculture*, 56.

9. Ruth Underhill, *Social Organization of the Papago Indians* (New York: Columbia University Press, 1939), 90.

10. Bernard L. Fontana and John Paul Schaefer, *Of Earth and Little Rain: The Papago Indians* (Tucson: University of Arizona Press, 1981), 50.

11. Ibid.

12. Underhill, *Papago Indian Religion*, 18.

13. Frank S. Crosswhite, "The Annual Saguaro Harvest and Crop Cycle of the Papago, with Reference to Ecology and Symbolism," *Desert Plants* 2, no. 1 (1980).

14. Ruth Underhill, *Singing for Power: The Song Magic of the Papago Indians of Southern Arizona* (Tucson: University of Arizona Press, 1993).

15. Underhill, *Papago Indian Religion*, 78.

16. Ibid., 109.

17. Ibid., 116.

18. Ibid., 124.

19. Ibid., 147.

20. Ibid., 165.

21. Fontana and Schaefer, *Of Earth and Little Rain*, 50–51.

22. Ruth Underhill, *Rainhouse & Ocean: Speeches for the Papago Year* (Tucson: University of Arizona Press, 1997), 7–8.

23. Deborah McGregor, "Traditional Ecological Knowledge and Sustainable Development: Towards Coexistence," In *In the Way of Development: Indigenous Peoples, Life Projects and Globalization*, ed. Mario Blaser, Harvey Feit, and Glenn McRae (London: Zed/IDRC, 2004), 79.

24. Gary Paul Nabhan, *The Desert Smells Like Rain: A Naturalist in O'Odham Country* (Tucson: University of Arizona Press, 2002), 41.

25. The 2000 and 2001 numbers are estimates by the authors based on field observations made at the time.

26. This analysis is based on Tohono O'odham Community Action reports found in the archives of the organization. An earlier version of this analysis also appears in Daniel Lopez, Tristan Reader, and Paul Buseck, "Community Attitudes toward Traditional Tohono O'Odham Foods" (2002), 9.

27. Fontana and Schaefer, *Of Earth and Little Rain*.

28. Ibid., 117–18.

29. The analysis in this section is based on field work/participatory research conducted by the authors between 1996 and the present.

30. Underhill, *Papago Woman*, 37.

31. Personal communication with the authors.

32. This section is based on the founding documents, various reports, and other items found in the archives of TOCA. More information can be found at www.tocaonline.org.

33. http://www.guidestar.org/profile/86-0883222 (accessed May 10, 2016).

Conclusion

Searching for Annapurna; or, Cultivating Earthbound Regenerative Abundance in the Anthropocene

Pramod Parajuli

I'd like to begin the conclusion to this book by quoting the leading agrarian essayist and champion of sustainable agriculture, Wendell Berry. "Wherever they live, if they eat, people have agricultural responsibilities just as they have cultural responsibilities. Eating without knowledge is the same as eating without gratitude. What's the use of thanking God for food that has come at an unbearable expense to the world and other people? Every eater has a responsibility to find out where food comes from and what its real costs are, and then to do something to reduce the costs."[1] Within the rubric of searching for Annapurna—a mountain range in the Nepalese Himalayas in whose shadow the author grew up—this chapter invites readers to imagine how we could create an abundant and resilient food system. Because climate disruption has already affected the current functioning of agriculture and food systems and its impact will only increase in the future, my urging in this concluding chapter is to all religious traditions and worldviews covered in this volume as well as those that lie outside it. The chapter also situates the discussion in a context in which we have overstepped planetary boundaries in some crucial areas of earthly processes. The earth is already feverish due to the rise in temperature. We have lost biodiversity to an extent that is affecting agriculture and food systems as well as the viability of regenerating life on earth. High levels of carbon, nitrogen, and phosphorous are floating in the wrong places in our environment. This has been proved to be detrimental to the well-being of the biosphere—the thin membrane that is our homeplace, our habitat. Warning signals also abound in global freshwater supply/use as well as the extent of changes in land use.[2] Equally serious is the collapse of the aquatic food chain due to ocean acidification, which is having a ripple effect, bringing pressures to terrestrial agriculture.

Yet I urge readers to consider how by practicing agriculture and growing our food in particular ways, we can not only feed ourselves and the world but also significantly address the issues of greenhouse gases, carbon dioxide emissions, carbon sequestration, soil depletion, and nitrogen runoff. I refer to those particular ways as "earthbounding." If we follow principles of earthbounding (more on this concept later), we can grow healthier food in smaller land areas, sequester carbon and other greenhouse gases, and also save water. In order to bring these new opportunities and priorities into practice, human social systems (which I refer to here as the ethnosphere) and natural systems (the biosphere) have to work in close collaboration and in mutually beneficial ways. As Wendell Berry urges us in the quote with which I opened, I am seeking to enhance mutualistic synergies between three spheres—biosphere, ethnosphere, and what I call the learningsphere.

Although modern life has created tremendous negative ecological footprints to support itself, through reimagination and redesign of our agriculture and food systems, could we make our food footprint beneficial to the biosphere as well as the ethnosphere? One way is by earthbounding carbon in soils, where it should be. By not letting carbon (and other greenhouse gases) get skybound, where it should not be, and bringing it to soil, forests, and water bodies, we could make our food and agricultural economy serve as a "sink" rather than a "source" of greenhouse gases. Yet this is not just a technical fix between carbon and soil; it requires shifts and transitions in the very equation between the human household and the earth household. This also requires that we pay close attention to some of the salient features of indigenous and peasant households, and some 570 million smallholder farmers and their agrarian practices that are based on "regenerative reciprocity."[3]

Remaking the Human Household within the Earth Household

Over twenty years ago, in 1994, I asked my mother, Parvati Parajuli (1923–2008), what "environment" meant to her. The official word used in Nepali development lingo for environment, *prayabaran,* did not resonate with her. She did not accept that there was environment as such but, she said, as a *kisan* (peasant) householder, she interacted with what is around her in three ways. First, she was supposed to use (*chalaam*) nature; it was the right and obvious thing to do. The second verb she used was *bachaam,* by which she meant to use nature prudently, carefully, and thoughtfully. As there was no refrigerator or electricity at her farm in a village in Chitwan District, my mother harvested her vegetables, fruits, and even some grains three times a day, for breakfast, lunch, and dinner. The idea was not to harvest from the fields and gardens unless it was necessary to meet her family's immediate needs. This way, she said, she could allow the plants and fruits to grow and ripen until the last minute. The third verb she used was *jogaam,* which to her meant that there are certain things that should never be harvested by humans—other species should have them. Examples are the sacred groves around water sources, or sacred trees such as *tulsi* (basil), which used to be planted in front of each house, and other trees, *peepal, bar,* and *swami,* which are planted in public places. In the ripe season, all these trees are visited by numerous birds, bats, and other pollinators. All these were outside the purview of her chalaam and bachaam.

Accordingly, my parents built at least five *chautaris* (resting places/platforms in public pathways) and planted peepal, bar, and swami trees on them. Except for rituals and celebrations, even in emergencies, these trees were spared being cut for fodder or harvested for timber. Such trees were the peasant householder's contribution to the well-being of the larger cosmos, including birds, insects, and a host of pollinators. However, collecting dry wood and fallen branches and leaves from these trees was allowed. For my mother, humans had to be beneficial to nature while actively interacting with it every day and every moment.

It is with some hesitation that I attempt to make sense of what aspects of my mother's culture and life informed such practices. I should mention that my mother's views represent and resonate with most of the Hindu peasants of the Himalayan foothills of Nepal. However, I do not want to universalize these as Hindu views, nor do I want to characterize them as merely Hindu views. I propose that they emanate from the particular situatedness of my mother: in a subsistence peasant economy. As several chapters in this volume also suggest, irrespective of what religions they espouse, subsistence peasant economies are deeply embedded in local ecologies.

As I closely observed my mother's daily life over eighteen months during 1993–1994, I found that she was not a silent witness of /to what unfolded in nature but an active partner in its living and reliving, life as well as death. She did not consider her acts of chalaam, bachaam, and jogaam as transgression into nature's domain but embraced them as her essential *dharma* to behave according to the right conduct. My mother was not acting alone in this peasant universe; in the Nepali peasant calendar, rituals were performed, festivities were organized, or celebrations were held at least one-third of the days in a year. The life cycle of trees was celebrated as much as the human life cycle of birth, coming of age, maturing, and death. As mentioned by Dr. Adhikari in chapter 5, family members of the dead disbursed hundreds of different kinds of seeds in a forest, on temple grounds, or around remote lakes in the Himalayas. Known as *satbeej charne*, this act allowed living family members to repay nature for providing all the requirements for the life of their dead kin. My parents had done satbeej charne four times in their lives, one dedicated to each of their parents. In what I call my mother's mode of "dharmic right livelihood," humans were not entitled to take more than they needed from nature. At the least, we must give back as much as we take. Better, we give back more than we take from nature. Yet this was not a gesture of Hindu nonmateriality or otherworldliness, as it is often understood or interpreted in Euro-American academia. I suggest, for lack of other words, that it was a sort of utilitarian outlook but one with an ethic of mutualistic reciprocity embedded in it. My mother believed such acts would pile up her *punya* (merit) and thus balance the potential *paap* (misconduct) she might have done due to negligence or ignorance. In other words, it was her way of negotiating with nature as well as an indication of how she considered herself a partner in the making and unmaking of life.

I am at a loss to accurately name such a practice, this delicate balancing act of my mother. I can simply say that in her peasant worldview/cosmovision, both humans and nature were alive, co-players in the cycle of life and death, rebirth and renewal. Both sides, nature and culture, were considered sacred in a continuum. Nature was not a given or inert but an equal—and often more powerful—entity with which we have to negotiate the terms of exchange. My mother once told me: "We were children of nature, for sure, but Mother Nature needed our caring and nurturing, too. It was not a one-way but a two-way process." As she said, "You cannot clap with one hand; it requires both hands." She added: "And both hands need to be willing to clap." In the Andes, they express it as *criar y criar* (nurture and be nurtured). I equate such worldviews and practices with what we would call in the present lingo being "regenerative."

What Is My Dharma in the Anthropocene Household?

Fast-forward to 2015 and let us look at the life of one of my mother's sons, Pramod, the author of this chapter. In the title of this chapter, I mean to evoke the notion of Annapurna to imagine an abundant and resilient/regenerative food system. As I remember it clearly, during my childhood, I was given the impression that the Annapurna mountain range was already saving and would save my life, when needed. Now, after the April 25, 2015, earthquake in the Nepal/Himalayas, I am also open to accepting that Annapurna's well-being is simultaneously dependent upon me—my conduct and my dharma. The melting of the mountain range's snow cover now depends on how I live my life, whether my life's ecological footprint is beneficial or harmful to the mountain.

My ecological footprint is pretty heavy. A professor at Prescott College in Prescott, Arizona, I live on a ten-acre mini-ranch where I raise one dozen Churro-Navajo sheep and half a dozen goats. I eat on the high end of mostly organic and locally grown foods. I identify myself as a locavore—a person who prefers to eat local food as much as possible. I drive a car, traveling on average nine thousand miles a year. I also travel by air. I use a laptop computer more than eight hours a day.

Such a lifestyle is a phenomenon of the Anthropocene epoch. Perhaps in my mother's lifetime, Annapurna was abundant while her modest needs only lightly impacted the life of Annapurna. But as for her son, as a result of participating in carbon-intensive lifeways, my presence is mostly harmful. I then ponder: compared to my mother, have not my duties as a householder increased in volume and intensity? Thus, it would not be enough for me to do just what my mother used to do; my responsibilities are multiplied several times in scope and scale. To what extent have they multiplied? I question myself and my readers: at this juncture of the Anthropocene epoch, will regenerating the traits of ancient religious traditions such as Hinduism, Islam, or Judaism suffice, or do we have additional thinking to do? Do we have additional responsibilities and obligations to uphold? In short, what does a dharmic life look like in the Anthropocene epoch?

Opinions vary widely about whether the human species is capable of changing the destructive course we are on, whether we can make the U-turn needed to again become resilient and abundant, as I am suggesting here. Some, including my coeditors of this volume, believe that overall, the human species would rather jump off a cliff than change its present course. Many estimate, perhaps accurately, that we have about thirty to fifty years left to make that fateful turn—or collapse. I tend to trust that as a species, humans are hardwired to learn, unlearn, and relearn. If that is the case, to borrow an apt term from Bill McDonough and Michael Braungart, I should be capable of "upcycling" my beneficial aspects while downsizing my harmful aspects vis-à-vis the wellbeing of Annapurna.[4] In this chapter, I am seeking to find some direction for myself and my fellow members of the Anthropocene epoch. What could be our dharma at this juncture, irrespective of the religion we may have been born into? We did not have a choice of what religious tradition we inherited, but should we not have the choice to find appropriate spiritual (or even secular) pathways suitable to the Anthropocene reality in which we live?

Whatever our religious origins might be, one crucial aspect of this new dharma for me is to deeply rethink the very idea of what food is, where it comes from, how it should be grown, how it should be consumed, and by whom, where, and in what ways. I call this a rethinking of every step in the continuum of "soil to supper and back to soil." As author Michael Pollan suggests, "What would happen, if we were to start thinking about food as less of a thing and more of a relationship?"[5] I would like to add that food constitutes a relationship not only between humans as food growers and eaters but also among all other species and entities that join us in this drama—the earthworms, the pollinators, the sun, the moon, the water cycles, the biogeochemical cycles, the soils, the seeds, and other numerous actors. It is time to bring along our companion species, which garden and farm themselves and help us humans do our farming and gardening.

Given how seriously planetary boundaries are already compromised, our designs for finding solutions must match the amount of damage we have done. If that means to create room for multiple theologies and even new theosophies, the author is ready to welcome them. As William McDonough and Michael Braungart suggest in their recent book, *The Upcycle,* humans do not have an energy problem. Energy is abundant. What we have is a "materials in the wrong place" problem. Could this be pertinent in assessing current dilemmas in agriculture and food systems as well? In other words, let me propose: we do not have a food problem; what we have is elements of our food systems in the wrong places. For example, the world currently produces sufficient calories per head to feed 12–14 billion people. Yet around 1 billion people suffer from chronic starvation; and another 2 billion plus are malnourished, poorly fed, or obese. Globally, now there are two and a half times more overweight people than undernourished ones. More than one in three adults are overweight. Ironically, some 70 percent of these starving and malnourished people are small farmers or agricultural laborers. But these same people still grow some 70 percent of the global food supply. This tragedy is largely due to the preference given to industrial agriculture, whose main tenets are monocultures; the significant use of external inputs in the form of industrially grown seeds, chemicals, and pesticides; biofuels; concentrated animal feed and feedlots; and excessively meat-based diets. Furthermore, some 40 percent of the food grown goes to waste between the farm and the dinner plate. This wasted food alone could feed people who lack the money to buy food. It has also been proposed that if people were to embrace a vegan diet, the impact on lands, soil, water, and human health would be immediate and far-reaching.[6] Changing to a vegan diet would entail a huge stretch in my own thinking about what I eat and how. Although the idea is tempting, I do not anticipate a majority of people switching to a vegan diet anytime soon.

Whatever path we take—vegan, vegetarian, or omnivore—I argue that we urgently need a system in which the top 10 percent is not able to hold the planet and the rest of humanity hostage. Just 11 percent of the global population (also the eaters at the top of the food chain) generate around 50 percent of carbon emissions, while the 50 percent at the bottom of the economic pyramid create only 11 percent. Nitrogen use and abuse is one example. Currently, humanity is using nitrogen at four times the globally sustainable rate. Nitrogen is overused to produce fertilizers for crops and animal feed, to manage manure and human sewage, and in the burning of fossil fuels and biomass. Not each of the 7 billion humans on earth is contributing equally to this. For example, the European Union—home to just 7 percent of the world's population—uses 33 percent of the globally sustainable nitrogen budget simply to grow and import animal feed.[7] High-income countries, comprising 16 percent of the global population, account for 64 percent of the world's spending on consumer products and use 57 percent of the world's electricity, while the people in the bottom one-third of the economic pyramid are hungry, live in the dark due to lack of electricity, and do not have access to goods and services. They are not the priority of the present global order. Most of the 7 billion people on our planet are denied opportunities to meaningfully participate in either the ethnosphere (human-made economic and social systems/institutions) or the biosphere (earthly assets and processes, ecosystem services).

As I will elaborate later, another irony is we are releasing greenhouse gases into the air, whereas these should be earthbound, sequestered in soils. Rather than building soil health and wealth by enriching carbon and organic matter, we have been eroding topsoil at an alarming rate. By some accounts, the United States has lost 75 percent of its topsoil. Global trends are not better. In one estimate, the world's cultivated soils have lost 50–70 percent of their carbon stock.[8]

It seems pertinent to note that we must rethink not only the science of agriculture and food but also our deeply held beliefs, worldviews, and religions. What if the new guiding principles for a sustainable, resilient, and abundant food system were a mix of the secular and sacred? What if we were to cultivate ritualistic agriculture in which there is regenerativity, reciprocity, mutuality, and reverential attitudes at every step in what I call the "soil to supper/sustenance and back to soil" continuum? What if, rather than decrying the overwhelming human influence on the biosphere, we turn that influence into a beneficial engagement, one that enhances instead of depleting it? What if we turn the presence of 7.2 billion (and growing) humans into a blessing rather than a curse? Could we imagine a scenario in which 10 billion humans (estimated for 2050), while feeding themselves, will also feed their fellow species through a food system that is efficient and ecologically effective? In such a system, everybody eats and everybody feeds. A significant share of humans will grow food suitable to local and regional environments. The system could be abundant, one that upcycles. The system could be permanent and regenerative, one in which there is no waste because the so-called waste becomes nutrients for the next cycle. As discussed below, in this system soil and carbon are earthbound and are enhanced to create abundance. Rather than relying on machines and petrochemicals to grow our food, the new system could be labor intensive and knowledge intensive for the primary food growers and eaters, such that everybody has meaningful work. This would be the most effective way to address rural poverty, unemployment, and the trend of small-scale peasants and farmers migrating to cities. Could we imagine a food system in which peasants, farmers, and a new generation of food growers have the means, resources, and needed support to practice beneficial agriculture? What if these primary food growers earned enough income from beneficial agriculture that they did not go hungry but could afford to eat nutritious and healthy foods? Why not? This can be assured if they eat what is grown on their own farms or locally. If there is surplus, they could sell and/or exchange it in the regional food markets. What incentives might young people need to embrace the idea that a labor-intensive and a knowledge-intensive agriculture is their way forward?

One encouraging trend is that the idea of who may be earthbound, who may live on and from the land, is moving to the city. I will refer to this development as the making of the urban *chacareros*. Chacareros are peasants in the Andes and the Amazon region who work on their agricultural plots, known as *chacras*. Imagine a scenario in which at least one-third of the food consumed in urban centers will be grown within the city limits. By some estimates, some 200 million city dwellers produce food for themselves and for the urban market, accounting for 15–20 percent of total global food production. What if the rest of the needed food could come from surrounding agricultural areas, forests, and water bodies? A recent report by Landscapes for People, Food and Nature, *City Regions as Landscapes for People, Food and Nature,* profiles over fifty initiatives that are trying to erase the traditional urban-rural divide. Instead, they are creating food and a land-based urban-rural continuum with mutually reinforcing, reciprocal relationships, characterized by flows of resources, people, and information.[9]

Additionally, imagine that in such a system soils are rich, dancing with life, while abundant pollinators go about their business. Building on the various religious and spiritual traditions of the food growers, a sacred thread could bind all elements of life in such a system. As we can see from the features mentioned, in such a system nature and culture are coupled and co-constitutive in the drama of food and agriculture. Moreover, with careful design, such a food system could achieve not only abundance but beauty and elegance as well.[10]

Definitely, what I have proposed is a tall order. There are valid reasons to proceed carefully, accepting such a vision with a degree of caution. Yet without a vision, we will be motionless, remaining stuck in a dysfunctional system that has proven to be unworkable.

From Unitary Dogmas to Pluriversity

As we recognize that the earth has become feverish due to increasing temperature and that we have entered the Anthropocene epoch, unitary notions of development or modernity are being challenged worldwide. So is the notion that markets are the only way to circulate food. Trying to articulate what my mother's cosmovision might have been, I prefer to talk about the human economy as an equation between two households—the earth household and the human household. How Wendell Berry, farmer, poet, and philosopher, defines economy resonates with me. He does not mean *economics* but *economy,* "the making of the human household upon the earth; the arts of adapting kindly the many, many human households to the earth's many eco-systems and human neighborhoods."[11] He adds:

> Of course, everything needed locally cannot be produced locally. But a viable neighborhood is a community; and a viable community is made up of neighbors who cherish and protect what they have in common. This is the principle of subsistence. A viable community, like a viable farm, protects its own production capacities. It does not import products that it can pro-

duce for itself. And it does not export local products until local needs have been met. The economic products of a viable community are understood either as belonging to the community's subsistence or as surplus, and only the surplus is considered to be marketable abroad.[12]

South American indigenous peoples and peasantry have also proposed the notion of *buen vivir* instead of development. The Spanish idea has many roots; to consider its expression a unitary voice would be another way of doing violence to the very plural ways this view has emerged among indigenous nations of the Amazon. While for the Aymara this means *suma qamaña*, for the Quechua *sumak kawsay* approximates it. For the Guarani *ñandereco* could be the word for an ideal life. There are fundamental differences between various indigenous languages, but it is interesting to note that they all share a concept of an ideal life.[13]

Realigning the Biosphere, Ethnosphere, and Learningsphere

My rather upbeat proposal to create a resilient and abundant food system is predicated upon the human species having the capacity to learn. But what kind of learning is required? I propose an idea of pluriversity, a pluriversity for resilience and abundance. For example, the Annapurna that I grew up looking at during my childhood refers to a mountain range in the Nepalese Himalayas that protected my village. As I was told by my elders in the Himalayan foothills of Nepal, the meaning of the word *Annapurna* is "Full Harvest of Grain/Food." I was also told that although a mountain is usually considered a male being in the Nepali cosmovision, Annapurna is thought of as a goddess. This goddess is especially endowed with qualities to ensure a full harvest of food and crops. I remember my father, a precise astrologer, looking up to the snow on the southern face of Annapurna to predict whether and what kind of harvest we would have that year. He could forecast a dry year or if there would be excessive rain or hailstorms. Accordingly, he also advised what kind of seeds to plant and where, when, and in what elevations.

I regret that I did not take the opportunity to learn from my father how he made these predictions. This is part of the loss my generation of Nepalese suffered in our pursuit of modern education. In my case, I detoured from my ancestors' vernacular knowledge to obtain a Ph.D. from Stanford University and seek a life in US academia. The time seems ripe for me to embark on an inquiry, asking not only what my fellow South Asians think about the Annapurna mountain range but what and how does Annapurna herself think? Do the snows and mountain also feel, think, and communicate? How can humans learn to listen to them? Is the recent earthquake in Nepal just a geological event in a preset cycle or is it the Himalayas' response to the stresses of the Anthropocene? What can the subjectivities and voices of the mountains, rocks, and glaciers tell us? I will be searching for answers to these questions in the years to come.

For my peasant community, the goddess Annapurna embodied the entanglement of the natural world, here the mountain, and the human collectivity living around it. The goddess Annapurna partakes both of the mountain and of the human ability to speak, act, feel, and reciprocate. Both goddess and mountain, Annapurna exemplifies the intraspecies entanglement of the biosphere and the ethnosphere. For Wade Davis, who proposed this term, *ethnosphere* is "the sum total of all thoughts, dreams, ideas, beliefs, myths, intuitions, and inspirations brought into being by the human imagination since the dawn of consciousness. It's a symbol of all that we've accomplished and all that we can accomplish. The ethnosphere is humanity's great legacy."[14] No wonder that Annapurna is worshiped throughout the South Asian subcontinent. Yet please notice the bioculturality of Annapurna as both a mountain and also as the commonly held notion of abundance in food and harvest. In such biocultural spaces, I imagine a new birthing of knowledge democracy that is circular, not hierarchical. Such a sphere of knowledge, learning, and leading is ever evolving, not fixed, or frozen. It has pluriverses and pluri-epicenters, not one.

Learningsphere for Earthbound Religiosity

In his 2013 Gifford Lecture, Bruno Latour indicates a deeper shift possible in our religious worldviews in the Anthropocene epoch. He asks: have we been misreading the Christian gospel "What good would it be to possess the world, if you forfeit your soul?" Now, ironically, perhaps the better question is "What use is it to save your soul, if you forfeit the earth?"[15] A new dilemma emerges in the Anthropocene epoch for humans, the dominant species of the biosphere. We have become too successful in our own mission—indeed, we have possessed the world. What new ethic of land, food, and agriculture do we need now that nourishes our bodies and souls but does not forfeit the earth? Would Aldo Leopold's *Land Ethic* or Sir Albert Howard's *New Agricultural Testament* suffice, or do we need another version more attuned to the current Anthropocene reality? Do Vandana Shiva's *Earth Democracy* (2005) or Francis Moore Lappe's *Eco-mind* (2013) approximate what we are looking for?

What role could current religious traditions play in this new epoch of food growing, food making, food learning, and eating? Will there be resilient and abundant Hindu food systems, Buddhist food systems, Christian food systems, Islamic food systems, indigenous peoples food systems, or nomadic food systems in the future? Will these religious traditions rise to the occasion and give prescriptions for sustainable and plentiful food systems? This is difficult to guess. Perhaps there will be many different configurations. How will each religion integrate as well as renew its traditions to embrace the new reality? As demonstrated amply in this volume, although each religious tradition demonstrates some deeply ecological roots, most have acquiesced in the emergence of an industrial epoch. From the Peruvian Andes to the Nepalese Himalayan foothills, the motions of markets and modernity have overwhelmed the local agroecological systems. Globally, a significant part of the land is under monoculture, aimed at exporting food. We have to admit that we entered the Anthropocene epoch despite the religious wisdoms espoused in churches, mosques, synagogues, and Hindu temples. For instance, in the Bible, the word *wilderness* appears three hundred times, and usually its meaning is derogatory.[16] As Jagannath Adhikari's chapter 5 and Pankaj Jain's chapter 6 in this volume suggest, the Hindu ethos and textual instructions that could inform current agroecological practices on the ground are few and far between. I am wondering if the time is ripe for writing new agricultural testaments. Indeed, people like Wes Jackson, Wendell Berry, Vandana Shiva, Bill Mollison, Janine Benyus, Gary Nabhan, and Masanoba Fukuoka have explored such new terrains.

Bruno Latour, in the same 2013 lecture, commented, "Earth should be understood as a historical, or better, as a geostorical adventure." *Geostori-*

cal is a term he proposes to express what it means to live at the epoch of the Anthropocene.[17] In other words, the next phase of human imagination, choices, and designs will have to be a mix of the anthropohistorical and the geostorical because we will be able to observe and feel the immediate impact of our actions on the earth: impacts on soil, water, carbon, climate, and/or food. The earth is no longer indifferent to our actions; it reacts—quickly and often furiously, through excessive and irregular rainfalls, floods, droughts, earthquakes, or hurricanes. Could these immediate feedback loops from nature be blessings in disguise? One of the brighter sides of us being in the Anthropocene epoch could be that the human species will become more cognizant of these reactions to our own actions. Will such feedback loops lead to correcting our harmful ways and seeking beneficial relationships and paths?

Obviously, the task ahead is overwhelming, requiring tremendous effort and strength of conviction. We have accumulated a mountain of "systems of problems." Will humanity rise to the occasion and find "systems of solutions" instead? For example, can we keep these planetary boundaries to where they should be and not exceed them? Through over 250,000 years of the history of Homo sapiens, we have shown maximum resiliency in learning, relearning, and unlearning the way we have imagined and lived our lives. We are always seeking to find more satisfying lives. In many cases, humans have played a role as the keystone species in enriching and enhancing the workings of the biosphere. Could playing such a role be even more pertinent now, when two-thirds of the earth is already somehow altered and dominated by humans? Or, as Wes Jackson asks, is settled and annual agriculture itself the root of the problem? Should we abandon annual monocultural agriculture as we know it? Would "perennial polyculture," with which they are experimenting at the Land Institute in Salina, Kansas, be a better substitute?

There is some bright news amid the dark reality of growing and eating food during the Anthropocene. A near-global consensus has been reached that a massive paradigm shift is needed in agriculture and food systems. This is evident even in current reports from the World Bank and the United Nations Conference in Trade and Development (UNCTAD), both of which were previously promoters of large-scale industrial food production and consumption. A recent UNCTAD report concludes that the world needs a paradigm shift in agriculture from a "green revolution" to an "ecological intensification" approach. The report recommends a shift "towards a mosaic of sustainable, regenerative production systems that also considerably improves the productivity of the small-scale farmers." The report further makes a significant proposal to "move from a linear to a holistic approach in agricultural management, which recognizes that a farmer is not only a producer of agricultural goods but also a manager of an agroecological system that provides quite a number of public goods and services (e.g. water, soil, landscape, energy, biodiversity and recreation)."[18] These findings are similar to those reached by an earlier United Nations Environmental Program report.[19] That report, based on nine hundred participants from 110 countries, also noted the urgent need to build local and national capacity for biodiverse, ecologically resilient farming to cope with increasing environmental stresses, including the rise in global temperature. The general tenor and recommendations of these reports are welcome. I would like to take such a corrective from dominant institutions as an opportunity to dig deeper directly into the emergent practices of peasants, farmers, gardeners and fisherfolk, nomadic herders, and desert dwellers.

In the middle of the realignment between the well-being of the biosphere and the ethnosphere, a crucial role is played by the learningsphere. As the world is turning serious attention to issues of food and agriculture, a new science and knowledge base of/for food systems is in the making. Knowing one's own food system from soil to supper/sustenance and back to soil, which I call the SOSuS loop, is already becoming a priority area for teaching and learning. This author was one of the pioneers who designed and developed learning gardens in Portland, Oregon, from 2003 to 2008. Unexpected places, such as asphalt areas in schools, are turning into ecosystems and gardens. I predict an increased connection between pedology (the science of soil) and pedagogy (the way we learn and teach). In the new learningsphere, peasants and farmers as well as children and youth will be co-learners, co-gardeners, and prosumers (both producers and consumers).

One of the most promising aspects of the learningsphere is that embedded within it are new designs to create abundance while benefiting the biosphere. Examples of humans aiding the earth can still be traced among indigenous peoples, nomads, forest dwellers, fishing folk, and as exemplified by my mother's story, even some peasant communities. I have referred to them as "ecological ethnicities"—today, these might comprise at least 40 percent of humanity.[20] Interestingly, new initiatives, such as Janine Benyus's "biomimicry" and "permaculture" design, have joined the ancient wisdom of how to live on the land, grow food, and support livelihoods while not depleting the earth but enriching it.[21] This author's own small experiment (since 1993) in creating a permaculture household in Chitwan, Nepal, precisely tries to blend the age-old peasant traditions of food and agriculture with principles of agroecology, bio-intensive agriculture, and permaculture.[22]

These very diverse, dynamic, and versatile emergent initiatives are worth mentioning, although they have yet to reach enough critical mass to alter the dominant food system. While some use the term *agroecology,* others call themselves proponents of diversified agriculture. Some champion the term *ecological agriculture,* while others think of what they are doing as climate-friendly, climate-smart agriculture. Some identify themselves as permaculturists, while others are agroforesters who also create edible forest gardens. Some call their efforts small-scale agriculture, while others pin their hopes on biodynamic agriculture. One among many, the Rodale Institute in the United States has a long history of promoting what it calls regenerative agriculture, while the Land Institute in Kansas is experimenting in growing perennial polyculture.[23] Joel Salatin's Polyface Farm in Virginia has taken the new agrarian community by storm with its closed-loop system of agro-animal-pastoral efficiency (see chapter 13 by Ragan Sutterfield in this volume).

The acequia system in the US Southwest shows promise in how to use gravity to regulate water and nutrients. Professor Devon Pena and a team have shown that the ecological services provided by acequias can be classified into five major categories: biological, hydrological, geochemical, topographical, and cultural. These services are not only important to the acequia agroecosystem but also yield benefits for the larger regional economy as well.[24] In Detroit, even while the city is experiencing bankruptcy, abandoned lots are turned into community gardens. As Wes Jackson, founder of the Land Institute, has aptly warned us: "If we don't get sustainability in agriculture first, sustainability will not happen." The time seems ripe to raise our goal: to achieve not only a sustainable food system but one that is abundant and resilient. Can diverse religious traditions embrace this idea?

Earthbounding Soil and Carbon

This author proposes that we can reach the goal of creating a regenerative, resilient, abundant agrarian order. Let us look at "earthbounding" soil and carbon as one of the ways to achieve that. As author Charles Mann and soil scientist Rattan Lal have pointed out, the future of food and agriculture lies under our feet. Our hopes of surviving and thriving on this earth depend on the soil and what we do with the soil—whether we use or abuse it, nourish or deplete it, enhance or degrade it.[25] The properties of soil in relation to food and climate are still a hidden secret. If we knew how to properly work with soil, carbon, water, and biodiversity, we could abundantly feed the world while retaining resiliency in the overall system. But we might have to first recognize that our initial step is to feed and grow the soil, not crops and vegetables. The good news is that this is also the way to design and practice a climate-friendly, climate-smart ecological-agricultural path. But how do we change the prevailing mindset, persuade people that we should shift our attention from growing crops, vegetables, and fruits to growing soil? Once we grow healthy soils, the soils will in turn grow all the crops, vegetables, and fruits we need. Notions about how to increase productivity in agriculture also need to be reexamined.

It is rather ironic that climate change luminaries like Al Gore and the Nobel Prize–winning Intergovernmental Panel on Climate Change present the facts as if humans and other species are better off in a "low-carbon" or even "carbon-free" world. Carbon has been vilified unfairly. In reality, carbon is the fifteenth most abundant element in the earth's crust, and the fourth most abundant element in the universe by mass, after hydrogen, helium, and oxygen. Carbon is present is all life forms, and in the human body it is the second most abundant element after oxygen.[26] The problem is not carbon, it is excess carbon in the atmosphere, which is beyond a threshold of 350 parts per million.

Peasants and indigenous economies around the world, including in the Andes, the Amazon, and the Himalayan region, show that carbon is as essential to life as soil, water, or oxygen. The problem is not the presence of carbon in our environment—soil or forest—but how the carbon is contained or used, stored or released. Global warming is not caused by having carbon in soil, vegetation cover, or forests as such; it is caused by the ways we are releasing four hundred years' worth (in one estimate) of naturally stored carbon (coal and petroleum, also known as ancient sunlight) into the atmosphere by burning it in a short period. However, when carbon is in its proper place—earthbound, forestbound, and soilbound—it becomes the source of life.

Among others, Rattan Lal, a prominent soil scientist at Ohio State University, shows that even devastated soils can be restored. And a restored soil, according to Professor Lal, is our chance to address not only world hunger but also problems like water scarcity and even global warming.[27] One question not yet fully explored is: could global warming be significantly slowed by using vast stores of carbon to reengineer the world's bad soils? This would be a smart strategy to pursue, argues Charles Mann.[28] First sequester carbon and restore soils. "Political stability, environmental quality, hunger, and poverty all have the same root," Lal notes. "In the long run, the solution to each is restoring the most basic of all resources, the soil."[29]

In other words, our future squarely depends on what we do with soil. And the vitality of soil rests on what happens with its carbon content. The amount of carbon in the soil is a function of the historical vegetative cover and productivity, which in turn are dependent in part upon climatic variables. Albert Bates calculates that before people began to use plows on soil and graze goats, there was perhaps 20 percent carbon in soils. Today, that has been reduced to 0.5 to 5 percent. Soil scientist Rattan Lal estimates that the ultimate soil carbon uptake is 1 gigaton (1 billion tons) per year.[30]

Among others, Vandana Shiva, Charles Mann, Rattan Lal, Gary Paul Nabhan, and Michael Pollan have shown we can actually create a "carbon-rich" world that can give us organic food and sustainable livelihoods. Evidence abounds demonstrating that through sustainable agroecological and regenerative agrarian technologies we can systematically build organic matter into the soil, capturing carbon from the atmosphere rather than releasing it into the air. In her foreword to Albert Bates's *Biochar Solution*, Vandana Shiva writes:

> Soil is a major store of carbon, containing three times as much carbon as the atmosphere and five times as much as forests. About 60% of this is in the form of organic matter in the soil. The principal component of soil carbon is humus, a stable form of organic carbon with the average lifetime of hundreds to thousands of years. We need to grow living soils because they are the very

source of life. Living soils provide multiple ecological services, including conservation of water and maintenance of hydrological cycle. They are the basis of food security and climate resilience.[31]

Addressing the latest drought in the US Southwest, Gary Paul Nabhan suggests a similar soil-building tactic: "One strategy would be to promote the use of locally produced compost to increase the moisture-holding capacity of fields, orchards and vineyards. In addition to locking carbon in the soil, composting buffers crop roots from heat and drought while increasing forage and food-crop yields. By simply increasing organic matter in their fields from 1 percent to 5 percent, farmers can increase water storage in the root zones from 33 pounds per cubic meter to 195 pounds."[32]

One of the promising stories of soil building comes from the Amazon region. Dating as far back as five thousand years, indigenous peoples seem to have created a very fertile soil, known as the *terra preta do Indio,* or Amazonian dark earth (ADE). The soil scientist Wim Sombroek had wondered if modern farmers might create their own terra preta, which he called *terra preta nova.* Now this technology is researched and championed by, among others, William Woods, soil scientist at the University of Kansas, and Johannes Lehmann of Cornell University. As elaborated in chapter 2 of this volume, Dr. Frédérique Apffel-Marglin, founder of the Sachamama Center for Biocultural Regeneration at Lamas, Peru, has begun to prepare and use terra preta in peasant fields, which she calls the Chacra-Huerto project. This author is involved in helping Sachamama Center to create school gardens where ADE is made and used. In this pedagogical project, the idea is to enable children and youth to "learn to garden" while they also "garden to learn."

Charles Mann describes the promise of terra preta not only in improving soils but also in sequestering carbon: "Key to terra preta is charcoal, made by burning plants and refuse at low temperatures. Adding and mixing terra preta in tropical soils seems to enhance fertility because such soils quickly lose microbial richness when converted to agriculture. Soil combined with Charcoal seems to provide habitat for microbes—making a kind of artificial soil within the soil—partly because nutrients bind to the charcoal rather than being washed away. Tests by a U.S.-Brazilian team in 2006 found that terra preta had a far greater number and variety of microorganisms than typical tropical soils—it was literally more alive."[33]

Wim Sombroek has argued that creating terra preta around the world would use so much carbon-rich charcoal that it could more than offset the release of soil carbon into the atmosphere. As Charles Mann comments, "Terra preta could unleash what the scientific journal *Nature* has called a 'black revolution' across the broad arc of impoverished soil from Southeast Asia to Africa. A black revolution might even help combat global warming."[34] William McDonough and Michael Braungart suggest that we do not burn the valuable carbon and make it a source of pollution but instead "create a high-carbon diet for the planet." Carbon should be considered not a waste to be minimized but an asset to be enhanced and multiplied. Responding to carbon-trading regimes, McDonough and Braungart also caution that carbon's real value is not in trade but in capturing in soil—because it is a nutrient, not merely a numerical asset to be turned into a derivative and exchanged by traders in an abstract numerical universe.[35]

Concluding Food for Thought

I am not suggesting that the new economy of soil and carbon is all about science, or about just fixing cycles of technical nutrients. Religious traditions have very significant and deep roles to play. Some religious traditions, such as those of peasants and indigenous communities, already embrace these rules of regenerative reciprocities such as giving thanks to soils and other natural elements. For example, peanuts were planted January 8–10, 2014, at the Sachamama Center for Biocultural Regeneration because these three days happened to be right after the first five days of the new moon. As powerful as Amazonian dark earth is, the planting also has to be aligned with the cycle of the moon. Similar lunar cycles were observed when harvesting the peanuts.

Another aspect I want to stress is that any solutions proposed will have to connect both sides of the coin: the biosphere and social justice/equity in the ethnosphere. There can be no abundant and resilient food system without social justice, and nor will social justice be achievable without making healthy and nourishing food available to everyone. Thus, creating fairness and equity at the local, regional, and global scale will be the test of our new ventures and the new knowledge and science associated with achieving sustainable agriculture. In essence, what I call the new earthbound secular religiosity also needs to be pro-justice, in favor of those who actually work on the land and grow food.

It is important to revisit, rethink, and even regenerate our religious traditions at this juncture because planetary boundaries and climate disruption are not merely natural events; they are equally the product of human-made social institutions and arrangements that reflect our worldviews and our cosmovisions.

Despite looming and potential ecological threats, this author is confident humanity will embark on a journey of transformation to find a more satisfying food system. As many examples above suggest, the need to shift to a more permanent and abundant agriculture and food system is now fully recognized. We have the requisite ideas, skills, and technologies among farmers, peasants, and scholars/activists. They are learning, unlearning, and relearning every day. The reasons for hope are personal as well as systemic: personal because Homo sapiens is hardwired to learn, unlearn, and relearn; systemic because food has entered the nervous system of the global economy through markets and nutritional networks. The issue is of deep concern because food directly impacts our health and well-being. Thus I have been speculating that our renewed engagement with soil, farms, forests, gardens, and food might be the gateway to larger transformations that are not only "deep" but also "delicious."

One insight that arises from the interplay of biosphere, ethnosphere, and learingsphere is that the whole is almost always more than the sum of its parts. How the whole will form and emerge will involve pleasant surprises, because living systems like the soils, carbon, plants, forests, and water bodies are not machines—living systems make and unmake themselves.[36]

Welcome to pleasant surprises!

Study Questions

1. Why is it important to rethink/revisit what we eat and how we grow food in the context of the Anthropocene epoch?
2. Will humanity rise to the occasion and create a resilient and abundant food system?
3. What other examples of regenerative agrarian food systems are you aware of?

4. How useful are the notions of the biosphere, ethnosphere, and learningsphere in reimagining a resilient and abundant food system?
5. Will existing religious traditions find a way to reform/update themselves so that, as Bruno Latour suggests, they do not "forfeit the earth?"
6. What new forms of religiosities can you imagine in the continuum of the sacred and secular by 2030 or 2050?

Notes

1. Wendell Berry, *Conversations with Wendell Berry* (Jackson: University Press of Mississippi, 2007), 116.

2. For the planetary boundaries, see Paul Crutzen, "Can We Survive the Anthropocene Period?" http://www.project-syndicate.org/commentary/crutzen1/English (accessed December 4, 2012); and Johan Rockstrom et al., "Planetary Boundaries: Exploring the Safe Operating Space for Humanity," http://www.stockholmresilience.org/download/18.8615c78125078c8d3380002197/ES-2009-3180.pdf (accessed December 4, 2012).

3. For example, see chapter 1 by Leonor Hurtado Paz y Paz and Cristóbal Cojtí García, chapter 5 by Jagannath Adhikari, chapter 8 by Alexander Kaufman, and chapter 15 by Tristan Reader and Terrol Dew Johnson in this volume.

4. William McDonough and Michael Braungart, *The Upcycle: Beyond Sustainability, Designing for Abundance* (New York: North Point, 2013).

5. Michael Pollan's prolific work on food and culinary traditions has been an inspiration to me. I have derived many insights in this chapter from his two recent books: *Cooked: A History of Ecological Transformation* (New York: Penguin, 2013) and *In Defense of Food* (New York: Penguin, 2008).

6. Among others, this argument is elaborated by Richard Oppenlander, *Comfortably Unaware: Global Depletion and Food Responsibility* (New York: Beaufort Books, 2012). See also Oppenlander's website, *Global Depletion and Food Responsibility*, http://www.comfortablyunaware.com (accessed August 20, 2014).

7. Kate Raworth, "A Safe and Just Space for Humanity," http://www.oxfam.org/en/grow/video/2012/introducing-doughnut-safe-and-just-space-humanity (accessed March 5, 2013).

8. Judith Schwartz, "Soil as Carbon Storehouse: New Weapons in the Climate Fight," http://e360.yale.edu/feature/soil_as_carbon_storehouse_new_weapon_in_climate_fight/2744/ (accessed April 7, 2014).

9. Thomas Forster and Arthur Getz Escudero, *City Regions as Landscapes for People, Food and Nature* (Washington, D.C.: Ecoagriculture Partners, 2014), http://landscapes.ecoagriculture.org/global_review/city_regions (accessed September 13, 2014).

10. McDonough and Braungart, *The Upcycle*.

11. Wendell Berry, *What Matters? Economics for a Renewed Commonwealth* (Berkeley, Calif.: Counterpoint, 2010).

12. Wendell Berry, "The Idea of a Local Economy" http://home2.btconnect.com/tipiglen/localecon.html (accessed October 12, 2012).

13. M. Lang and D. Mokrani, *Beyond Development: Alternative Visions from Latin America,* http://rosalux-europa.info/userfiles/file/Beyond_Development_RLS_TNI_2013.pdf (accessed January 2, 2014).

14. Wade Davis, *Light at the Edge of the World* (Washington, D.C.: National Geographic Society, 2007).

15. Bruno Latour, "Facing Gaia: Six Lectures on the Political Theology of Nature," Gifford Lecture, 2013, http://www.bruno-latour.fr/sites/default/files/downloads/GIFFORD-SIX-LECTURES_1.pdf (accessed November 4, 2013).

16. Rene Dubos, "Wooing of the Earth," in McDonough and Braungart, *The Upcycle*, 27.

17. Latour, "Facing Gaia."

18. UNCTAD, *Wake Up before It Is Too Late* (United Nations, 2013).

19. United Nations Environmental Program, *International Assessment of Agricultural Knowledge, Science and Technology for Development,* 2009, http://www.unep.org/dewa/agassessment/reports/IAASTD/EN/Agriculture%20at%20a%20Crossroads_Synthesis%20Report%20(English).pdf (accessed November 4, 2013).

20. Pramod Parajuli, "How Can Four Trees Make a Jungle?" in *The World and the Wild* (Tucson: University of Arizona Press, 2002), 3–20.

21. David Holmgren, *Future Scenarios: How Communities Can Adapt to Peak Oil and Climate Change* (White River Junction, Vt.: Chelsea Green, 2009).

22. The Ajamvari Permaculture Farm is designed to meet a Nepali peasant household's needs of food, fiber, fertilizer, fodder, and pharmaceuticals. See www.ajamvarifarm.org.

23. Rodale Institute has been pioneering organic and regenerative methods of agriculture and soil improvement for the last several decades. Its recent white paper, "Regenerative Agriculture and Climate Change," summarizes many years of research and can be found at http://rodaleinstitute.org/assets/WhitePaper.pdf (accessed September 3, 2014).

24. The prolific writings of Devon Pena on food justice and the acequia system can be found at www.ejfood.blogspot.com. I also recommend his book, *Mexican Americans and the Environment* (Tucson: University of Arizona Press, 2005).

25. Charles Mann, "Our Good Earth: The Future Rests on the Soil beneath Our Feet," http://ngm.nationalgeographic.com/2008/09/soil/mann-text/1 (accessed November 2, 2013); Rattan Lal, "Carbon Sequestration," *Philosophical Transactions of the Royal Society,* http://rstb.royalsocietypublishing.org/content/363/1492/815.full.pdf (accessed March 5, 2014).

26. Albert Bates, *The Biochar Solution* (Gabriola Island, British Columbia: New Society, 2011).

27. Lal, "Carbon Sequestration."
28. Mann, "Our Good Earth."
29. Lal, "Carbon Sequestration."
30. Ibid.
31. Vandana Shiva, foreword to Bates, *The Biochar Solution*.
32. Gary Paul Nabhan, "Our Coming Food Crisis," *New York Times*, July 21, 2013, http://www.nytimes.com/2013/07/22/opinion/our-coming-food-crisis.html?_r=1& (accessed September 5, 2013).
33. Mann, "Our Good Earth," 8.
34. Ibid.
35. McDonough and Braungart, *The Upcycle*, 39.
36. Peter Senge et al., "Awakening Faith in an Alternative Future," *Reflections: The SOL Journal on Knowledge, Learning and Change,* http://www.ai.wu.ac.at/~kaiser/birgit/awakening-faith-scharmer-et-al.pdf (accessed November 15, 2014).

Acknowledgments

The editors would like to express their heartfelt thanks to all the contributors. The insights and knowledge the chapters contain make this a unique contribution to the human understanding of agriculture. We appreciate the contributors' hard work as well as their patience with this project. The editors also thank the staff at the University Press of Kentucky for help with the life history of this book, and extend special thanks to Allison Webster and Ila McEntire. They also would like to thank Vandana Shiva for taking time from her busy schedule to write a very rich foreword. Lastly, each editor expresses sincere thanks to all the sustainable, regenerative, and ethical farmers, peasants, and indigenous peoples on our planet who are working on creating a resilient, equitable (for both human and nonhuman organisms), sustainable agriculture. It is from the roots of their humble, sacred actions that a healthy future will grow.

Contributors

Maximilian Abouleish-Boes is responsible for SEKEM's Department for Sustainable Development, which fosters research, development, and innovation in the company's economic, cultural, societal, and ecological life. Abouleish-Boes is also involved in the Social Innovation Center of Heliopolis University for Sustainable Development, an integrated part of SEKEM's innovation ecosystem.

Jagannath Adhikari is a social scientist, scholar, and consultant who works and writes extensively on issues of ecoagriculture, food systems, and food security in South Asia. He is also an adjunct faculty member at Curtin University of Technology, Perth, Australia. He has authored several books on the political ecology, agriculture, and food systems of Nepal, including *Land Reform in Nepal, Pokhara: Biography of a Town, Food Crisis in Nepal,* and *Farm Management Strategies in Midhills of Nepal.*

Frédérique Apffel-Marglin is professor emerita in the Department of Anthropology at Smith College in Massachusetts. She is the founder and director of the Sachamama Center in Lamas, San Martin, Peru.

Cristóbal Cojtí García is recognized as the day keeper of the Mayan calendar and cosmology. The sacred calendars kept by the Mayans have long predicted the shifts that we are experiencing in modern times. As depicted in the prophecy of sacred birds, the *nawales* Eagle, Quetzal, and Condor, Mayan people foresaw a time when the North and South would come together, combining the higher knowledge of both in an opportunity to create a better life for all, honoring Mother Earth, our community, and ourselves.

Yigal Deutscher is an educator, farmer, and permaculture designer. He has participated in the Adamah Fellowship and trained with the Center for Agroecology & Sustainable Food Systems (University of California, Santa Cruz) as well as the Permaculture Research Institute in Australia. From 2006 to 2010 he was the farm manager and permaculture educator at the Chava v'Adam Farm in Israel, where he also founded the Shorashim/Eco-Israel apprenticeship program. He currently runs 7Seeds, an educational project weaving together Jewish wisdom traditions and permaculture design strategies, activating the foundations for sacred Jewish agriculture. He writes and teaches regularly on the subjects of Jewish spirituality, ecology, and culture. He is the author of *Envisioning Sabbatical Culture: A Shmita Manifesto.*

Mark H. Dixon is an associate professor of philosophy at Ohio Northern University. He is coeditor, with Forrest Clingerman, of *Placing Nature on the Borders of Religion, Philosophy and Ethics.* His research interests are Asian philosophies and environmental philosophy and ethics.

Leonor Hurtado Paz y Paz, Gutemalan by birth, earned her Ph.D. from the Universidad LaSalle, Costa Rica. She serves as senior advisor to the Oakland, California-based Food First/Institute for Food and Development Policy. Over the last fifteen years she has worked extensively with local and international service organizations and social movements in Guatemala in the areas of human rights, indigenous rights, community health, anti-mining, and cultural survival. During the 1980s, she worked in Nicaragua on adult literacy, farmworker training, and international solidarity and information campaigns. As both a scholar and an educator, Dr. Hurtado is involved in issues of food sovereignty, social movements, action research, and community education.

Pankaj Jain is the author of *Dharma and Ecology of Hindu Communities: Sustenance and Sustainability,* which won the 2012 DANAM Book Award and the 2011 Uberoi Book Award. He is an assistant professor in the Department of Anthropology and the Department of Philosophy and Religion at the University of North Texas, where he teaches courses on religions, cultures, ecologies, and films of India and Asia. He received the Fulbright-Nehru Environmental Leadership Fellowship in 2012 and the Wenner-Gren Grant in 2014. He is also director of the Eco-Dharma and Bhumi-Seva Project and is working with the Hindu and Jain temples in North America on their "greening" efforts. Currently, he is editing a volume on Indian philosophical theories and writing books on Jain philosophy and dharma and science in India.

Terrol Dew Johnson is an award-winning Tohono O'odham basket weaver whose work appears in a variety of permanent collections. He provides consultations to museums and is an active organizer within his community.

Alexander Harrow Kaufman grew up in the United States and lived in Spain, Brazil, and Japan before settling in Thailand in 1996. Through his consultancy work, Kaufman has carried out social audits of factories in Cambodia, China, Indonesia, Macau, Mongolia, Korea, Taiwan, Thailand, and Vietnam. His research focuses on the use of organic agriculture as a rural development strategy in northeast Thailand. He serves as an advisor to the Moral Rice Network at the Dharma Garden Temple in Yasothon Province, Thailand.

Michael Lemons received his Ph.D. from the Department of Anthropology at the University of Florida. He is a lecturer in anthropology at Hawaii Community College in Hilo, Hawaii, where he researches permaculture, theories of deviance, and cultural and ecological anthropology.

Todd LeVasseur is a visiting assistant professor at the College of Charleston, where he works in the religious studies department and is also program director of the environmental and sustainability studies program.

Pramod Parajuli serves as the core faculty for the Ph.D. program in sustainability education at Prescott College, Prescott, Arizona. At Prescott, Pramod was also active in designing new graduate programs and bioregional and international studies, including developing learning gardens for Masaai schools in Masaailand, Kenya, and in his homeland of Nepal/Himalayas. An interdisciplinary/applied scholar from Stanford University (1983–1990), Parajuli has thirty years of teaching experience in US higher education. At Stanford, Syracuse University, Portland State University, and currently at Prescott College, he has designed and developed award-winning graduate programs in sustainability studies, sustainable food systems, farm- and garden-based ecological literacy, and learning gardens.

Eston Dickson Pembamoyo, a reverend of the Anglican Church, is from Malindi in Mangochi in Malawi, where he is active in a variety of permaculture-based initiatives.

Raymond F. Person Jr. is professor of religion at Ohio Northern University. He has authored numerous books, articles, and chapters, including *Deuteronomy and Environmental Amnesia*. He and his wife, Elizabeth, live on a twenty-acre organic farm operated by a cooperative he founded of nineteen families.

Anna Peterson is professor in the Department of Religion at the University of Florida, where she teaches social, environmental, and animal ethics, researches sustainability and social change, and is a leading scholar in religion and nature theory.

Tristan Reader is a practitioner of community-based systems change, specializing in the theory and practice of food systems development, Native American community regeneration, health and wellness promotion, cultural revitalization, and grassroots community organizing. He earned his Ph.D. in agroecology and food security from Coventry University, England.

A. Whitney Sanford is associate professor of religion in the Department of Religion at the University of Florida in Gainesville, where she researches and teaches about religion and nature, South Asian religions, South Asian and Global South environmentalisms, and issues of food sustainability. Her books include *Growing Stories from India: Religion and the Fate of Agriculture*.

Elaine Solowey is director of the Center for Sustainable Agriculture at the Arava Institute in Israel. Her Ph.D. work focused on land reclamation, and her current research explores crop suitability for both arid and saline lands.

Ragan Sutterfield is a writer, ordained minister in the Episcopal Church, and onetime livestock farmer. His books include a theology of agriculture, *Cultivating Reality,* and a spiritual memoir, *This Is My Body.*

Norman Wirzba is professor of theology, ecology, and rural life at Duke Divinity School. His research and teaching interests reside at the intersections of theology, philosophy, ecology, and agrarian and environmental studies. His current research is centered on a recovery of the doctrine of creation and a restatement of humanity in terms of its creaturely life. Professor Wirzba's most recent books are *Food and Faith: A Theology of Eating* and (with Fred Bahnson) *Making Peace with the Land: God's Call to Reconcile with Creation.* He also has edited several books, including *The Essential Agrarian Reader: The Future of Culture, Community, and the Land* and *The Art of the Commonplace: The Agrarian Essays of Wendell Berry.* Professor Wirzba serves as general editor for the book series Culture of the Land: A Series in the New Agrarianism, published by the University Press of Kentucky, and is cofounder and executive committee member of the Society for Continental Philosophy and Theology.

Index

Ab', 25, 26
Abouleish, Dr. Ibrahim, 19, 295–96, 298, 301–2, 303, 306, 307, 311
Agarwal, Anil, 148
agency of nonhuman others, 50–52, 53–54, 55, 64–65, 123
agrarianism, 279–80, 290
Agricultural Revolution, environmental critique of, 241
agricultural sustainability, 10–11, 86, 345, 352; and religion, 12–20, 168, 244–47, 337–57; attributes of, 11; Gary Fick, 10; gift exchange, 50; holistic, 10; Mayan/Guatemalan practice of, 25–27, 28–29, 32, 37, 38–41; new agrarianism, 168; peasant farming, 36, 50; regenerative, 10, 12, 20, 51, 64–65, 223, 312, 337–57; resilience, 123; traditional ecological knowledge, 16, 20, 74, 82–83, 126–27, 207, 210, 325
agriculture, 1, 5, 51, 196, 209, 212–13, 222, 229; critiques of Green Revolution/industrial agriculture, vii–viii, 3–4, 11, 32–34, 36, 39, 42, 51, 56, 74–75, 76, 79, 80–81, 123, 128, 133, 135, 175, 177, 199, 241–42, 247, 256, 267, 300, 301, 309, 343, 349, 350; food ecology, 222, 223; formation of/Fertile Crescent, 215–16, 218, 222, 225; Green Revolution, start of, 5–6; industrial, 3, 6–10, 14, 17, 29; organic, 77; perennial, 16, 223; sustainable, 2, 3, 4, 10, 77, 206, 207, 240–41, 242–44, 246–47, 290; three levels of relations involved with, 257

agroecology, 28, 38, 40–41, 351, 352
Alexander Pancho Memorial Farm, 315, 333
Amazon, 12, 45, 60; agriculture in, 59–62; deforestation of, 55, 61; Peruvian, history of, 61–62. *See also* Lamas, Peru
Amazonian dark earth, 12, 20, 45, 46–47, 50, 57–60, 62–63, 64, 355, 356; biochar, 62–63; Terra Preta Nova project, 62–63, 64
Amish. *See* Old Order Amish
Anabaptist(s), 18, 233–34, 244–47, 251–52, 272
animism, 124, 134, 173
Annapurna, 20, 125, 130, 337, 341, 342, 347–48
Anthropocene, the, 4, 20, 337, 341, 342, 346, 348, 349, 350; dharmic life of, 342; environmental metrics of, 4–5, 350; geostorical, 349–50; greenhouse gasses, 344; nitrogen, 344
anthropocentrism, 244; Christian dualism, 251
Asiatic water buffalo, 16, 174; *Su Khwan Kwai* ritual, 175
Athavale, Guru Pandurang Shastri, 15, 139, 141–43, 147, 149–50
axis mundi, 106, 107, 118

back-to-the-land, 99, 100, 104
Bailey-Dick, Matthew, 257–58
being/*Dasein,* 18, 253–57
Berry, Wendell, 279, 282–83, 287, 337, 346; and economy, 346–47
biblical/traditional agriculture, 16, 195–96, 198, 199, 204, 206–9, 212, 213–31; *adam,* 213; agricultural

biblical/traditional agriculture *(cont.)* celebrations/calendar, 196, 197, 198, 200, 203, 224–25, 226, 228, 231; Cain and Abel, 216–17, 219; the Diaspora and end of in Israel, 204–5, 229; *eved*, 215; Garden of Eden, 216–19, 223, 227, 230; Hebrew Bible and passages about agriculture, 195–200, 202, 203, 213–31; *Kil-i-yim*, 207; *la'avod/avodah*, 213–15, 217, 218, 219, 224, 226, 228, 229, 231; *migrash*, 222; Speaking Place/Midbar, 218, 220, 221, 228

biochar, 12

biosphere, 20, 338, 344, 347, 348, 349, 351, 356

Bhagavad Gita, 15, 156, 162, 165, 166, 170

Bhave, Vinoba, 15, 153–54, 155, 157, 158, 162, 164, 168–69; ashrams, 15, 153; Brahma Vidya, 159; Brahma Vidya Mandir (BVM), 15, 153, 154, 155, 158–64, 167–69; critiques of the Green Revolution/industrial agriculture, 154, 158, 162, 166; Gandhian values of, 153, 163–64; *Gita Pravachan*, 156–57; karma yogi, 156–57; Nilayam Nivedita, 15, 154, 158, 164–69; Paramdham Ashram, 154; Paunar, Maharashtra, 153, 158; Samvad Farm, 15, 154, 158, 164–69; social workers, 159; sustainable agricultural practices and Gandhian/religious influences on, 154, 157, 158–64, 164–69

Boonserng, Wijit, 183–85

Buddhism, 14, 124, 134, 173–93; environmental values of, 175–77, 186; Pali canon, 176; precepts, 15–16, 176–77, 178, 182, 184, 186, 189, 192; Thailand, history of, 15, 173, 179; Theravada Buddhism, 173, 190

"Buddhist agriculture," 176, 185; Buddhadasa, 176; *dhammachart*, 176; Khwan Khao (Rice Soul), 16, 173, 174; Mae Khongka (River Goddess/Mother), 16, 173, 187; Mae Phosop (Rice Mother), 16, 173, 174, 187; Mae Thoranee (Earth Mother), 16, 173, 187; *pong khao*, 174; religious significance of, 173; rice, 15, 173–74; rituals of, 16, 173–74; *singwaedlom*, 176; Suanmokkh Temple, 176; traditional farming practices, 173–74, 175; Prawes Wasi, 176

carbon, 21, 61–63, 353–56

Catholicism. *See* Roman Catholicism

Chachapoyas cultural area, map of, 48

Chacra-Huerto project, 59, 63–65, 355

Chol q'ij, 25, 26

Christianity, 277–94; blessing/grace, 19, 278, 291, 292; views of creation, 19, 277, 278, 283, 287, 291, 292

colonialism, 27–28, 31

Columbian Exchange, 7

communitarian, 104, 106

Consultative Group for International Agricultural Research, 29

Cooperation for Rural Development in the Western Region of Guatemala, 37–39

cosmocentric economy, 53, 54, 65

cosmocentrism, 13, 21, 45, 50–51

cosmovision, 20, 346, 347, 357

Creoles, 29

Declaration of the Rights of Indigenous Peoples, 36

Deshpande, Mohan Shankar, 148

dharma. *See under* Hinduism

Dharma Garden Community and Temple, 16, 178–84, 189, 190;

Dharma Garden Foundation, 178; ecocentric values of, 190; Khun Grasehboon, 180–81; marketing of Moral Rice, 189; Monk Khammak/Luang Poh Thammachart, 178, 184; Poh Nikom, 180–81, 186; Poh Songkran, 180–81
dharmic ecology, 14
dramaturgy, 13, 103, 116–17

earthbounding, 20, 337, 338, 345–46, 352, 353–57; ecological and agricultural benefits of, 338; religiosity of, 348–52, 356–57; synergies between bio-, ethno-, and learningsphere, 338, 347–48, 351, 356–57
ecological ethnicities, 134
egalitarianism, 105, 110, 118–19
Egypt, 19, 295, 298; agriculture in, 296–98, 301, 304; Bedouins, 303–4; Belbes, 299; religion in, 301; "Revolution of the Hungry," 297
Egyptian Biodynamic Association, 300
El Salvador, 17, 237–38, 239–40, 241, 243, 244, 246, 248; Catholicism, 238, 239; civil war, 238; postwar, 233
Ethnic Council of the Kichwa People of Amazonia, 57
ethnosphere, 20, 338, 344, 347, 348, 351, 356

Fick, Gary, 10
food justice, 21
food security, 21, 74, 77, 182, 297
food sovereignty, 40–41, 57, 64, 135
food system, 337, 343, 357; human-caused stresses on, 337–38, 343; SOSuS loop (soil to supper/sustenance back to soil), 342, 344, 351

fossil fuels in agriculture, 6, 17, 29
free trade agreements and neoliberal economics, 34; Central American Free Trade Agreement, 34; criticisms of, 34, 36, 39, 52
Fukuoka, Masanoba, 148, 349

Gandhi, Mahatma, 15, 153–58, 161–63, 165–70; *ahimsa* (nonviolence), 155, 156, 166, 167; *aparigraha*, 156; appropriate technologies, 155, 161, 163; food democracy, 153, 169; Gandhism, 167; *Hind Swaraj*, 168; Jain influences, 156; karma yoga, 156; *sarvodaya*, 155, 165, 167; Sevagram, 158; *swadeshi*, 167; village India, 155; Wardha, 158
Garifuna, 26
Gaud, William, 31
genetically modified organisms (GMOs), 23, 36, 81, 157, 166, 207–8, 210, 300
gift economy, 52–53
globalization, 3, 14, 122, 134; trade agreements, criticisms of, 7
Global North, 36
Global South, 36, 155, 237; environmental concerns, 1148, 155, 240–41, 246
green economy, 52
Green Revolution, 8–10, 13, 20, 29–34, 80–81, 134, 135, 195; chemicals used, 7–8, 29; Consultative Group for International Agricultural Research, 29; Food and Agriculture Organization, 29; William Gaud, 31; US Agency for International Development, 29, 35
Guatemala, 26, 28; colonial history, 28–29, 34, 39; contemporary political history, 31–32, 33–36, 39; experience with the Green Revolution, 34–36, 39; Latifundio-

Guatemala *(cont.)*
  minifundio agricultural system, 28–29; map of, 30; race relations in, 28–29, 31, 39–40

Haberman, David, 149
Hawaii, 13, 99; Big Island, 13, 99; Puna, 13, 99, 101, 105–7
health: animals in agriculture, 8–10, 40, 175, 207; environmental, 8–10, 11, 20, 40, 134, 300; human, 9–10, 11, 20, 40, 175; religious, 10, 20, 40; social, 7, 8–10, 11, 20, 40, 134
Heidegger, Martin, 18, 252–57, 262–63, 272, 273–74; being/*Dasein*, 18, 253–57; building, 18, 253–57; dwelling, 18, 253–57, 262–63, 274; environmental and spiritual ethics derived from, 253–54, 262, 272; Christine Swanton, 253–54
himdag, 19, 315, 316, 319–25, 328–33; agri/culture, 19, 315, 316, 319–23; decline of, 328–30; revitalization of, 331–33; songs, 321–22; Tohono O'odham Community Action, 330, 331–33, 336
Himalayas, 121, 337, 341, 347
Hindu epics, 139
Hinduism, vii–x, 14, 121, 123–24, 128–29, 151, 171; ashrams, 154; *Bhagavad Gita*, 15, 156, 162, 165, 166, 170; Brahmanic rituals, 173; *dharma*, viii, 14, 15, 121, 124, 128, 130, 134–35, 136, 341–42; *karma*, 14, 121, 124, 128, 131, 134; *Krisi-Parashar*, 122; *Mahabharata*, ix, 156; *paap*, 14, 121, 124, 130, 134; religion and farming, 122, 129–33, 134–36; Vedas, vii, 122; Vedic, 14; Vedic model of agriculture, 124, 136
Howard, Sir Albert, vii, 11

India, vii, ix, 15, 139, 151, 155, 159, 160, 168; map of, 160

Indwelling God, 142, 143
International Federation of Organic Agriculture, 178, 182, 184, 185, 188, 192; definition of "organic," 192
International Labor Organization, 36
International Monetary Fund, 33
Islam, 295, 302–11; *adl*, 19, 310–11; Allah, 303, 305, 306–7; *amanah*, 309; *amin*, 309; *fitra*, 19, 305–6; *ihsan*, 19, 307–8; *ilm*, 19, 306–7; *khalifa/khulafa*, 19, 309–10, 311; *mizan*, 305; Prophet Muhammad, 306–7; Qur'an and passages from, 303, 306, 309, 310–11; *tawhid*, 19, 303–5; *ummah*, 308
Israel, 16, 195–210, 212, 221, 226, 228, 229; Canaan, 218; renewal of agriculture, 205–6; *Tsena*, 205

Jackson, Wes, 350; Land Institute, the, 350, 352
Jewish social health, 227–28; *hefker*, 227; *leket*, 227; *peah*, 227
Judaism, 16, 195, 212; Hebrew Bible, 16, 195, 197, 201, 207, 213; jubilee/*yovel*, 221, 231; Middle East, 195, 202, 205; Mishnah, 231; *shabbat*/Sabbath, 219–20, 222–23; *shmita*, 220, 223, 227–28, 229, 231; *siach*, 214; Torah, 17, 219, 228, 230

Kichwa-Lamista(s), 12, 45–65; agricultural practices, 45, 47, 50–51, 52, 53, 57, 64; *chakra*, 12, 13, 45, 47, 49–50, 57, 64; colonization and history of, 45–46, 47–49, 55, 56; Ethnic Council of the Kichwa People of Amazonia, 57; gift economy and exchange, 50–52, 53–54, 55; Mama Allpa, 12, 13, 45, 47, 51–52, 54; Mama Qilla, 12, 13, 47, 54; mestizos, 47, 56; Native

communities, 56–57, 63–64; Quechua, 56, 57; reform of, 54–55; religion/spirituality, 45, 47, 50–54, 57; Tayta Inti, 12, 47; Yakumama, 12, 47
kisan dharma, 14
Kumar, Satish, 167

Lamas, Peru, 55, 57, 64
Land Institute, the, 350, 352
learningsphere, 20, 337, 347, 348, 351
liberation theology, 17, 32, 238
Locke, John: influence on stewardship views in America, 284–85, 287

Malawi, 13, 71–97; and the Anglican Church, 89; Chichewa, 85; Christianity/Genesis, 90, 96, 97; colonialism/postcolonialism, and agriculture, 75–76, 87, 88, 93, 94, 96; food sovereignty, 71, 72–73, 77, 86; indigenous farming knowledge and practices, 76, 80, 82–83, 85, 87–88, 92, 93, 96–97; Islam, 89–90; Malawi Association for Christian Support, 91; *Malimidwe a makalo,* 13; map of, 73; *mwanaalirenji,* 71, 77, 92, 93; permaculture, 71–97; Permaculture Network in, 85; religion and sustainable agriculture, 72, 75, 82–83, 87, 88, 90, 91, 96; westernization of, 71, 74–75; women, 89
Malawi Association for Christian Support, 91
materialist/materialism, 111, 114
Mayans (Quiché), 12, 25; calendars and calendrical system, 25–27; colonialism/colonization of, 27, 28, 31–32, 34; ethnocide of, 34–35; experience with the Green Revolution, 29, 32–34, 39; Kaqchikel, 37; missionization of, 31; religious principles and spirituality, 25, 28, 37–41, 42
Mennonites, 18, 234, 251–52; embodied theology, 18, 252; farm-social environment relation, 255, 269–70; First Mennonite Church, Bluffton, 266; Mennonite homestead (Ohio) and farming as an example of, 264–72; simple living, 252, 256, 258–63, 265, 271–72
Mennonite Central Committee, 252, 257–58, 262, 266; Matthew Bailey-Dick, 257–58; cookbooks, 18, 257–58, 266; *Extending the Table,* 257, 258–62; *More-with-Less,* 257, 258–62, 265; relationship to the Christian gospel, 257, 259; *Simply in Season,* 257, 258–62; Rebekah Trollinger, 259–60, 263
millenarianism, 117
milpa, 12, 37
missionization, 42, 49, 96
monoculture, 6
Monsanto Quit India, 157
Moral Rice Network, 16, 177, 178, 181–85, 186–90; Agri-Nature Foundation, 177; Wijit Boonserng, 183–85; civil society organizations (CSOs), 177, 178; Green Net Cooperative/Earth Net Foundation, 177; Luang Poh Supatto, 181; New Theory Agriculture (NTA), 177; Poh Suvit, 185; religion and/seven vows of, 182, 184–89; Santi Asoke Group, 178; "Sufficiency Economy," 177, 192
mysticism, 105

natural resources, 52
nature/culture dualism, vii–ix, 13, 51–54, 58, 65, 99, 102, 103, 107, 110, 115–16,

nature/culture dualism *(cont.)*
  123, 226, 346, 348; dwelling in
  holiness, 263; in Nepal, 132–36,
  339–41; Ultimate Sacred Postulate,
  102
nature spirituality, 105, 107, 110, 114
Nepal, 14, 20, 121, 337, 339, 341, 347;
  Baidam, Pokhara Valley, 121,
  124, 136–37; Himalayas, 121,
  337, 341, 347; impacts of Green
  Revolution/industrial farming,
  123, 125–26, 133, 134; *prayabaran*,
  339; traditional farming practices,
  123–27, 131–33
Nepalese/Hindu peasants, 128–34,
  339, 348; *bachaam*, 339, 340;
  chalaam, 339, 340; *jogaam*, 339,
  340; *paap*, 340; *punya*, 340;
  regenerative farming practices of,
  339–40; *satbeej charne*, 14, 340
Network for Security and Sovereignty
  for Food in Guatemala, 38

Old Order Amish, 17, 233–37, 242,
  243–44; farming methods,
  233–37, 242–44; North American
  agricultural history, 234–25; social
  organization/religion, 233, 237,
  242, 244–47; values, 252, 257, 272,
  273

patents, 7
Paunar, Maharashtra, 153, 158
peasants, 3, 14, 18, 20, 50, 158, 303,
  338, 351–53; ecological ethnicities,
  352; El Salvador, 233; *chacareros*,
  345; Guatemalan/Mayan, 29,
  32, 33, 36, 39–40; regenerative
  reciprocity of, 338, 341, 356;
  relation with local ecologies, 340.
  See also Nepalese/Hindu peasants
permaculture, 13, 16, 77, 84–86, 103,
  110, 114–15, 117, 352; Ajamvari

Permaculture Farm, Nepal,
  359n22; Big Island, Hawaii,
  99–120; ethics of, 95; Global North
  vs. Global South varieties of, 13;
  growth of, 103; indigenous, 69;
  Malawi, 71–97; principles of, 95,
  103
Permaculture Network in Malawi, 85
phenomenology, 18
Pokhara Valley, 14
Polyface Farm, 19, 288–91
postcolonialism, 17, 20
postmaterialism, 105, 111–12, 114
priesthood, 19, 279, 287–92, 294
Promised Land, 16

Quiché Maya people. *See* Mayans
  (Quiché)
Qur'an, the, 303, 306, 309, 310–11

religion, 3, 67, 233–49, 337–57; Mayan,
  25–27; religious values, 4
religious agrarianism, 2
repopulators, 17, 237–44, 248; farming
  practices, 240–44; religion and
  social organization, 244–47
Research Foundation for Science,
  Technology, and Ecology (RFSTE),
  157
Rodale Institute, 352, 359
Roman Catholicism, 17, 29, 244–47,
  248, 324–25; Second Vatican
  Council, 238

Sachamama Center for Biocultural
  Regeneration, 12, 45, 59, 63–65,
  355, 356; Chacra-Huerto project,
  59, 63–65, 355; *yana allpa*, 12, 13
Salatin, Joel, 19, 288, 289, 291, 352;
  farming practices of, 288–89;
  Polyface Farm, 19, 288–91
saguaro cacti, 20, 315
Save, Bhaskar, 157–58; Kalpavriksha, 157

seeds, 9, 35, 40, 82–83, 126, 127, 128, 130, 132–33, 135, 166–67; seed democracy, 157
SEKEM, 19, 295–96, 298–311, 313; Dr. Ibrahim Abouleish, 19, 295–96, 298, 301–3, 306, 307, 311; business and development model of, 301, 308; Demeter, 304; Egyptian Biodynamic Association, 300; Fairtrade, 310; Islamic and Muslim influences on, 301–11; mission of, 299; regenerative/sustainable agriculture, 299–300; SEKEM Development Foundation, 301
Seven Species (*Shivat Minim*), 16, 195, 196, 200, 202, 203, 205, 207–9, 222; barley, 195, 197–98, 205; date palm, 195, 200–201, 205; Emmer wheat, 195, 196–97, 198, 205; fig, 195, 201–2, 205; olive, 195, 198–200, 205; pomegranate, 195, 202–3; vine (grape), 195, 201–2, 205
shamanism, 124, 134
Shiva, Vandana, 157, 166; earth democracy, 157; Gandhian influence on, 157; Monsanto Quit India, 157; Research Foundation for Science, Technology, and Ecology (RFSTE), 157; seed democracy, 157
soil, 212, 213–14, 299–300, 352, 353–56; health and fertility, 11, 28, 50, 175, 344
Sonoran Desert, 20, 315, 317, 318, 327
stewardship, 19, 278–79, 284, 285, 286, 287, 291–92; industrialism, 280, 290
structural adjustment programs, 33; Guatemalan critiques of, 33
Suanmokkh Temple, 176
subsidies, 7
Su Khwan Kwai ritual, 175
sustainability, 72, 100, 104

Swadhyaya (Swadhyayis), 15, 139–51; Anil Agarwal, 148; Guru Pandurang Shastri Athavale, 15, 139, 141, 142, 143, 147, 149–50; Mohan Shankar Deshpande, 148; *dharma*, 139, 340; earth dharma, 143–47, 149, 150; ethnosociological, 141; environmental activism, 149; Masanoba Fukuoka, 148; Gujarat/Gujarati, 15, 139, 140, 141, 142, 143, 146, 147, 149; David Haberman, 149; Hindu epics, 139; Indwelling God, 142, 143; *karma*, 141; *Kriśipārāśara*, 148; *krtibhakti*, 150; *Mahābhārata*, 139; Maharashtra, 139, 140, 147, 149; *moksa*, 139–40; Muslims, 145–46; *prasāda*, 148; *prayog*, 145–46, 151; rainwater/water ethic, 143–47; rural Swadhyayis' religious devotional and agriculture/environment interactions, 139, 141, 142, 143, 146, 147–50; *Śridarśanam*, 147; sustainability, 141, 143; vermiculture, 148; *Yogeśvara Krsi*, 15, 147–50
Swanton, Christine, 253–54
Steiner, Rudolph, 295, 299; biodynamics and farming practices of, 295, 299–300, 303, 304, 306, 307, 310

Terra Preta Nova project, 62–63, 64
Thai Alternative Agriculture Network (AAN). *See* Moral Rice Network
Thailand, 15, 16, 190; history of, 15, 173, 179; map of, 179
theocentric/ism, 244–45; theocentric ecological ethic, 245
Tohono O'odham, 19, 291, 315–36; Alexander Pancho Memorial Farm, 315, 333; *bahidaj*, 315;

Tohono O'odham *(cont.)*
  damage to traditional food system of, 326–28; I'itoi, 315, 318, 326; impact of Christianity, 324–25; *jujkida,* 315–16, 319, 320–25, 329, 334; *kukuipad,* 315; *mamakai,* 320, 324; rituals of, and relation to agriculture, 315–16; saguaro cacti, 20, 315; *sitol,* 315; Sonoran Desert, 20, 315, 317, 318, 327; traditional food system of, 316–19
Tohono O'odham Community Action, 330, 331–33, 336
Torah, 17, 219, 228, 230
Trollinger, Rebekah, 259–60, 263
Tyson, John, 280, 283, 286, 291–92; religious views of, 283–84

Tyson Foods: chicken farming practices, 19, 280–83, 286, 293

United Nations, 36; Food and Agricultural Organization, 59
US Agency for International Development, 29, 35

Wasi, Prawes, 176
water buffalo, Asiatic, 16, 174
World Bank, 33

Xinca, 26

Yasothon Province, 178; map of, 179
"you are my other self" principle, 25, 27

Zizioulas, John, 290

CULTURE OF THE LAND: A SERIES IN THE NEW AGRARIANISM

This series is devoted to the exploration and articulation of a new agrarianism that considers the health of habitats and human communities together. It demonstrates how agrarian insights and responsibilities can be worked out in diverse fields of learning and living: history, science, art, politics, economics, literature, philosophy, religion, urban planning, education, and public policy. Agrarianism is a comprehensive worldview that appreciates the intimate and practical connections that exist between humans and the earth. It stands as our most promising alternative to the unsustainable and destructive ways of current global, industrial, and consumer culture.

SERIES EDITOR
Norman Wirzba, Duke University, North Carolina

ADVISORY BOARD
Wendell Berry, Port Royal, Kentucky
Ellen Davis, Duke University, North Carolina
Patrick Holden, Soil Association, United Kingdom
Wes Jackson, Land Institute, Kansas
Gene Logsdon, Upper Sandusky, Ohio
Bill McKibben, Middlebury College, Vermont
David Orr, Oberlin College, Ohio
Michael Pollan, University of California at Berkeley, California
Jennifer Sahn, *Orion* magazine, Massachusetts
Vandana Shiva, Research Foundation for Science, Technology & Ecology, India
Bill Vitek, Clarkson University, New York